# Galois Theory

Since 1973, Galois theory has been educating undergraduate students on Galois groups and classical Galois theory. In **Galois Theory, Fifth Edition**, mathematician and popular science author Ian Stewart updates this well-established textbook for today's algebra students.

**New to the Fifth Edition**

- Reorganised and revised Chapters 7 and 13
- New exercises and examples
- Expanded, updated references
- Further historical material on figures besides Galois: Omar Khayyam, Vandermonde, Ruffini, and Abel
- A new final chapter discussing other directions in which Galois theory has developed: the inverse Galois problem, differential Galois theory, and a (very) brief introduction to $p$-adic Galois representations

This bestseller continues to deliver a rigorous, yet engaging, treatment of the subject while keeping pace with current educational requirements. More than 200 exercises and a wealth of historical notes augment the proofs, formulas, and theorems.

**Ian Stewart** is an emeritus professor of mathematics at the University of Warwick and a fellow of the Royal Society. Dr. Stewart has been a recipient of many honors, including the Royal Society's Faraday Medal, the IMA Gold Medal, the AAAS Public Understanding of Science and Technology Award, and the LMS/IMA Zeeman Medal. He has published more than 210 scientific papers and numerous books, including several bestsellers co-authored with Terry Pratchett and Jack Cohen that combine fantasy with nonfiction.

# Galois Theory

## Fifth Edition

Ian Stewart
University of Warwick, UK

CRC Press
Taylor & Francis Group
Boca Raton London New York

CRC Press is an imprint of the
Taylor & Francis Group, an **informa** business
A CHAPMAN & HALL BOOK

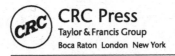

Fifth edition published 2023
by CRC Press
6000 Broken Sound Parkway NW, Suite 300, Boca Raton, FL 33487-2742

and by CRC Press
4 Park Square, Milton Park, Abingdon, Oxon, OX14 4RN

*CRC Press is an imprint of Taylor & Francis Group, LLC*



First edition published by CRC Press 1973
Second edition published by CRC Press 1989
Third edition published by CRC Press 2004
Fourth edition published by CRC Press 2015



ISBN: 978-1-032-10159-0 (hbk)
ISBN: 978-1-032-10158-3 (pbk)
ISBN: 978-1-003-21394-9 (ebk)

DOI: 10.1201/9781003213949

Typeset in CMR10 font
by KnowledgeWorks Global Ltd.

*Publisher's note*: This book has been prepared from camera-ready copy provided by the author.

Access the Support Material: www.Routledge.com/9781032101590

Portrait of Évariste Galois, age 15.

# Contents

# *Acknowledgements*

The following illustrations are reproduced, with permission, from the sources listed.

Frontispiece and Figures 4–6, 23 from Robert Bourgne and J.-P. Azra *Écrits et Mémoires Mathématiques d'Évariste Galois*, Gauthier-Villars, Paris 1962.

Figure 1 (left) from B.L. van der Waerden, *Erwachende Wissenschaft 2: Die Anfänge der Astronomie*, Birkhäuser, Basel 1968.

Figures 1 (right), 2 (right) from David M. Burton, *The History of Mathematics: an Introduction*, Allyn and Bacon, Boston 1985.

Figure 27 from Carl Friedrich Gauss, *Werke*, Vol. X, Georg Olms, Hildesheim and New York 1973.

The quotations in Chapter 25 are reproduced with permission from Peter M. Neumann, *The Mathematical Writings of Évariste Galois*, European Mathematical Society, Zürich 2011.

# Preface to the Fifth Edition

The first edition of *Galois Theory* appeared in 1973. New editions introduced various changes, especially the third, which began with a more concrete treatment using subfields of the complex numbers. This change was made so that students could appreciate the main ideas of Galois theory without becoming involved in a large amount of background material from abstract algebra. Anyone preferring the abstract treatment could omit this introductory material. This decision was a compromise and had a mixed reception. After wide consultation with mathematicians who have been teaching courses using the book, I have retained this two-stage structure. Since five prefaces seem excessive, this one begins with edited extracts from previous prefaces. Then I summarise the changes made for this edition.

## First and Second Editions

Galois theory is a showpiece of mathematical unification, bringing together several different branches of the subject and creating a powerful machine for the study of problems of considerable historical and mathematical importance. The central theme is the application of the Galois group to the quintic equation. As well as the traditional approach by way of the 'general' polynomial equation, I have included a direct approach that demonstrates the insolubility by radicals of a specific quintic polynomial with integer coefficients, a more convincing result. Other topics include the impossibility of duplicating the cube, trisecting the angle, and squaring the circle; the construction of regular polygons; the solution of cubic and quartic equations; the structure of finite fields; and the 'Fundamental Theorem of Algebra'.

In order to make the treatment as self-contained as possible, and to bring together all the relevant material in a single volume, I have included several digressions. The most important of these is a proof of the transcendence of $\pi$, which all mathematicians should see at least once in their lives. There is a discussion of Fermat numbers, to emphasise that the problem of regular polygons, although reduced to a simple-looking question in number theory, is by no means completely solved. A construction for the regular 17-gon is given, on the grounds that such an unintuitive result requires more than just an existence proof.

Much of the motivation for the subject is historical, and I have taken the opportunity to weave historical comments into the body of the book where appropriate. There are two sections of purely historical matter: a short sketch of the history of polynomials, and a biography of Évariste Galois. The latter is culled from several sources, listed in the references.

The text includes many examples that illustrate the general theory, and there are around two hundred exercises, with twenty harder ones for the more advanced or more ambitious student.

## Third and Fourth Editions

The first two editions of *Galois Theory* followed the fashion of the time, which favoured generality and abstraction. But educational fashions change, and at many institutions, the presentation of mathematics veered back towards specific examples and a preference for more concrete presentations.

Since both approaches are valuable, the story in these editions starts with polynomials over the complex numbers, and the central quest is to understand when such polynomials have solutions that can be expressed by radicals—algebraic expressions involving nothing more sophisticated than $n$th roots. Only after this tale is complete is any serious attempt made to generalise the theory to arbitrary fields, and to exploit the language and thought-patterns of rings, ideals, and modules.

## Fifth Edition

The main issue was whether to reverse the decision made in the third edition: to begin with a concrete treatment before introducing a slicker abstract version. After widespread consultation—eleven expert reviewers, sometimes with differing opinions—I decided not to make any significant changes to the overall structure. However, I have tried, as far as possible, to follow their advice on almost everything else.

I have brought the notation up to date, the main changes being $L/K$ rather than $L{:}K$ for a field extension, and $\mathbb{F}_q$ rather than $\mathbb{GF}(q)$ for the finite field with $q$ elements. The references have been updated, known typos have been corrected, and all mathematical figures have been redrawn. Chapter 7 has been reorganised, and a minor gap has been filled. I have added several more examples to Chapter 13, which calculates the Galois groups for various polynomials and works through the details of the Galois correspondence. Some standard results that were previously omitted are now included, especially the Theorem of the Primitive Element and the construction of the algebraic closure of a field. I have streamlined the proof that cyclotomic polynomials are irreducible and included a proof that there is a unique simple group of order 60 using only methods that Galois might have known—that is, not using Sylow's Theorem.

I have also included a new final chapter discussing other directions in which Galois Theory has developed: the inverse Galois problem, differential Galois theory, and a (very) brief introduction to $p$-adic Galois representations. I have shortened some of the historical remarks, while retaining the historical viewpoint that most reviewers found appealing. I have also included extra material about key figures other than Galois: Omar Khayyam, Vandermonde, Ruffini, and Abel. I have added new exercises intended to encourage students to develop their own ideas rather than just learning pre-prepared material. Some of them require a computer algebra package—any of the standard ones will suffice.

Many people have offered helpful advice, which has improved this book in many ways. In particular, I thank Justin Archer, George Bergman, Owen Brison, Tom Brissenden, Ronnie Brown, Thiago Castilho de Mello, Ian Iscoe, Gerry Myerson, Frans Oort, Gianfranco Prini, Miles Reid, José Carlos Santos, David Tall, F. Javier Trigos, Martin Trimmel—and eleven anonymous reviewers.

*University of Warwick* Ian Stewart
*Coventry, UK, May 2022*

# *Historical Introduction*

The prehistory of Galois theory can be traced back at least to 1600 BC, where among the mudbrick buildings of exotic Babylon, some priest/mathematician worked out how to solve a quadratic equation, and he or one of his students inscribed it in cuneiform on a clay tablet. Many such tablets survive to this day, along with others ranging from tax accounts to observations of the motion of the planet Jupiter, Figure 1 (left).

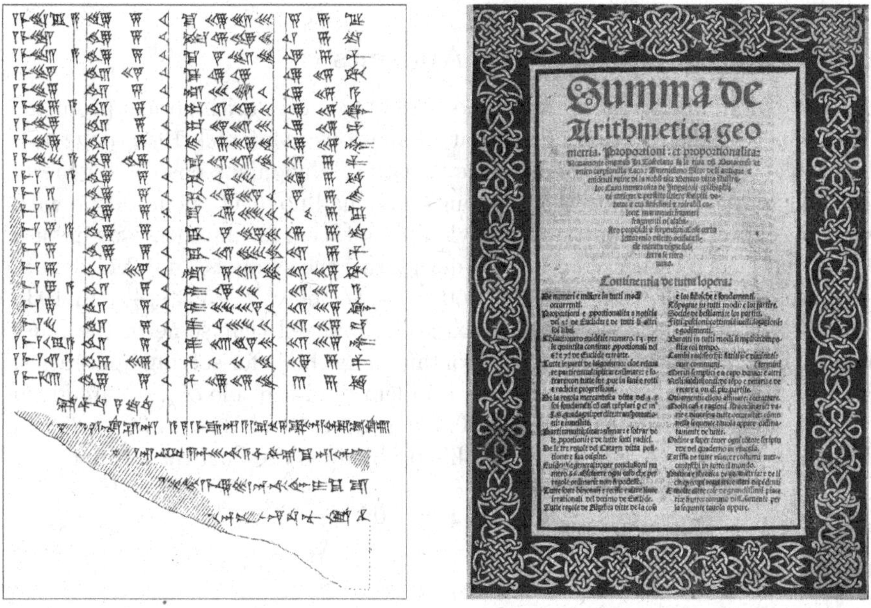


FIGURE 1: *Left*: A Babylonian clay tablet recording the motion of Jupiter. *Right*: A page from Pacioli's *Summa di Arithmetica*.

Adding to this rich historical brew, the problems that Galois theory solves, positively or negatively, have an intrinsic fascination—squaring the circle, duplicating the cube, trisecting the angle, constructing the regular 17-sided polygon, solving the quintic equation. If the hairs on your neck do not prickle at the very mention of these age-old puzzles, you need to have your mathematical sensitivities sharpened.

If those were not enough: Galois himself was a colourful and tragic figure—a youthful genius, one of the greatest mathematicians who has ever lived, but also a political revolutionary during one of the most turbulent periods in the history of France. At the age of 20, he was killed in a duel, ostensibly over a woman and quite possibly with a close friend, and his work was virtually lost to the world. Only some smart thinking by Joseph Liouville, probably encouraged by Galois's brother Alfred, rescued it. Galois's story is one of the most memorable among the lives of the great mathematicians, even when the more excessive exaggerations and myths are excised.

Our tale, therefore, has two heroes: a mathematical one—the humble polynomial equation—and a human one—the tragic genius. We take them in turn, along with a large supporting cast.

---

## Polynomial Equations: the Ancients

A Babylonian clay tablet from about 1600 BC poses arithmetical problems that reduce to the solution of quadratic equations (Midonick 1965, page 48). The tablet also provides firm evidence that the Babylonians possessed general methods for solving quadratics, although they had no algebraic notation with which to express their solutions. Babylonian notation for numbers was in base 60, so that (when transcribed into modern form) the symbols 7,4,0;3,11 denote the number $7 \times 60^2 + 4 \times 60 + 3 \times 60^{-1} + 11 \times 60^{-2} = 25440\frac{191}{3600}$. In 1930 the historian of science Otto Neugebauer announced that some of the most ancient Babylonian problem tablets contained methods for solving quadratics. For instance, one tablet contains this problem: find the side of a square given that the area minus the side is 14,30. Bearing in mind that 14,30 = 870 in decimal notation, we can formulate this problem as the quadratic equation

$$x^2 - x = 870$$

The Babylonian solution reads:

> Take half of 1, which is 0;30, and multiply 0;30 by 0;30, which is 0;15. Add this to 14,30 to get 14,30;15. This is the square of 29;30. Now add 0;30 to 29;30. The result is 30, the side of the square.

Although this description applies to one specific equation, it is laid out so that similar reasoning can be applied in greater generality, and this was clearly the Babylonian scribe's intention. The method is the familiar procedure of completing the square, which nowadays leads to the usual formula for the solution of a quadratic. See Joseph (2000) for more on Babylonian mathematics.

The ancient Greeks, in effect, solved quadratics by geometric constructions, but there is no sign of an algebraic formulation until at least AD 100 (Bourbaki

1969 page 92). The Greeks also possessed methods for solving cubic equations, which involved the points of intersection of conics. Again, algebraic solutions of the cubic were unknown, and in 1494 Luca Pacioli ended his *Summa di Arithmetica* (Figure 1, right) with the remark that (in his archaic notation) the solution of the equations $x^3 + mx = n$ and $x^3 + n = mx$ was as impossible at the existing state of knowledge as squaring the circle.

Pacioli was thinking of algebraic solutions. Geometric methods for solving cubics had already been found—notably by Omar Khayyam. To most of us, this name immediately calls to mind his long poem, the *Rubaiyat*, made familiar in its English translation by Edward Fitzgerald. Historians of mathematics know that Omar was a brilliant mathematician, who among other things used methods of Greek geometry to solve cubic equations. His full name was Ghiyath al-Din Abu'l-Fath Umar ibn Ibrahim Al-Nisaburi al-Khayyami. The word 'al-Khayyami' means 'tent-maker', probably his father Ibrahim's trade. Omar was born in 1047 and lived mainly at Naishapur in Persia—now Neyshabur, near Masshad in Iran, near the Turkmenistan border. When his childhood friend Nizam became Administrator of Affairs to the Sultan Alp Arslan, Omar was given a government salary to free his time for study. The Persian calendar was a solar calendar, so the date of New Year was subject to change. He was appointed to a commission to reform the calendar, applying his knowledge of mathematics and astronomy to calculate the date of New Year's Day in any given year.

Omar developed geometric solutions for all cubic equations, and explained them in his *Algebra* of 1079. Not recognising negative numbers, he arranged the equations so that all terms are positive. This convention led to a huge number of case distinctions, which nowadays we would consider to be essentially the same except for the signs of the numbers. Omar distinguished fourteen different types of cubic, depending on which (positive) terms appear on each side of the equation.

His geometric constructions for solutions necessarily went beyond the ruler and compass constructions of Euclidean geometry but not far beyond. They relied on conic sections, building on earlier Greek methods. Previous mathematicians, he noted, had discovered solutions of various cases, but these methods were all very special and each case was tackled by a different construction; no one before him had worked out the whole extent of possible cases, let alone found solutions to them. 'Me, on the contrary—I have never ceased to wish to make known, with exactitude, all of the possible cases, and to distinguish among each of the cases the possible and impossible ones.' By 'impossible' he meant 'having no positive solution'.

The next major advance would take more than four centuries.

## Renaissance Italy

The impasse on algebraic solutions of cubic equations was dramatically broken as new knowledge from the Middle and Far East swept across Europe and the Christian Church's stranglehold on intellectual innovation began to weaken. Italian Renaissance mathematicians at Bologna discovered that the solution of the cubic can be reduced to that of three basic types: $x^3 + px = q$, $x^3 = px + q$, and $x^3 + q = px$. Like Omar, they distinguished these cases because they did not recognise the existence of negative numbers. It is thought, on good authority (Bortolotti 1925), that Scipio del Ferro solved all three types; he certainly passed on his method for one type to a student, Antonio Fior. News of the solution leaked out, and others were encouraged to try their hand. Solutions for the cubic equation were rediscovered by Niccolo Fontana (nicknamed Tartaglia, 'The Stammerer'; Figure 2, left) in 1535.

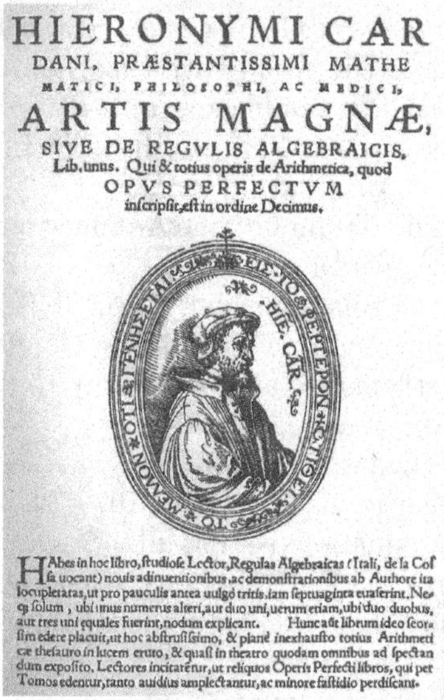
HIERONYMI CAR
DANI, PRÆSTANTISSIMI MATHE
MATICI, PHILOSOPHI, AC MEDICI,
ARTIS MAGNÆ,
SIVE DE REGVLIS ALGEBRAICIS,
Lib. unus. Qui & totius operis de Arithmetica, quod
OPVS PERFECTVM
inſcripſit, eſt in ordine Decimus.

HAbes in hoc libro, ſtudioſe Lector, Regulas Algebraicas (Itali, de la Coſſa uocant) nouis adinuentionibus, ac demonſtrationibus ab Authore ita locupletatas, ut pro pauculis antea uulgò tritis, iam ſeptuaginta euaſerint. Neq; ſolum, ubi unus numerus alteri, aut duo uni, uerum etiam, ubi duo duobus, aut tres uni equales fuerint, nodum explicant. Hunc aũt librum ideo ſeorſim edere placuit, ut hoc abſtruſiſsimo, & planè inexhauſto totius Arithmeticæ theſauro in lucem eruto, & quaſi in theatro quodam omnibus ad ſpectandum expoſito, Lectores incitarentur, ut reliquos Operis Perfecti libros, qui per Tomos edentur, tanto auidius amplectantur, ac minore faſtidio perdiſcant.

FIGURE 2: *Left*: Niccolo Fontana (Tartaglia), who discovered how to solve cubic equations. *Right*: Title page of Girolamo Cardano's *Ars Magna*.

One of the more charming customs of the period was the public mathematical contest, in which mathematicians engaged in mental duels using computational expertise as their weapons. Mathematics was a kind of performance art. Fontana demonstrated his methods in a public competition with Fior, but refused to reveal the details. Finally, he was persuaded to tell them to the physician Girolamo Cardano, having first sworn him to secrecy. Cardano, the 'gambling scholar', was a mixture of genius and rogue, and when his *Ars Magna* (Figure 2, right) appeared in 1545 it contained a complete discussion of Fontana's solution. Although Cardano claimed motives of the highest order (see the modern translation of his *The Book of My Life*, 1931), and fully acknowledged Fontana as the discoverer, Fontana was justifiably annoyed. In the ensuing wrangle, the history of the discovery became public knowledge. The *Ars Magna* also contained a method, due to Cardano's student Ludovico

Ferrari, for solving the quartic equation by reducing it to a cubic. All the formulas discovered had one striking property, which can be illustrated by Fontana's solution of $x^3 + px = q$, which in modern notation is:

$$x = \sqrt[3]{\frac{q}{2} + \sqrt{\frac{p^3}{27} + \frac{q^2}{4}}} + \sqrt[3]{\frac{q}{2} - \sqrt{\frac{p^3}{27} + \frac{q^2}{4}}}$$

This expression, usually called Cardano's formula because he was the first to publish it, is built up from the coefficients $p$ and $q$ by repeated addition, subtraction, multiplication, division, and—crucially—extraction of roots. Such expressions became known as *radicals*.

## Symmetric Functions

Since all equations of degree $\leq 4$ were now solved by radicals, it was natural to ask how to solve the quintic equation by radicals. Ehrenfried Walter von Tschirnhaus claimed a solution in 1683, but Gottfried Wilhelm Leibniz correctly pointed out that it was fallacious. Leonhard Euler failed to solve the quintic, but found new methods for the quartic, as did Etienne Bézout in 1765.

Further progress came with the work of Alexandre-Théophile Vandermonde, born in Paris in 1735. Initially he intended to become a musician, as his father wished, and he became an accomplished violinist. By 1770, however, he turned to mathematics. By then it was known that the coefficients of a polynomial are symmetric functions of its zeros. Vandermonde published a paper using symmetric functions to prove that the equation $x^n - 1 = 0$, associated with the regular $n$-gon, can be solved by radicals if $n = 11$. At the time, this was the smallest $n$ for which no such solution was known. We now know that this equation is soluble by radicals for any $n$—see Chapter 21. According to the great French analyst Augustin-Louis Cauchy, Vandermonde was the first mathematician to realise that symmetric functions can be applied to the solution of equations by radicals.

At much the same time the great French algebraist Joseph-Louis Lagrange took a major step forward in his magnum opus *Réflexions sur la Résolution Algébrique des Équations* of 1770–1771, when he unified the separate tricks used for the equations of degree $\leq 4$. He showed that they all depend on finding functions of the roots of the equation that are unchanged by certain permutations of those roots, and he showed that this approach *fails* when it is tried on the quintic. In modern terms, these permutations form subgroups of the symmetric group $\mathbb{S}_n$ of all permutations of the $n$ roots, and Lagrange was aware of this idea. His results did not prove that the quintic is insoluble by radicals, because other methods might succeed where this particular one did not. But the failure of such a general method was, to say the least, suspicious. It was starting to dawn on the mathematical community that the quintic might not be soluble by radicals.

## Paolo Ruffini

The first person to publish a largely correct impossibility proof was Paolo Ruffini, Figure 3 (left). He was born in Velantano in the Papal States, now Italy, in 1765. He studied literature, medicine, mathematics, and philosophy at the University of Modena, learning geometry from Luigi Fantini and calculus from Paolo Cassiani. In 1788 he was awarded a degree in philosophy, medicine, and surgery, and in 1789 he added a mathematics degree. Soon afterwards, he took over Fantini's professorship.

Meanwhile, in the wider world, Napoleon Bonaparte was embarking on a series of military campaigns. In 1796 he defeated the armies of Austria and Sardinia and captured Milan. Soon he had occupied Modena. In order to retain his professorship, Ruffini was required to swear allegiance to the French Republic, but he refused on religious grounds. Out of a job, he turned his attention to the quintic, and became convinced that there was no solution by radicals.

In 1799 Ruffini published a two-volume book, *Teoria Generale delle Equazioni*, whose 516 pages constituted an attempt to prove the insolubility of the quintic, writing: 'The algebraic solution of general equations of degree greater than four is always impossible. Behold a very important theorem which I believe I am able to assert (if I do not err): to present the proof of it is the main reason for publishing this volume. The immortal Lagrange, with his sublime reflections, has provided the basis of my proof.' He sent Lagrange a copy in 1801 but received no reply. A second copy a few months later was also ignored, as was a third attempt in 1802. After several years without any recognition, Ruffini decided that his proof was too complicated. In 1803 he simplified it, but on the whole, the new proof fared no better. There were a few exceptions. In 1821 Cauchy wrote to Ruffini: 'Your memoir on the general resolution of equations is a work which has always seemed to me worthy of the attention of mathematicians and which, in my judgment, proves completely the impossibility of solving algebraically equations of higher than the fourth degree.' But this praise came far too late.

In 1814, after the fall of Napoleon, Ruffini became rector of the University of Modena, simultaneously holding the chairs of applied mathematics, practical medicine, and clinical medicine. In 1817 there was a typhus epidemic. Ruffini continued to treat his patients until he caught the disease. He survived but never fully regained his health. In 1819 he gave up his chair of clinical medicine, and he died in 1822, barely a year after Cauchy had written to praise his work on the quintic.

Ruffini's impossibility proof was based on conditions that must be satisfied by any quintic whose roots can be expressed by radicals, and he realised that symmetric functions of the roots held the key to the problem. He showed that if the roots can be expressed by radicals, the set of 120 permutations of the roots must possess certain structural features. He then proved that these features are absent, implying that there is no such formula. Although his work sank into obscurity, it did change the prevailing wisdom. Before Ruffini, mathematicians generally believed that the quintic could be solved, but the solution would be very complicated. After Ruffini, there was a widespread feeling that the quintic is not soluble by radicals. As a side effect, most mathematicians lost interest: a solution would have been a big breakthrough, but an impossibility proof seemed to lead nowhere.

Ironically, it later emerged that Ruffini's work had gaps, which no one had spotted at the time. The most significant omission was the assumption that all of the radicals in a hypothetical formula must be based on the same rational functions of the roots (see Section 8.7). Nonetheless, Ruffini had made a big step forward, even though it was not appreciated at the time. The real breakthrough was his method.

FIGURE 3: *Left*: Paolo Ruffini. *Right*: Niels Henrik Abel.

## Niels Henrik Abel

As far as the mathematical community of the period was concerned, the quintic question was finally settled by Niels Henrik Abel in 1824, Figure 3 (right). Abel proved conclusively that the general quintic equation is insoluble by radicals. He was born in 1802 to Anne Marie (*née* Simonsen) and her husband Sören Abel, a pastor. He was a sickly child, and his mother had to spend a lot of time looking after him. Abel and his brother Hans Mathias attended the Cathedral School in Oslo in 1815. The mathematics teacher, Peter Bader, used to beat his pupils as a (misguided) incentive to learning—a common 'educational' technique at the time. In 1818, this tendency came to a head when he gave one pupil such a severe beating that the boy died, and Bader was replaced by Bernt Holmboe. Holmboe allowed his pupils to tackle interesting problems outside the usual syllabus, and he allowed Abel to borrow classic textbooks, among them some by Euler. 'From now on,' Holmboe later wrote, 'Abel devoted himself to mathematics with the most fervent eagerness and progressed in his science with the speed characteristic of a genius.' Shortly before finishing at school, Abel thought that he had solved the quintic equation. Ferdinand Degen, a prominent Danish mathematician, was unable to find any mistakes, but wisely suggested testing the calculations on specific examples. Abel quickly realised that something was wrong. He was disappointed, but relieved that he had not published an erroneous result.

His family life disintegrated: his father drank himself to death and his mother got drunk at the funeral and disgraced herself with a servant. His mathematics fared better. In 1821 he took the entrance examination to the University of Christiania (now Oslo), obtaining high grades in mathematics and terrible ones in everything else. Some of the professors, recognising his unusual talent, created a fellowship for him. Thus provided for, Abel devoted himself to mathematics, especially the quintic. He worked in other areas too, and in 1823 he studied elliptic integrals, where he made his best-known discoveries. He tried without success to prove Fermat's Last Theorem, though he did show that any counterexample must involve gigantic numbers. Meeting his future wife Christine Kemp, universally known as 'Crelly', at a ball, improved his confidence and gave his mathematics a boost. In 1823, unaware of Ruffini's work, he devised an impossibility proof for the quintic.

Impressed by this discovery, the University of Christiania granted money for a research visit to Paris, where Abel would meet some of the world's leading mathematicians. In preparation for the trip, he had his work on the theory of equations printed privately in French. Its title, translated, was 'Memoir on algebraic equations, wherein one proves the impossibility of solving the general equation of the fifth degree'. To reduce printing costs, Abel distilled the proof down to its essentials, and the printed version ran to just six pages. This was probably a mistake, because he had to leave out many crucial details, making the paper so compressed that most recipients probably found it incomprehensible. Later, Abel produced two longer versions of his proof. Having by this time heard of Ruffini, he wrote: 'The first to attempt a proof of the impossibility of an algebraic solution of the general equation was the mathematician Ruffini; but his memoir is so complicated that it is difficult to judge the correctness of the argument. It appears to me that his reasoning is not always satisfactory.' Like everyone else, Abel did not specify the precise gap in Ruffini's proof. Later it turned out that Abel's method was just what was needed to bridge that gap. Abel's proof, it turned out, was unnecessarily lengthy, and it also contained a minor error, which, fortunately, did not invalidate the method. In 1879 Leopold Kronecker published a simple, rigorous proof that tidied up Abel's ideas.

By Christmas 1828 Abel was very ill with a serious lung condition, probably tuberculosis. His friend August Crelle had persuaded the Department of Education to create a mathematical institute in Berlin, and wanted to appoint Abel. On the morning of his planned departure Abel was coughing violently and spitting blood. The family doctor prescribed bed rest, and his wife acted as nurse. Within a few weeks Abel could sit in a chair for short periods but was forbidden to do mathematics.

Adrien-Marie Legendre wrote to say how impressed he was with Abel's work on elliptic functions, and his reputation was growing fast, but it all came too late. Crelle sent what he thought would be good news. 'The Education Department has decided to call you to Berlin for an appointment... For your future you need no longer have any concern; you belong to us and are secure.' By then, however, Abel was too ill to travel. He died on the morning of 6 April 1829.

Abel had proved that the 'general' quintic is insoluble by radicals, but this left open other questions. In particular, special quintic equations might still be soluble. Some definitely are, for instance, $x^5 - 2 = 0$; less trivial examples are discussed in Section 1.4. In fact, for all Abel's methods could prove, *every* particular quintic equation might be soluble, with a special formula for each equation. So a new problem now arose: to find conditions that determine whether any particular equation can be solved by radicals. Abel was working on this question just before he died. Now, the baton would pass to another.

---

## Évariste Galois

In 1832 a young Frenchman, Évariste Galois, was killed in a duel. He had for some time sought recognition for his mathematical theories, submitting three memoirs to the Academy of Sciences in Paris. They were all rejected, and his work appeared to be lost to the mathematical world. Then, on 4 September 1843, Liouville addressed the Academy. He opened with these words:

> I hope to interest the Academy in announcing that among the papers of Évariste Galois I have found a solution, as precise as it is profound, of this beautiful problem: whether or not there exists a solution by radicals...

The most accessible account of Galois's troubled life, Bell (1965), is often unreliable, and distorts the events surrounding his death. The best sources I know are Rothman (1982a, 1982b). For Galois's papers and manuscripts, consult Bourgne and Azra (1962) for the French text and facsimiles of manuscripts and letters, and Neumann (2011) for English translation and parallel French text. Scans of the entire body of his work can be found on the web at

www.bibliotheque-institutdefrance.fr

FIGURE 4: Portrait of Évariste Galois drawn from memory by his brother Alfred, 1848.

Galois (Figure 4) was born at Bourg-la-Reine near Paris on 25 October 1811. His father Nicolas-Gabriel Galois was a Republican (Kollros 1949)—that is, he favoured the abolition of the monarchy. He was head of the village liberal party, and after Louis XVIII returned to the throne in 1814, Nicolas became town mayor. For the first twelve years of his life Galois was educated by his mother Adelaide-Marie (*née* Demante), a fluent reader of Latin, and his childhood appears to have been a happy one. In October 1823 he entered a preparatory school, the College de Louis-le-Grand. There he got his first taste of revolutionary politics: during his first term the students rebelled and refused to chant in chapel. He also witnessed heavy-handed retribution, for a hundred of the students were expelled for their disobedience.

Galois performed well during his first two years at school, obtaining first prize in Latin, but then boredom set in. As refuge from the tedium, he began to take a serious interest in mathematics. He came across a copy of Legendre's *Éléments de Géométrie*, a classic text which broke with the Euclidean tradition of school geometry. He read the original memoirs of Lagrange and Abel, and by the age of fifteen he was reading material intended only for professional mathematicians. But his classwork remained uninspired, and he seemed to have lost all interest in it. Even his own family considered him rather strange at that time.

Galois did make life difficult for himself. For a start, he was an untidy worker, as can be seen from some of his manuscripts (Bourgne and Azra 1962). Figure 5 shows two examples. Worse, he tended to work in his head, committing only the results of his deliberations to paper. Without adequate preparation, and a year early, he took the competitive examination for entrance to the École Polytechnique. A pass would have ensured a successful mathematical career, for the Polytechnique was the breeding-ground of French mathematics. Of course, he failed.

In 1828 Galois enrolled in an advanced mathematics course offered by Louis-Paul-Émile Richard, who recognised his ability and was very sympathetic towards him. The following year saw the publication of Galois's first research paper (Galois 1897) on continued fractions; though competent, it held no hint of genius. Meanwhile, Galois had been making fundamental discoveries in the theory of polynomial equations, and he submitted some of his results to the Academy of Sciences. The referee was Augustin-Louis Cauchy, who had already published work on the behaviour of functions under permutations of the variables, a central theme in Galois's theory.

As Rothman (1982a) says, 'We now encounter a major myth.' Many sources state that Cauchy lost the manuscript, or even deliberately threw it away. But René Taton (1971) found a letter written by Cauchy in the archives of the Academy. Dated 18 January 1830, it reads in part:

> I was supposed to present today to the Academy first a report on the work of the young Galoi [spelling was not consistent in those days] and second a memoir on the analytic determination of primitive roots [by Cauchy]... Am indisposed at home. I regret not being able to attend today's session, and I would like you to schedule me for the following session for the two indicated subjects.

So Cauchy still possessed the manuscript six months after Galois had submitted it, and he had found the work sufficiently interesting to want to draw it to the Academy's attention. However, at the next session of the Academy, on 25 January, Cauchy presented only his own paper. What had happened to the paper by Galois? Taton suggests that Cauchy was actually very impressed by Galois's research, because he advised Galois to prepare a new (no doubt improved) version, and to submit it for the Grand Prize in Mathematics, which he did in February.

The same year held two major disasters. On 2 July 1829 Galois's father committed suicide after a bitter political dispute in which the village priest forged Nicolas's signature on malicious epigrams aimed at his own relatives. A few days later Galois again sat for entrance to the Polytechnique—his final chance—but again he failed. Faced with the prospect of the École Normale, then called the École Preparatoire, which at that time was far less prestigious than the Polytechnique, Galois belatedly prepared for his final examinations.

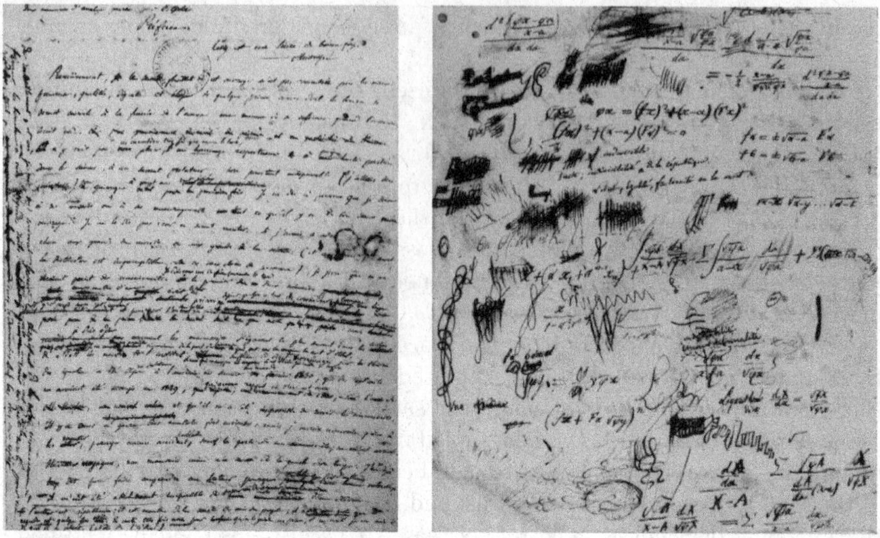

FIGURE 5: *Left*: First page of preface written by Galois when in jail. *Right*: Doodles left on the table before departing for the fatal duel. *'Une femme'*, with the second word scribbled out, can be seen near the lower-left corner.

His performance in mathematics and physics was excellent, in literature less so; he obtained both the Bachelor of Science and Bachelor of Letters on 29 December 1829.

In February 1830 Galois presented a new version of his research to the Academy of Sciences in competition for the Grand Prize. The manuscript reached the secretary Joseph Fourier, who took it home for perusal, but he died before reading it and the manuscript could not be found among his papers. The loss was probably an accident, but according to Dupuy (1896) Galois was convinced that the repeated losses of his papers were not just bad luck. He saw them as the inevitable effect of a society in which genius was condemned to an eternal denial of justice in favour of mediocrity, and he blamed the politically oppressive Bourbon regime. He may well have had a point, accident or not.

At that time, France was in political turmoil. King Charles X succeeded Louis XVIII in 1824. In 1827 the liberal opposition made electoral gains; in 1830 more elections were held, giving the opposition a majority. Charles, faced with abdication, attempted a *coup d'état*. On 25 July he issued his notorious *Ordonnances* suppressing the freedom of the press. The populace was in no mood to tolerate such repression, and revolted. The uprising lasted three days, after which as a compromise the Duke of Orléans, Louis-Philippe, was made king. During these three days, while the students of the Polytechnique were making history in the streets, Galois and his fellow students were locked in by Guigniault, Director of the École Normale. Galois was incensed, and subsequently wrote a blistering attack on the Director in the *Gazette des Écoles*, signing the letter with his full name. The editor removed the signature, the Director was not amused, and Galois was expelled because of his 'anonymous' letter (Dalmas 1956).

Galois promptly joined the Artillery of the National Guard, a branch of the militia composed almost entirely of Republicans. On 21 December 1830 the Artillery was stationed near the Louvre, awaiting the verdict of the trial of four ex-ministers. The public wanted these functionaries executed, and the Artillery was planning to rebel if they received only life sentences. Just before the verdict was announced, the Louvre was surrounded by the full National Guard, plus other troops who were far more trustworthy. When the verdict of a jail sentence was heralded by a cannon shot, the revolt failed to materialise. On 31 December, the king abolished the Artillery of the National Guard on the grounds that it constituted a serious security threat.

Galois was now faced with the urgent problem of making a living. On 13 January 1831 he set up as a private teacher of mathematics, offering a course in advanced algebra. Forty students enrolled, but the class soon petered out, probably because Galois was too involved in politics. On 17 January he submitted a third version of his memoir to the Academy: *On the Conditions of Solubility of Equations by Radicals*. After two months he had heard no word. He wrote to the President of the Academy, but received no reply.

During the spring of 1831, Galois's behaviour became more and more extreme, verging on the paranoid. On April 18 Sophie Germain, one of the few women mathematicians of the time, who studied with Carl Friedrich Gauss, wrote to Guillaume Libri about Galois's misfortunes: 'They say he will go completely mad, and I fear this is true.' See Henry (1879). Also in April, 19 members of the Artillery of the National Guard, arrested after the events at the Louvre, were put on trial charged with attempting to overthrow the government. The jury acquitted them, and on 9 May a celebratory banquet was held. About 200 Republicans were present, all extremely hostile to the government of Louis-Philippe. The proceedings became more and more riotous, and Galois was seen with a glass in one hand and a dagger in the other. His companions allegedly interpreted this as a threat to the king's life, applauded mightily, and ended up dancing and shouting in the street.

Next day, Galois was arrested. At his subsequent trial, he admitted everything, but claimed that the toast proposed was actually 'To Louis-Philippe, *if he turns traitor*,' and that the uproar had drowned the last phrase. The jury acquitted him and he was freed on 15 June. On 4 July he heard the fate of his memoir. Poisson declared it 'incomprehensible'. The report (reprinted in full in Taton, 1947) ended as follows:

> We have made every effort to understand Galois's proof. His reasoning is not sufficiently clear, sufficiently developed, for us to judge its correctness, and we can give no idea of it in this report. The author announces that the proposition which is the special object of this memoir is part of a general theory susceptible of many applications. Perhaps it will transpire that the different parts

> of a theory are mutually clarifying, are easier to grasp together rather than in isolation. We would then suggest that the author should publish the whole of his work in order to form a definitive opinion. But in the state which the part he has submitted to the Academy now is, we cannot propose to give it approval.

The report may well have been entirely fair. Tignol (1988) points out that Galois's entry 'did not yield any workable criterion to determine whether an equation is solvable by radicals.' The referees' report was explicit:

> [The memoir] does not contain, as [its] title promised, the condition of solubility of equations by radicals; indeed, assuming as true M. Galois's proposition, one could not derive from it any good way of deciding whether a given equation of prime degree is soluble or not by radicals, since one would first have to verify whether this equation is irreducible and next whether any of its roots can be expressed as a rational function of two others.

The final sentence here refers to a beautiful criterion for solubility by radicals of equations of prime degree that was the climax of Galois's memoir. It is indeed unclear how it can be applied to any specific equation. What the referees wanted was some kind of condition on the *coefficients* that determined solubility; what Galois gave them was a condition on the *roots*. No simple criterion based on the coefficients has ever been found, nor is one remotely likely, but that was unclear at the time. See Chapter 25 for further discussion.

On 14 July, Bastille Day, Galois and his friend Ernest Duchâtelet were at the head of a Republican demonstration. Galois was wearing the uniform of the disbanded Artillery and carrying a knife, several pistols, and a loaded rifle. It was illegal to wear the uniform, and even more so to be armed. Both men were arrested, and Galois was charged with the lesser offence of illegally wearing a uniform. They were sent to the jail at Sainte-Pélagie to await trial. On 23 October Galois was tried and convicted, and his appeal was turned down on 3 December. By this time he had spent more than four months in jail. Now he was sentenced to six months there. He worked for a while on his mathematics (Figure 5 left); then in the cholera epidemic of 1832 he was transferred to a hospital. Soon he was put on parole.

Along with his freedom he experienced his first and only love affair, with a certain Mlle. 'Stéphanie D.' Until recently, the lady's surname was unknown, adding to the romantic image of the *femme fatale*. The full name appears in one of Galois's manuscripts, but the surname has been deliberately scribbled over, no doubt by Galois. Some forensic work by Carlos Infantozzi (1968), deciphering the name that Galois had all but obliterated, led to the suggestion that the lady was Stéphanie-Felicie Poterin du Motel, the entirely respectable daughter of Jean-Louis Auguste Poterin du Motel. Jean-Louis was resident physician at the Sieur Faultrier, where Galois spent the last few months of his life. The identification is plausible, but it relies on extracting a sensible name from beneath Galois's scribbles, so naturally there is a some controversy about it.

In general, much mystery surrounds this interlude, which has a crucial bearing on subsequent events. Apparently Galois was rejected and took it very badly. Not long afterwards, he was challenged to a duel, ostensibly because of his unwanted advances towards the young lady. Again, the circumstances are veiled in mystery. One school of thought (Bell, 1965; Kollros, 1949) asserts that Galois's infatuation with Mlle. du Motel was used by his political opponents, who found it the perfect excuse to eliminate their enemy on a trumped-up 'affair of honour'. There are even suggestions that Galois was in effect assassinated by a police spy. But in his *Mémoires*, Alexandre Dumas says that Galois was killed by Pescheux D'Herbinville, a fellow Republican, see Dumas (1967). Dalmas (1956) cites evidence from the police report, suggesting that the other duellist was one of Galois's revolutionary comrades, and the duel was exactly what it appeared to be. This theory is largely borne out by Galois's own words on the matter (Bourgne and Azra, 1962):

> I beg patriots and my friends not to reproach me for dying otherwise than for my country. I die the victim of an infamous coquette. It is in a miserable brawl that my life is extinguished. Oh! why die for so trivial a thing, for something so despicable! ... Pardon for those who have killed me, they are of good faith.

Figure 5 right shows a doodle by Galois with the words 'Une femme' partially crossed out. It does appear that Stéphanie was at least a proximate cause of the duel, but very little else is clear.

On 29 May, the eve of the duel, Galois wrote a famous letter to his friend Auguste Chevalier, outlining his mathematical discoveries, sketching the connection between groups and polynomial equations and stating that an equation is soluble by radicals provided its group is soluble. But he also mentioned other ideas about elliptic functions and the integration of algebraic functions, and other things too cryptic to be identifiable.

The duel was with pistols. The post-mortem report (Dupuy 1896) states that they were fired at 25 paces, but the truth may have been even nastier. Dalmas reprints an article from the 4 June 1832 issue of *Le Precursor*, which reports:

> Paris, 1 June—A deplorable duel yesterday has deprived the exact sciences of a young man who gave the highest expectations, but whose celebrated precocity was lately overshadowed by his political activities. The young Évariste Galois ... was fighting with one of his old friends ... It is said that love was the cause of the combat ... At point-blank range they were each armed with a pistol and fired. Only one pistol was charged. Galois was pierced through and through by a ball from his opponent; he was taken to the hospital Cochin where he died in about two hours. His age was 22. L.D., his adversary, is a bit younger.

Who was 'L.D.'? Does the initial 'D' refer to d'Herbinville? Perhaps. 'D' is acceptable because of the variable spelling of the period; the 'L' may have been a mistake. The article is unreliable on details: it gets the date of the duel wrong, and also the day Galois died and his age. So the initial might also be wrong. Rothman has a more convincing theory. The person who best fits the description is not d'Herbinville, but Duchâtelet. Bourgne and Azra (1962) give his Christian name as 'Ernest', but that might be wrong, or again the 'L' may be wrong. To quote Rothman: 'we arrive at a very consistent and believable picture of two old friends falling in love with the same girl and deciding the outcome by a gruesome version of Russian roulette.'

This theory is also consistent with a final horrific twist to the tale. Galois was hit in the stomach, a particularly serious wound that was almost always fatal. If indeed the duel was at point-blank range, this is no great surprise. If at 25 paces, he was unlucky. He did not die two hours later, as *Le Precursor* says, but a day later on 31 May, of peritonitis; he refused the office of a priest. On 2 June 1832 he was buried in the common ditch at the cemetery of Montparnasse. His letter to Chevalier ended with these words (Figure 6 right):

> Ask Jacobi or Gauss publicly to give their opinion, not as to the truth, but as to the importance of these theorems. Later there will be, I hope, some people who will find it to their advantage to decipher all this mess ...

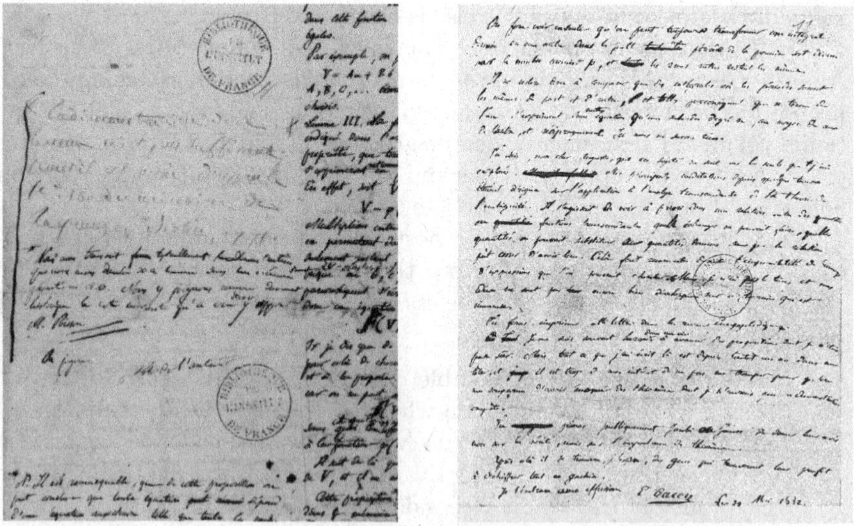

FIGURE 6: *Left*: Marginal comment by Poisson. *Right*: The final page written by Galois before the duel. 'To decipher all this mess' (*déchiffrer tout ce gâchis*, is the next to last line).

# Chapter 1

# Classical Algebra

In the first part of this book, Chapters 1–15, we present a (fairly) modern version of Galois's ideas in the same setting that he used, namely, the complex numbers. Later, from Chapter 16 onwards, we generalise the setting, but the complex numbers are familiar and concrete. By initially restricting ourselves to complex numbers, we can focus on the main ideas that Galois introduced, with the additional advantage that the material is more accessible. Moreover, we appreciate the abstract viewpoint when it finally makes its appearance, and we understand where it comes from. However, the concrete setting is sometimes clumsy compared to the elegance of an axiomatic approach, and we have to check which proofs in the complex case go through unchanged to more general fields and which do not. It turns out that most proofs remain unchanged, but a few key results require extra conditions before they apply in general. We point out these exceptions, and the extra conditions involved, as they arise.

We assume familiarity with the basic theory of real and complex numbers, but to set the scene we recall some of the concepts involved. We begin with a brief discussion of complex numbers and introduce two important ideas. Both relate to subsets of the complex numbers that are closed under the usual arithmetic operations. A subring of the complex numbers is a subset that contains 1 and is closed under addition, subtraction, and multiplication; a subfield is a subring that is also closed under division by any non-zero element. Both concepts were formalised by Richard Dedekind in 1871, though the ideas go back to Peter Gustav Lejeune-Dirichlet and Kronecker in the 1850s.

We then show that the historical sequence of extensions of the number system, from natural numbers to integers to rationals to reals to complex numbers, can with hindsight be interpreted as a quest to make more and more equations have solutions. We are thus led to the concept of a polynomial, which is central to Galois theory because it determines the type of equation that we wish to solve. And we appreciate that the existence of a solution depends on the kind of number that is permitted.

Throughout, we use the standard notation $\mathbb{N}, \mathbb{Z}, \mathbb{Q}, \mathbb{R}, \mathbb{C}$ for the natural numbers, integers, rationals, real numbers, and complex numbers. These systems sit inside each other:

$$\mathbb{N} \subseteq \mathbb{Z} \subseteq \mathbb{Q} \subseteq \mathbb{R} \subseteq \mathbb{C}$$

DOI: 10.1201/9781003213949-1

and each $\subseteq$ symbol hints at a lengthy historical process in which 'new numbers' were proposed for mathematical reasons—usually against serious resistance on the grounds that they might well be new, but they were not numbers and therefore did not exist.

---

## 1.1 Complex Numbers

A *complex number* has the form

$$z = x + \mathrm{i}y$$

where $x, y$ are real numbers and $\mathrm{i}^2 = -1$. Therefore $\mathrm{i} = \sqrt{-1}$, in some sense.

Throughout, we use the Roman letter i for $\sqrt{-1}$. This frees up italic $i$ for other uses.

The easiest way to define what we mean by $\sqrt{-1}$ is to consider $\mathbb{C}$ as the set $\mathbb{R}^2$ of all pairs of real numbers $(x, y)$, with algebraic operations

$$\begin{aligned}(x_1, y_1) + (x_2, y_2) &= (x_1 + x_2, y_1 + y_2) \\ (x_1, y_1)(x_2, y_2) &= (x_1x_2 - y_1y_2, x_1y_2 + x_2y_1)\end{aligned} \tag{1.1}$$

Then we identify $(x, 0)$ with the real number $x$ to arrange that $\mathbb{R} \subseteq \mathbb{C}$, and define $\mathrm{i} = (0, 1)$. In consequence, $(x, y)$ becomes identified with $x + \mathrm{i}y$. The formulas (1.1) imply that $\mathrm{i}^2 = (0, 1)(0, 1) = (-1, 0)$, which is identified with the real number –1, so i is a 'square root of minus one'. Observe that $(0, 1)$ is not of the form $(x, 0)$, so i is not real, which is as it should be, since $-1$ has no real square root.

This approach seems first to have been published by the Irish mathematician William Rowan Hamilton in 1837, but in that year, Gauss wrote to the geometer Wolfgang Bolyai that the same idea had occurred to him in 1831. This was probably true, because Gauss usually worked things out before anybody else did, but he set himself such high standards for publication that many of his more important ideas never saw print under his name. Moreover, Gauss was somewhat conservative and shied away from anything potentially controversial.

Once we see that complex numbers are just pairs of real numbers, the previously mysterious status of the 'imaginary' number $\sqrt{-1}$ becomes much more prosaic. In fact, to the modern eye, it is the 'real' numbers that are mysterious, because their rigorous definition involves analytic ideas such as sequences and convergence, which lead into deep philosophical waters and axiomatic set theory. In contrast, the step from $\mathbb{R}$ to $\mathbb{R}^2$ is essentially trivial—except for the peculiarities of human psychology.

## 1.2 Subfields and Subrings of the Complex Numbers

Abstract algebra courses usually introduce (at least) three basic types of algebraic structure, defined by systems of axioms: groups, rings, and fields. Linear algebra adds a fourth, vector spaces, which are so important that they usually warrant a separate lecture course. For the first half of this book, we steer clear of abstract rings and fields, but we do assume the basics of finite group theory and linear algebra.

Recall that a *group* is a set $G$ equipped with an operation of 'multiplication' written $(g, h) \mapsto gh$. If $g, h \in G$, then $gh \in G$. The associative law $(gh)k = g(hk)$ holds for all $g, h, k \in G$. There is an identity $1 \in G$ such that $1g = g = g1$ for all $g \in G$. Finally, every $g \in G$ has an inverse $g^{-1} \in G$ such that $gg^{-1} = 1 = g^{-1}g$. The classic example here is the *symmetric group* $\mathbb{S}_n$, consisting of all permutations of the set $\{1, 2, \ldots, n\}$ under the operation of composition. We assume familiarity with these axioms, and with subgroups, isomorphisms, homomorphisms, normal subgroups, and quotient groups; see Humphreys (1996), Neumann, Stoy and Thompson (1994), or any other introductory group theory text.

Rings are sets equipped with operations of addition, subtraction, and multiplication; fields also have a notion of division. The formal definitions were supplied by Heinrich Weber in 1893. The axioms specify the formal properties assumed for these operations—for example, the commutative law $ab = ba$ for multiplication.

In the first part of this book, we do not assume familiarity with abstract rings and fields. Instead, we restrict attention to subrings and subfields of $\mathbb{C}$, or polynomials and rational functions over such subrings and subfields. Informally, we assume that the terms 'polynomial' and 'rational expression' (or 'rational function') are familiar, at least over $\mathbb{C}$, although for safety's sake we define them when the discussion becomes more formal, and redefine them when we make the whole theory more abstract in the second part of the book. There were no formal concepts of 'ring' or 'field' in Galois's day, and linear algebra was in a rudimentary state. He had to invent groups for himself. So we are still permitting ourselves a more extensive conceptual toolkit than his.

**Definition 1.1.** A *subring* of $\mathbb{C}$ is a subset $R \subseteq \mathbb{C}$ such that $1 \in R$, and if $x, y \in R$ then $x + y, -x$, and $xy \in R$.

(The condition that $1 \in R$ is required here because we use 'ring' as an abbreviation for what is often called a 'ring-with-1' or 'unital ring'.)

A *subfield* of $\mathbb{C}$ is a subring $K \subseteq \mathbb{C}$ with the additional property that if $x \in K$ and $x \neq 0$ then $x^{-1} \in K$.

Here $x^{-1} = 1/x$ is the reciprocal. As usual, we often write $x/y$ for $xy^{-1}$.

It follows immediately that every subring of $\mathbb{C}$ contains $1 + (-1) = 0$ and is closed under the algebraic operations of addition, subtraction, and multiplication. A subfield of $\mathbb{C}$ has all of these properties and is also closed under division by any nonzero element. Because $R$ and $K$ in Definition 1.1 are subsets of $\mathbb{C}$, they inherit the usual rules for algebraic manipulation.

**Examples 1.2.** (1) The set of all $a + b\text{i}$, for $a, b \in \mathbb{Z}$, is a subring of $\mathbb{C}$, but not a subfield.

Since this is the first example we outline a proof. Let

$$R = \{a + b\text{i} : a, b \in \mathbb{Z}\}$$

Since $1 = 1 + 0\text{i}$, we have $1 \in R$. Let $x = a + b\text{i}, y = c + d\text{i} \in R$. Then

$$\begin{aligned} x + y &= (a + c) + (b + d)\text{i} \in R \\ -x &= -a - b\text{i} \in R \\ xy &= (ac - bd) + (ad + bc)\text{i} \in R \end{aligned}$$

and the conditions for a subring are valid. However, $2 \in R$ but its reciprocal $2^{-1} = \frac{1}{2} \notin R$, so $R$ is not a subfield.

(2) The set of all $a + b\text{i}$, for $a, b \in \mathbb{Q}$, is a subfield of $\mathbb{C}$.

Let

$$K = \{a + b\text{i} : a, b \in \mathbb{Q}\}$$

The proof is just like case (1), but now a straightforward calculation shows that

$$(a + b\text{i})^{-1} = \frac{a}{a^2 + b^2} - \frac{b}{a^2 + b^2}\text{i} \in K$$

so $K$ is a subfield.

(3) The set of all polynomials in $\pi$, with integer coefficients, is a subring of $\mathbb{C}$, but not a subfield.

(4) The set of all polynomials in $\pi$, with rational coefficients, is a subring of $\mathbb{C}$. We can appeal to a result proved in Chapter 24 to show that this set is not a subfield. Suppose that $\pi^{-1} = f(\pi)$ where $f$ is a polynomial over $\mathbb{Q}$. Then $\pi f(\pi) - 1 = 0$, so $\pi$ satisfies a nontrivial polynomial equation with rational coefficients, contrary to Theorem 24.5 of Chapter 24.

(5) The set of all rational expressions in $\pi$ with rational coefficients (that is, fractions $p(\pi)/q(\pi)$ where $p, q$ are polynomials over $\mathbb{Q}$ and $q(\pi) \neq 0$) is a subfield of $\mathbb{C}$.

(6) The set $2\mathbb{Z}$ of all even integers is not a subring of $\mathbb{C}$, because (by our convention) it does not contain 1.

(7) The set of all $a + b\sqrt[3]{2}$, for $a, b \in \mathbb{Q}$, is not a subring of $\mathbb{C}$ because it is not closed under multiplication. However, it *is* closed under addition and subtraction.

**Definition 1.3.** Suppose that $K$ and $L$ are subfields of $\mathbb{C}$. An *isomorphism* between $K$ and $L$ is a map $\phi : K \to L$ that is one-to-one and onto and satisfies the conditions

$$\phi(x + y) = \phi(x) + \phi(y) \qquad \phi(xy) = \phi(x)\phi(y) \tag{1.2}$$

for all $x, y \in K$.

If $\phi$ satisfies (1.2) and is one-to-one but not necessarily onto, it is a *monomorphism*. An isomorphism of $K$ with itself is called an *automorphism* of $K$.

**Proposition 1.4.** *If $\phi : K \to L$ is a monomorphism, then:*

$$\begin{aligned}\phi(0) &= 0\\ \phi(1) &= 1\\ \phi(-x) &= -\phi(x)\\ \phi(x^{-1}) &= (\phi(x))^{-1}\end{aligned}$$

*Proof.* Since $0 + 0 = 0$, we have $\phi(0) + \phi(0) = \phi(0)$. Therefore $\phi(0) = 0$.

Since $1.1 = 1$, we have $\phi(1)\phi(1) = \phi(1)$. Therefore $\phi(1) = 0$ or $\phi(1) = 1$. But $0 \neq 1$ and $\phi$ is a monomorphism, so $\phi(1) = 1$.

Since $x + (-x) = 0$ for all $x \in K$, we have $\phi(x) + \phi(-x) = \phi(0) = 0$. Therefore $\phi(-x) = -\phi(x)$.

Since $x.x^{-1} = 1$ for all $x \in K$, we have $\phi(x).\phi(x^{-1}) = \phi(1) = 1$. Therefore $\phi(x^{-1}) = (\phi(x))^{-1}$. □

It is easy to see that a necessary and sufficient condition for $\phi$ satisfying (1.2) to be a monomorphism is that $\phi(x) = 0$ implies $x = 0$. Another such condition is that $\phi$ does not map the whole of $K$ to 0.

Throughout the book, we make extensive use of the following terminology:

**Definition 1.5.** A *primitive* $n$th root of unity is an $n$th root of 1 that is not an $m$th root of 1 for any proper divisor $m$ of $n$.

For example, i is a primitive fourth root of unity, and so is $-$i. Since $(-1)^4 = 1$, the number $-1$ is also a fourth root of unity, but it is not a primitive fourth root of unity because $(-1)^2 = 1$.

Over $\mathbb{C}$ the standard choice for a primitive $n$th root of unity is

$$\zeta_n = \mathrm{e}^{2\pi \mathrm{i}/n} = \cos\frac{2\pi}{n} + \mathrm{i}\sin\frac{2\pi}{n}$$

We omit the subscript $n$ when this causes no ambiguity.

The next result is standard, but we include proof for completeness.

**Proposition 1.6.** *Let $\zeta = \mathrm{e}^{2\pi\mathrm{i}/n}$. Then $\zeta^k = \mathrm{e}^{2k\pi\mathrm{i}/n}$ is a primitive $n$th root of unity if and only if $k$ is prime to $n$.*

*Proof.* We prove the equivalent statement: $\zeta^k = \mathrm{e}^{2k\pi\mathrm{i}/n}$ is not a primitive $n$th root of unity if and only if $k$ is not prime to $n$.

Suppose that $\zeta^k$ is not a primitive $n$th root of unity. Then $(\zeta^k)^m = 1$, where $m$ is a proper divisor of $n$. That is, $n = mr$ where $r > 1$. Therefore $\zeta^{km} = 1$, so $mr = n$ divides $km$. This implies that $r|k$, and since also $r|n$ we have $(n, k) \geq r > 1$, so $k$ is not prime to $n$.

Conversely, suppose that $k$ is not prime to $n$, and let $r > 1$ be a common divisor. Then $r|k$ and $n = mr$ where $m < n$. Now $km$ is divisible by $mr = n$, so $(\zeta^k)^m = 1$. That is, $\zeta^k$ is not a primitive $n$th root of unity. □

**Examples 1.7.** (1) Complex conjugation $x + iy \mapsto x - iy$ is an automorphism of $\mathbb{C}$. Indeed, if we denote this map by $\alpha$, then:

$$\begin{aligned}\alpha((x+iy)+(u+iv)) &= \alpha((x+u)+i(y+v))\\ &= (x+u)-i(y+v)\\ &= (x-iy)+(u-iv)\\ &= \alpha(x+iy)+\alpha(u+iv)\\ \alpha((x+iy)(u+iv)) &= \alpha((xu-yv)+i(xv+yu))\\ &= xu-yv-i(xv+yu)\\ &= (x-iy)(u-iv)\\ &= \alpha(x+iy)\alpha(u+iv)\end{aligned}$$

(2) Let $K$ be the set of complex numbers of the form $p+q\sqrt{2}$, where $p, q \in \mathbb{Q}$. This is a subfield of $\mathbb{C}$ because

$$(p+q\sqrt{2})(p-q\sqrt{2}) = p^2 - 2q^2$$

so

$$(p+q\sqrt{2})^{-1} = \frac{p}{p^2-2q^2} - \frac{q}{p^2-2q^2}\sqrt{2}$$

if $p$ and $q$ are non-zero. (Observe that $p^2 - 2q^2 = 0$ implies $p = q = 0$, since $\sqrt{2}$ is irrational.) The map $p + q\sqrt{2} \mapsto p - q\sqrt{2}$ is an automorphism of $K$.
(3) Let $\alpha = \sqrt[3]{2} \in \mathbb{R}$, and let

$$\omega = -\frac{1}{2} + i\frac{\sqrt{3}}{2}$$

be a primitive cube root of unity in $\mathbb{C}$. The set of all numbers $p + q\alpha + r\alpha^2$, for $p, q, r \in \mathbb{Q}$, is a subfield of $\mathbb{C}$, see Exercise 1.5. The map

$$p + q\alpha + r\alpha^2 \mapsto p + q\omega\alpha + r\omega^2\alpha^2$$

is a monomorphism onto its image, but not an automorphism, Exercise 1.6.

**Remark 1.8.** Working over the complex numbers has the advantage of making the *numbers* more concrete. For example, in the concrete approach, the field $\mathbb{Q}(i)$ is defined to be the set of all complex numbers $p + qi$ for $p, q \in \mathbb{Q}$. Compare this to the abstract approach, in which $\mathbb{Q}(i)$ is defined to be the quotient of the ring of all complex polynomials $p(t)$ in an indeterminate $t$ by the ideal consisting of all multiples of $t^2 + 1$, that is, all polynomials of the form $(t^2 + 1)p(t)$. (See Chapter 16 for the abstract definitions of 'ring' and 'ideal'.)

However, the natural geometry of $\mathbb{C}$ as a plane is usually not very helpful. For example, $\mathbb{Q}(i)$ is a dense subset of $\mathbb{C}$. The smallest subfield containing the $n$th roots of unity is also dense in $\mathbb{C}$ when $n = 5$ or $n \geq 7$, but all of these fields are different. Here the geometry does little to aid intuition.

However, the geometry can be helpful for some purposes. For example, the complex $n$th roots of unity are the vertices of a regular $n$-gon inscribed in the unit circle. The integer combinations $a+bi$ for $a, b \in \mathbb{Z}$ form a square lattice in $\mathbb{C}$. So the moral of this tale is that we can use the geometry as a thinking aid whenever it *is* helpful, but ignore it when it is not. Usually it quickly becomes obvious which is the case.

---

## 1.3 Solving Equations

We tend to think of the great problems of mathematics as being things like Fermat's Last Theorem, the Poincaré Conjecture, or the Riemann Hypothesis: problems of central importance that remained unsolved for decades or even centuries. But the really big problems of mathematics are more general. A problem that runs like an ancient river through the middle of the territory we are going to explore is: *Find out how to solve equations.* Or, as often as not, prove that it cannot be done with specified methods. What sort of equations? There are many kinds: Diophantine equations, differential equations (ordinary, partial, or delay), difference equations, integral equations, operator equations ... For Galois, it was polynomial equations. We work up to those in easy stages.

Historically, new kinds of number like $\sqrt{2}$ or i were introduced because the old ones were inadequate for solving some important problems. Most such problems can be formulated using equations, though it must be said that this is a modern interpretation, and the ancient mathematicians did not think in quite those terms.

For example, the step from $\mathbb{N}$ to $\mathbb{Z}$ is needed because although some equations, such as

$$t + 2 = 7$$

can be solved for $t \in \mathbb{N}$, others, such as

$$t + 7 = 2$$

cannot. However, such equations *can* be solved in $\mathbb{Z}$, where $t = -5$ makes sense. (The symbol $x$ is more traditional than $t$ here, but it is convenient to standardise on $t$ for the rest of the book, so we may as well start straight away.)

Similarly, the step from $\mathbb{Z}$ to $\mathbb{Q}$ (historically, it was initially from $\mathbb{N}$ to $\mathbb{Q}^+$, the positive rationals) makes it possible to solve the equation

$$2t = 7$$

because $t = \frac{7}{2}$ makes sense in $\mathbb{Q}$.

In general, an equation of the form

$$at + b = 0$$

where $a, b$ are specific numbers and $t$ is an unknown number, or 'variable', is called a *linear equation*. In a subfield of $\mathbb{C}$, any linear equation with $a \neq 0$ can be solved, with the unique solution $t = -b/a$.

The step from $\mathbb{Q}$ to $\mathbb{R}$ is related to a different kind of equation:

$$t^2 = 2$$

As the ancient Greeks understood (in their own geometric manner; they did not possess algebraic notation and thought in a very different way from modern mathematicians), the 'solution' $t = \sqrt{2}$ is an *irrational* number — it is not in $\mathbb{Q}$. (See Exercise 1.2 for a proof. An entirely different proof of a more general theorem is outlined in Exercise 1.3.)

Similarly, the step from $\mathbb{R}$ to $\mathbb{C}$ centres on the equation

$$t^2 = -1$$

which has no real solutions since the square of any real number is positive or zero. The remedy, which took at least 400 years to gain wide acceptance, is to augment the real numbers with new number i, which leads naturally to the complex numbers.

More generally, equations of the form

$$at^2 + bt + c = 0$$

are called *quadratic equations*. The classic formula for their solutions (there can be 0, 1, or 2 of these) is of course

$$t = \frac{-b \pm \sqrt{b^2 - 4ac}}{2a}$$

and this gives all the solutions $t$ provided the formula makes sense. For a start, we need $a \neq 0$. (If $a = 0$, then the equation is actually linear, so this restriction is not a problem.) Over the real numbers, the formula makes sense if $b^2 - 4ac \geq 0$, but not if $b^2 - 4ac < 0$. Over the complex numbers, it makes sense for all $a, b, c$. Over the rationals, it makes sense only when $b^2 - 4ac$ is a perfect square — the square of a rational number.

---

## 1.4 Solution by Radicals

We begin by reviewing the state of the art regarding solutions of polynomial equations as it was just before the time of Galois. We consider linear, quadratic, cubic, quartic, and quintic equations in turn. In the case of the quintic, we also describe some ideas that were discovered after Galois. Throughout, we make the default assumption of the period: the coefficients of the equation are complex numbers.

## Linear Equations

Let $a, b \in \mathbb{C}$ with $a \neq 0$. The general *linear* equation is

$$at + b = 0$$

and the solution is clearly

$$t = -\frac{b}{a}$$

## Quadratic Equations

Let $a, b, c \in \mathbb{C}$ with $a \neq 0$. The general *quadratic* equation is

$$at^2 + bt + c = 0$$

Dividing by $a$ and renaming the coefficients, we can consider the equivalent equation

$$t^2 + at + b = 0$$

The standard way to solve this equation is to rewrite it in the form

$$\left(t + \frac{a}{2}\right)^2 = \frac{a^2}{4} - b$$

Taking square roots,

$$t + \frac{a}{2} = \pm\sqrt{\frac{a^2}{4} - b}$$

so that

$$t = -\frac{a}{2} \pm \sqrt{\frac{a^2}{4} - b}$$

which is the usual quadratic formula except for a change of notation. The process used here is called *completing the square*; as remarked in the Historical Introduction, it goes back to the Babylonians 3600 years ago.

## Cubic Equations

Let $a, b, c \in \mathbb{C}$ with $a \neq 0$. The general *cubic* equation can be written in the form

$$t^3 + at^2 + bt + c = 0$$

where again we have divided by the leading coefficient to avoid unnecessary complications in the formulas.

The first step is to change the variable to make $a = 0$. This is achieved by setting $y = t + \frac{a}{3}$, so that $t = y - \frac{a}{3}$. Such a move is called a Tschirnhaus transformation, after the person who first made explicit and systematic use of it. The equation becomes

$$y^3 + py + q = 0 \tag{1.3}$$

where

$$\begin{aligned} p &= \frac{-a^2 + 3b}{3} \\ q &= \frac{2a^3 - 9ab + 27c}{27} \end{aligned}$$

To find the solution(s), we try (rabbit out of hat) the substitution

$$y = \sqrt[3]{u} + \sqrt[3]{v}\ .$$

Now

$$y^3 = u + v + 3\sqrt[3]{u}\sqrt[3]{v}(\sqrt[3]{u} + \sqrt[3]{v})$$

so that (1.3) becomes

$$(u + v + q) + (\sqrt[3]{u} + \sqrt[3]{v})(3\sqrt[3]{u}\sqrt[3]{v} + p) = 0$$

We now choose $u$ and $v$ to make *both* terms vanish:

$$u + v + q = 0 \tag{1.4}$$

$$3\sqrt[3]{u}\sqrt[3]{v} + p = 0 \tag{1.5}$$

which imply

$$u + v = -q \tag{1.6}$$

$$uv = -\frac{p^3}{27} \tag{1.7}$$

Multiply (1.6) by $u$ and subtract (1.7) to get

$$u(u + v) - uv = -qu + \frac{p^3}{27}$$

which can be rearranged to give

$$u^2 + qu - \frac{p^3}{27} = 0$$

which is a quadratic.

The solution of quadratics now tells us that

$$u = -\frac{q}{2} \pm \sqrt{\frac{q^2}{4} + \frac{p^3}{27}}$$

Since $u + v = -q$, we have

$$v = -\frac{q}{2} \mp \sqrt{\frac{q^2}{4} + \frac{p^3}{27}}$$

Changing the sign of the square root just permutes $u$ and $v$, so we can set the sign to $+$. Thus we find that

$$y = \sqrt[3]{-\frac{q}{2} + \sqrt{\frac{q^2}{4} + \frac{p^3}{27}}} + \sqrt[3]{-\frac{q}{2} - \sqrt{\frac{q^2}{4} + \frac{p^3}{27}}} \tag{1.8}$$

which (by virtue of publication, not discovery) is usually called *Cardano's formula.* (This version differs from the formula in the Historical Introduction because Cardano worked with $x^2 + px = q$, so $q$ changes sign.) Finally, remember that the solution $t$ of the original equation is equal to $y - a/3$.

## Peculiarities of Cardano's Formula

An old warning, which goes back to Aesop's *Fables*, is: 'Be careful what you wish for: you might get it'. We have wished for a formula for the solution, and we have got one. It has its peculiarities.

First: recall that over $\mathbb{C}$, every nonzero complex number $z$ has *three* cube roots. If one of them is $\alpha$, then the other two are $\omega\alpha$ and $\omega^2\alpha$, where

$$\omega = -\frac{1}{2} + \mathrm{i}\frac{\sqrt{3}}{2}$$

is a primitive cube root of 1. Then

$$\omega^2 = -\frac{1}{2} - \mathrm{i}\frac{\sqrt{3}}{2}$$

The expression for $y$ therefore appears to lead to *nine* solutions of the form

$$\begin{array}{ccc} \alpha + \beta & \alpha + \omega\beta & \alpha + \omega^2\beta \\ \omega\alpha + \beta & \omega\alpha + \omega\beta & \omega\alpha + \omega^2\beta \\ \omega^2\alpha + \beta & \omega^2\alpha + \omega\beta & \omega^2\alpha + \omega^2\beta \end{array}$$

where $\alpha, \beta$ are specific choices of the cube roots.

However, not all of these expressions are zeros. Equation (1.5) implies (1.7), but (1.7) implies (1.5) only when we make the correct choices of cube roots. If we choose $\alpha, \beta$ so that $3\alpha\beta + p = 0$, then the solutions are

$$\alpha + \beta \qquad \omega\alpha + \omega^2\beta \qquad \omega^2\alpha + \omega\beta$$

Another peculiarity emerges when we try to solve equations whose solutions we already know. For example,

$$y^3 + 3y - 36 = 0$$

has the solution $y = 3$. Here $p = 3, q = -36$, but Cardano's formula gives

$$y = \sqrt[3]{18 + \sqrt{325}} + \sqrt[3]{18 - \sqrt{325}}$$

which seems a far cry from 3. However, further (rather ingenious) algebra converts it to 3: see Exercise 1.4. As Cardano observed in his book, it gets worse: if his formula is applied to

$$t^3 - 15t - 4 = 0 \tag{1.9}$$

it leads to

$$t = \sqrt[3]{2 + \sqrt{-121}} + \sqrt[3]{2 - \sqrt{-121}} \tag{1.10}$$

in contrast to the obvious solution $t = 4$. This is very curious even today and must have seemed even more so in the Renaissance period.

Cardano had already encountered such baffling expressions when trying to solve the quadratic $t(10 - t) = 40$, with the apparently nonsensical solutions $5+\sqrt{-15}$ and $5-\sqrt{-15}$, but there it was possible to see the puzzling form of the 'solution' as expressing the fact that no solution exists. However, Cardano was bright enough to spot that if you ignore the question of what such expressions *mean*, and just manipulate them as if they are ordinary numbers, then they do indeed satisfy the equation. 'So,' Cardano commented, 'progresses arithmetic subtlety, the end of which is as refined as it is useless.'

However, this shed no light on why a cubic could possess a perfectly reasonable solution when the formula (more properly, the equivalent numerical procedure) could not find it. Around 1560 Raphael Bombelli observed that $(2 \pm \sqrt{-1})^3 = 2 \pm \sqrt{-121}$, and recovered (see Exercise 1.7) the solution $t = 4$ of (1.9) from the formula (1.10), again assuming that such expressions can be manipulated just like ordinary numbers. But Bombelli, too, expressed scepticism that such manoeuvres had any sensible meaning. In 1629 Albert Girard argued that such expressions are valid as *formal* solutions of the equations, and should be included 'for the certitude of the general rules'. Girard was influential in making negative numbers acceptable, but he was way ahead of his time when it came to their square roots.

In fact, Cardano's formula is pretty much useless whenever the cubic has three *real* roots. This is called the 'irreducible case' of the cubic. (This is a different use of 'irreducible' from that in Definition 3.12 below.) The traditional escape route is to use trigonometric functions, Exercise 1.8. All this rather baffled the Renaissance mathematicians, who did not even have effective algebraic notation, and were wary of negative numbers, let alone imaginary ones.

Using Galois theory, it is possible to prove that the cube roots of complex numbers that arise in the irreducible case of the cubic equation cannot be avoided. That is, there are no formulas in real radicals for the real and imaginary parts. See Van der Waerden (1953) volume 1 page 180, and Isaacs (1985).

## Quartic Equations

An equation of the fourth degree

$$t^4 + at^3 + bt^2 + ct + d = 0$$

is called a *quartic* equation (an older term is *biquadratic*). To solve it, start by making the Tschirnhaus transformation $y = t + a/4$, to get

$$y^4 + py^2 + qy + r = 0 \tag{1.11}$$

where

$$\begin{aligned} p &= b - \frac{3a^2}{8} \\ q &= c - \frac{ab}{2} + \frac{a^3}{8} \\ r &= d - \frac{ac}{4} + \frac{a^2b}{16} - \frac{3a^4}{256} \end{aligned}$$

Rewrite this in the form

$$\left(y^2 + \frac{p}{2}\right)^2 = -qy - r + \frac{p^2}{4}$$

Introduce a new term $u$, and observe that

$$\begin{aligned} \left(y^2 + \frac{p}{2} + u\right)^2 &= \left(y^2 + \frac{p}{2}\right)^2 + 2\left(y^2 + \frac{p}{2}\right)u + u^2 \\ &= -qy - r + \frac{p^2}{4} + 2uy^2 + pu + u^2 \end{aligned}$$

We choose $u$ to make the right-hand side a perfect square. If it is, it must be the square of $\sqrt{2u}y - \frac{q}{2\sqrt{2u}}$, and then we require

$$-r + \frac{p^2}{4} + pu + u^2 = \frac{q^2}{8u}$$

Provided $u \neq 0$, this becomes

$$8u^3 + 8pu^2 + (2p^2 - 8r^2)u - q^2 = 0 \tag{1.12}$$

which is a cubic in $u$. Solving by Cardano's method, we can find $u$. Now

$$\left(y^2 + \frac{p}{2} + u\right)^2 = \left(\sqrt{2u}y - \frac{q}{2\sqrt{2u}}\right)^2$$

so

$$y^2 + \frac{p}{2} + u = \pm\left(\sqrt{2u}y - \frac{q}{2\sqrt{2u}}\right)$$

Finally, we can solve the above two quadratics to find $y$.

If $u = 0$ we do not obtain (1.12), but if $u = 0$ then $q = 0$, so the quartic (1.11) is a quadratic in $y^2$ and can be solved using only square roots.

Equation (1.12) is called the *resolvent cubic* of (1.11). Explicit formulas for the roots can be obtained if required. Since they are complicated, we shall not give them here.

An alternative approach to the resolvent cubic, not requiring a preliminary Tschirnhaus transformation, is described in Exercise 1.13.

## Quintic Equations

So far, we have a series of special tricks, different in each case. We can approach the general *quintic* equation

$$t^5 + at^4 + bt^3 + ct^2 + dt + e = 0$$

in a similar way. A Tschirnhaus transformation $y = t + a/5$ reduces it to

$$y^5 + py^3 + qy^2 + ry + s = 0$$

However, all variations on the tricks that we used for the quadratic, cubic, and quartic equations grind to a halt.

In 1770–1771 Lagrange analysed all of the above special tricks, showing that they can all be 'explained' using general principles about symmetric functions of the roots. When he applied this method to the quintic, however, he found that it 'reduced' the problem to a sextic—an equation of degree 6. Instead of helping, the method made the problem *worse*.

Lagrange observed that all methods for solving polynomial equations by radicals involve constructing rational functions of the roots that take a small number of values when the roots $\alpha_j$ are permuted. Prominent among these is the expression

$$\delta = \prod_{1 \leq j < k \leq n} (\alpha_j - \alpha_k) \tag{1.13}$$

where $n$ is the degree. This takes just two values, $\pm\delta$: plus for even permutations and minus for odd ones. Therefore $\Delta = \delta^2$ (known as the *discriminant* because it is nonzero precisely when the roots are distinct, so it 'discriminates' among the roots) is a rational function of the coefficients. This gets us started, and it yields a complete solution for the quadratic, but for cubics upwards it does not help much unless we can find other expressions in the roots with similar properties under permutation.

Lagrange worked out what these expressions look like for the cubic and the quartic, and noticed a pattern. For example, if a cubic polynomial has roots $\alpha_1, \alpha_2, \alpha_3$, and $\omega$ is a primitive cube root of unity, then the expression

$$u = (\alpha_1 + \omega\alpha_2 + \omega^2\alpha_3)^3$$

takes exactly two distinct values. In fact, even permutations leave it unchanged, while odd permutations transform it to

$$v = (\alpha_1 + \omega^2\alpha_2 + \omega\alpha_3)^3$$

It follows that $u+v$ and $uv$ are fixed by all permutations of the roots and must therefore be expressible as rational functions of the coefficients. So $u + v = a$, $uv = b$ where $a, b$ are rational functions of the coefficients. Therefore $u$ and $v$ are the solutions of the quadratic equation $t^2 - at + b = 0$, so they can be expressed using square roots. But now the further use of cube roots expresses $\alpha_1 + \omega\alpha_2 + \omega^2\alpha_3 = \sqrt[3]{u}$ and $\alpha_1 + \omega^2\alpha_2 + \omega\alpha_3 = \sqrt[3]{v}$ by radicals. Since we also know that $\alpha_1+\alpha_2+\alpha_3$ is minus the coefficient of $t^2$, we have three independent linear equations in the roots, which are easily solved.

Something very similar works for the quartic, with expressions like

$$(\alpha_1 - \alpha_2 + \alpha_3 - \alpha_4)^2$$

But when we try the same idea on the quintic, an obstacle appears. Suppose that the roots of the quintic are $\alpha_1, \alpha_2, \alpha_3, \alpha_4, \alpha_5$. Let $\zeta$ be a primitive fifth root of unity. Following Lagrange's lead, it is natural to consider

$$w = (\alpha_1 + \zeta\alpha_2 + \zeta^2\alpha_3 + \zeta^3\alpha_4 + \zeta^4\alpha_5)^5$$

There are 120 permutations of 5 roots, and they transform $w$ into 24 distinct expressions. Therefore $w$ is a root of a polynomial of degree 24—a big step in the wrong direction, since we started with a mere quintic.

The best that can be done is to use an expression derived by Arthur Cayley in 1861, based on an idea of Robert Harley in 1859. This expression is

$$x = (\alpha_1\alpha_2+\alpha_2\alpha_3+\alpha_3\alpha_4+\alpha_4\alpha_5+\alpha_5\alpha_1-\alpha_1\alpha_3-\alpha_2\alpha_4-\alpha_3\alpha_5-\alpha_4\alpha_1-\alpha_5\alpha_2)^2$$

It turns out that $x$ takes precisely six values when the variables are permuted in all 120 possible ways. Therefore $x$ is a root of a sextic equation. The equation is very complicated and has no obvious roots; it is, perhaps, better than an equation of degree 24, but it is still no improvement on the original quintic. Except when the sextic happens, by accident, to have a root whose square is rational, in which case the quintic *is* soluble by radicals. Indeed, this is a necessary and sufficient condition for a quintic to be soluble by radicals, see Berndt, Spearman and Williams (2002). For instance, as they explain in detail, the equation

$$t^5 + 15t + 12 = 0$$

has the solution

$$t = \sqrt[5]{\frac{-75 + 21\sqrt{10}}{125}} + \sqrt[5]{\frac{-75 - 21\sqrt{10}}{125}} + \sqrt[5]{\frac{225 + 72\sqrt{10}}{125}} + \sqrt[5]{\frac{225 - 72\sqrt{10}}{125}}$$

with similar expressions for the other four roots.

Lagrange's general method, then, fails for the quintic. This does not prove that the general quintic is not soluble by radicals, because for all Lagrange or anyone else knew, there might be other methods that do not make the problem worse. But it does suggest that there is something very different about the quintic. Suspicion began to grow that *no* method would solve the quintic by radicals. Mathematicians stopped looking for such a solution, and a few of them started looking for an impossibility proof instead.

---

## EXERCISES

1.1 Use (1.1) to prove that multiplication of complex numbers is commutative and associative. That is, if $u, v, w$ are complex numbers, then $uv = vu$ and $(uv)w = u(vw)$.

1.2 Prove that $\sqrt{2}$ is irrational, as follows. Assume for a contradiction that there exist integers $a, b$, with $b \neq 0$, such that $(a/b)^2 = 2$.

1. Show that we may assume $a, b > 0$.
2. Observe that if such an expression exists, then there must be one in which $b$ is as small as possible.
3. Show that
$$\left(\frac{2b-a}{a-b}\right)^2 = 2$$
4. Show that $2b - a > 0, a - b > 0$.
5. Show that $a - b < b$, a contradiction.

1.3 Prove that if $q \in \mathbb{Q}$, then $\sqrt{q}$ is rational if and only if $q$ is a perfect square; that is, it can be written in the form $q = p_1^{a_1} \cdots p_n^{a_n}$ where the integers $a_j$, which may be positive or negative, are all even.

1.4* Prove without using Cardano's formula that
$$\sqrt[3]{18+\sqrt{325}} + \sqrt[3]{18-\sqrt{325}} = 3$$

1.5 Let $\alpha = \sqrt[3]{2} \in \mathbb{R}$. Prove that the set of all numbers $p + q\alpha + r\alpha^2$, for $p, q, r \in \mathbb{Q}$, is a subfield of $\mathbb{C}$.

1.6 Let $\omega$ be a primitive cube root of unity in $\mathbb{C}$. With the notation of Exercise 1.5, show that the map
$$p + q\alpha + r\alpha^2 \mapsto p + q\omega\alpha + r\omega^2\alpha^2$$
is a monomorphism onto its image but not an automorphism.

1.7 Use Bombelli's observation that $(2 \pm \sqrt{-1})^3 = 2 \pm \sqrt{-121}$ to show that (for suitable choices of values of the cube roots)

$$\sqrt[3]{2+\sqrt{-121}} + \sqrt[3]{2-\sqrt{-121}} = 4$$

1.8 Use the identity $\cos 3\theta = 4\cos^3\theta - 3\cos\theta$ to solve the cubic equation $t^3+pt+q=0$ when $27q^2+4p^3 < 0$. You may use the inverse trigonometric function $\cos^{-1}$ (or arccos) as well as cos.

1.9 Find radical expressions for all three roots of $t^3 - 15t - 4 = 0$.

1.10 When $27q^2 + 4p^3 < 0$ it seems reasonable to try to make sense of Cardano's formula by generalising Bombelli's observation; that is, to seek $\alpha, \beta$ such that

$$\left[\alpha \pm \beta\sqrt{\frac{q^2}{4}+\frac{p^3}{27}}\right]^3 = \frac{q}{2} \pm \sqrt{\frac{q^2}{4}+\frac{p^3}{27}}$$

Why is this usually pointless?

1.11* Let $P(n)$ be the number of ways to arrange $n$ zeros and ones in a row, given that ones occur in groups of three or more. Show that

$$P(n) = 2P(n-1) - P(n-2) + P(n-4)$$

and deduce that as $n \to \infty$ the ratio $\frac{P(n+1)}{P(n)} \to x$, where $x > 0$ is real and $x^4 - 2x^3 + x^2 - 1 = 0$. Factorise this quartic as a product of two quadratics, and hence find $x$.

1.12* The largest square that fits inside an equilateral triangle can be placed in any of three symmetrically related positions. Eugenio Calabi noticed that there is exactly one other shape of triangle in which there are three equal largest squares, Figure 7. Prove that in this triangle, the ratio $x$ of the longest side to the other two is a solution of the cubic equation $2x^3 - 2x^2 - 3x + 2 = 0$, and find an approximate value of $x$ to three decimal places.

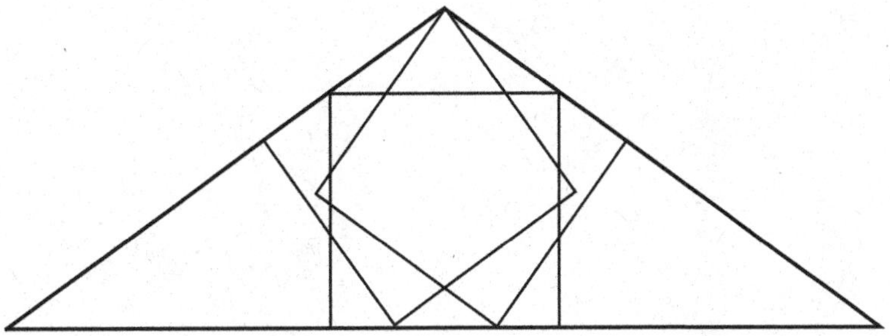

FIGURE 7: Calabi's triangle.

1.13 Investigate writing the general quartic $t^4 + at^3 + bt^2 + ct + d$ in the form

$$(t^2 + pt + q)^2 - (rt + s)^2$$

which, being a difference of two squares, factorises into two quadratics

$$(t^2 + pt + q + rt + s)(t^2 + pt + q - rt - s)$$

and can thus be solved in radicals if $p, q, r, s$ can be expressed in terms of the original coefficients $a, b, c, d$.

Show that doing this leads to a cubic equation.

1.14 Mark the following true or false.

(a) $-1$ has no square root.

(b) $-1$ has no real square root.

(c) $-1$ has two distinct square roots in $\mathbb{C}$.

(d) Every subring of $\mathbb{C}$ is a subfield of $\mathbb{C}$.

(e) Every subfield of $\mathbb{C}$ is a subring of $\mathbb{C}$.

(f) The set of all numbers $p + q\sqrt[7]{5}$ for $p, q \in \mathbb{Q}$ is a subring of $\mathbb{C}$.

(g) The set of all numbers $p + q\sqrt[7]{5}$ for $p, q \in \mathbb{C}$ is a subring of $\mathbb{C}$.

(h) Cardano's formula always gives a correct answer.

(i) Cardano's formula always gives a sensible answer.

(j) A quintic equation over $\mathbb{Q}$ can never be solved by radicals.

# Chapter 2

# The Fundamental Theorem of Algebra

At the time of Galois, the natural setting for most mathematical investigations was the complex number system. The real numbers were inadequate for many questions, because $-1$ has no real square root. The arithmetic, algebra, and—decisively—analysis of complex numbers were richer, more elegant, and more complete than the corresponding theories for real numbers.

In this chapter we establish one of the key properties of $\mathbb{C}$, known as the Fundamental Theorem of Algebra. This theorem asserts that every polynomial equation with coefficients in $\mathbb{C}$ has a solution in $\mathbb{C}$. This theorem is, of course, false over $\mathbb{R}$—consider the equation $t^2+1=0$. It was fundamental to classical algebra, but the name is somewhat archaic, and modern algebra bypasses $\mathbb{C}$ altogether, preferring greater generality. Because we find it convenient to work in the same setting as Galois, the theorem is fundamental for us.

All rigorous proofs of the Fundamental Theorem of Algebra require quite a lot of background. Here, we give a proof that uses a few simple ideas from algebra and trigonometry, estimates of the kind that are familiar from any first course in analysis, and one simple basic result from point-set topology. Later, we give an almost purely algebraic proof, but the price is the need for much more machinery: see Chapter 23. Ironically, that proof uses Galois theory to prove the Fundamental Theorem of Algebra, the exact opposite of what Galois did. The logic is not circular, because the proof in Chapter 23 rests on the abstract approach to Galois theory described in the second part of this book, which makes no use of the Fundamental Theorem of Algebra.

## 2.1 Polynomials

Linear, quadratic, cubic, quartic, and quintic equations are examples of a more general class: polynomial equations. These take the form

$$p(t) = 0$$

where $p(t)$ is a polynomial in $t$. Mathematics is littered with polynomial equations, arising in a huge variety of contexts.

DOI: 10.1201/9781003213949-2

A polynomial is an algebraic expression involving the powers of a 'variable' or 'indeterminate' $t$. We are used to thinking of such a polynomial as the function that maps $t$ to the value of the expression concerned, so that the polynomial $t^2 - 2t + 6$ represents the function $f$ defined by $f(t) = t^2 - 2t + 6$. This 'function' viewpoint is familiar, and it causes no problems when we are thinking about polynomials with complex numbers as their coefficients. Later (Chapter 16) we will see that when more general fields are permitted, it is not such a good idea to think of a polynomial as a function. So it is worth setting up the concept of a polynomial so that it extends easily to the general context.

We therefore define a *polynomial over* $\mathbb{C}$ *in the indeterminate* $t$ to be an expression

$$r_0 + r_1 t + \cdots + r_n t^n$$

where $r_0, \ldots, r_n \in \mathbb{C}$, $0 \leq n \in \mathbb{Z}$, and $t$ is undefined. What, though, is an 'expression', logically speaking? For set-theoretic purity we can replace such an expression by the sequence $(r_0, \ldots, r_n)$. In fact, it is more convenient to use an infinite sequence $(r_0, r_1, \ldots)$ in which all entries $r_j = 0$ when $j > n$ for some finite $n$: see Exercise 2.2. In such a formalism, $t$ is just a symbol for the sequence $(0, 1, 0 \ldots)$.

The elements $r_0, \ldots, r_n$ are the *coefficients* of the polynomial. In the usual way, terms $0t^m$ may be omitted or written as 0, and $1t^m$ can be replaced by $t^m$. To distinguish the zero polynomial (all coefficients = 0) from the number 0, we write it as

$$\mathbf{0} = 0 + \cdots + 0.t^m = (0, 0, 0 \ldots)$$

In practice it is often convenient write polynomials in descending order

$$r_n t^n + r_{n-1} t^{n-1} + \cdots + r_1 t + r_0$$

and from now on we use either ordering of the terms without further comment.

Two polynomials are defined to be equal *if and only if* the corresponding coefficients are equal, with the understanding that coefficients of powers of $t$ that do not occur in the polynomial are zero. To define the sum and the product of two polynomials, write

$$\sum r_i t^i$$

instead of

$$r_0 + r_1 t + \cdots + r_n t^n$$

where the summation is considered as being over all integers $i \geq 0$, and $r_k$ is defined to be 0 if $k \geq n$. Then, if

$$r = \sum r_i t^i \qquad s = \sum s_i t^i$$

we define

$$r + s = \sum (r_i + s_i) t^i \tag{2.1}$$

and

$$rs = \sum q_j t^j \quad \text{where} \quad q_j = \sum_{h+i=j} r_h s_i \tag{2.2}$$

It is now easy to check directly from these definitions that the set of all polynomials over $\mathbb{C}$ in the $t$ obeys all of the usual algebraic laws (Exercise 2.3). We denote this set by $\mathbb{C}[t]$, and call it the *ring of polynomials over* $\mathbb{C}$ *in the indeterminate* $t$.

We can also define polynomials in several indeterminates $t_1, t_2, \ldots, t_n$, obtaining the ring of $n$-variable polynomials

$$\mathbb{C}[t_1, t_2, \ldots, t_n]$$

in an analogous way.

An element of $\mathbb{C}[t]$ will usually be denoted by a single letter, such as $f$, whenever it is clear which indeterminate is involved. If there is ambiguity, we write $f(t)$ to emphasise the role played by $t$. Although this looks like function notation, technically it is not. However, polynomials over $\mathbb{C}$ can usefully be interpreted as functions, see Proposition 2.3 below.

Next, we introduce a simple but very useful concept, which quantifies how complicated a polynomial is.

**Definition 2.1.** If $f$ is a polynomial over $\mathbb{C}$ and $f \neq 0$, then the *degree* of $f$ is the highest power of $t$ occurring in $f$ with non-zero coefficient.

For example, $t^2 + 1$ has degree 2, and $723t^{1101} - 9111t^{55} + 43$ has degree 1101.

More generally, if $f = \sum r_i t^i$ and $r_n \neq 0$ and $r_m = 0$ for $m > n$, then $f$ has degree $n$. We write $\partial f$ for the degree of $f$. To deal with the case $f = 0$ we adopt the convention that $\partial 0 = -\infty$. This symbol is endowed with the following properties: $-\infty < n$ for any integer $n$, $-\infty + n = -\infty$, $-\infty \times n = -\infty$, $(-\infty)^2 = -\infty$. We do *not* set $(-\infty)^2 = +\infty$ because $0.0 = 0$.

The following result is immediate from this definition:

**Proposition 2.2.** *If $f$, $g$ are polynomials over $\mathbb{C}$, then*

$$\partial(f+g) \leq \max(\partial f, \partial g) \qquad \partial(fg) = \partial f + \partial g$$

□

The inequality in the first line is due to the possibility of the highest terms 'cancelling', see Exercise 2.4.

The $f(t)$ notation makes $f$ appear to be a function, with $t$ as its 'independent variable', and in fact we can identify each polynomial $f$ over $\mathbb{C}$ with the corresponding function. Specifically, each polynomial $f \in \mathbb{C}[t]$ can be considered as a function from $\mathbb{C}$ to $\mathbb{C}$, defined as follows: if $f = \sum r_i t^i$ and $\alpha \in \mathbb{C}$, then $\alpha$ is mapped to $\sum r_i \alpha^i$. The next proposition proves that when the coefficients lie in $\mathbb{C}$, it causes no confusion if we use the same symbols $f$ to denote a polynomial and the function associated with it.

**Proposition 2.3.** *Two polynomials $f, g$ over $\mathbb{C}$ define the same function if and only if they are equal as polynomials; that is, they have the same coefficients.*

*Proof.* Equivalently, by taking the difference of the two polynomials, we must prove that if $f(t)$ is a polynomial over $\mathbb{C}$ and $f(t) = 0$ for all $t$, then the coefficients of $f$ are all 0. Let $P(n)$ be the statement: If a polynomial $f(t)$ over $\mathbb{C}$ has degree $n$, and $f(t) = 0$ for all $t \in \mathbb{C}$, then $f = 0$. We prove $P(n)$ for all $n$ by induction on $n$.

Both $P(0)$ and $P(1)$ are obvious. Suppose that $P(n-1)$ is true. Write

$$f(t) = a_n t^n + \cdots + a_0$$

In particular, $f(0) = 0$, so $a_0 = 0$ and

$$\begin{aligned} f(t) &= a_n t^n + \cdots + a_1 t \\ &= t(a_n t^{n-1} + \cdots + a_1) \\ &= t g(t) \end{aligned}$$

where $g(t) = a_n t^{n-1} + \cdots + a_1$ has degree $n-1$. Now $g(t)$ vanishes for all $t \in \mathbb{C}$ except, perhaps, $t = 0$. However, if $g(0) = a_1 \neq 0$, then $g(t) \neq 0$ for $t$ sufficiently small. (This follows by continuity of polynomial functions, but it can be proved directly by estimating the size of $g(\varepsilon)$ when $\varepsilon$ is small.) Therefore $g(t)$ vanishes for all $t \in \mathbb{C}$. By induction, $g = 0$. Therefore $f = 0$, so $P(n)$ is true and the induction is complete. □

Proposition 2.3 implies that we can safely consider a polynomial over a subfield of $\mathbb{C}$ as either a formal algebraic expression or a function. It is easy to see that sums and products of polynomials agree with the corresponding sums and products of functions. Moreover, the same notational flexibility allows us to 'change the variable' in a polynomial. For example, if $t$, $u$ are two indeterminates and $f(t) = \sum r_i t^i$, then we may define $f(u) = \sum r_i u^i$. It is also clear what is meant by such expressions as $f(t-3)$ or $f(t^2+1)$.

---

## 2.2 Fundamental Theorem of Algebra

In Section 1.3 we saw that the development of the complex numbers can be viewed as the culmination of a series of successive extensions of the natural number system. At each step, equations that cannot be solved within the existing number system become soluble in the new, extended system. For example, $\mathbb{C}$ arises from $\mathbb{R}$ by insisting that $t^2 = -1$ should have a solution.

The question then arises: why stop at $\mathbb{C}$? Why not find an equation that has no solutions over $\mathbb{C}$, and enlarge the number system still further to provide a solution?

The answer is that no such equation exists, at least if we limit ourselves to polynomials. *Every* polynomial equation over $\mathbb{C}$ has a solution in $\mathbb{C}$. This proposition was a matter of heated debate around 1700. In a paper of 1702, Leibniz disputes that it can be true, citing the example

$$x^4 + a^4 = \left(x + a\sqrt{\sqrt{-1}}\right)\left(x - a\sqrt{\sqrt{-1}}\right)\left(x + a\sqrt{-\sqrt{-1}}\right)\left(x - a\sqrt{-\sqrt{-1}}\right)$$

and presumably thinking that $\sqrt{\sqrt{-1}}$ is not a complex number.

However, in 1676 Isaac Newton had already observed the factorisation into two real quadratics:

$$x^4 + a^4 = (x^2 + a^2)^2 - 2a^2x^2 = (x^2 + a^2 + \sqrt{2}ax)(x^2 + a^2 - \sqrt{2}ax)$$

and Nicholas Bernoulli published the same formula in 1719. In effect, the resolution of the dispute rests on observing that $\sqrt{\mathrm{i}} = \frac{1 \pm \mathrm{i}}{\sqrt{2}}$, which is in $\mathbb{C}$. In fact, every complex number has a complex square root:

$$\sqrt{a + b\mathrm{i}} = \sqrt{\frac{a + \sqrt{a^2 + b^2}}{2}} + \mathrm{i}\sqrt{\frac{-a + \sqrt{a^2 + b^2}}{2}} \tag{2.3}$$

(together with minus the same formula), as can be checked by squaring the right-hand side. Here the square root of $a^2 + b^2$ is the positive one, and the signs of the other two square roots are chosen to make their product equal to $b$. Observe that

$$a + \sqrt{a^2 + b^2} \geq 0 \qquad -a + \sqrt{a^2 + b^2} \geq 0$$

because $a^2 + b^2 \geq a^2$, so both of the main square roots on the right-hand side are real.

In 1742 Euler asserted, without proof, that every real polynomial can be decomposed into linear or quadratic factors with real coefficients. This is a way to say, without explicit mention of conmplex numbers, that it factorises into linear factors over $\mathbb{C}$, by combining complex conjugate pairs $x + \alpha, x + \bar{\alpha}$ into their product

$$(x + \alpha)(x + \bar{\alpha}) = x^2 + (\alpha + \bar{\alpha})x + \alpha\bar{\alpha}$$

which has real coefficients. After some lively discussion with Bernoulli, who thought he had a counterexample, Euler wrote to his friend Christian Goldbach, saying that he had proved the theorem for polynomials of degree $\leq 6$. Later, Euler and Jean Le Rond d'Alembert gave incomplete but plausible proofs for any degree. The first genuine proof was given by Gauss in his doctoral thesis of 1799. It involved the manipulation of complicated trigonometric series to derive a contradiction, and was far from transparent. Later he gave three other proofs, all based on different ideas. Other classical proofs use deep results in complex analysis. William Kingdon Clifford gave a proof based on induction on the power of 2 that divides the degree $n$, which is most easily explained using Galois theory. We present this in Chapter 23, Corollary 23.13.

All of these proofs are quite sophisticated. But there's an easier way, using a few ideas from elementary point-set topology and estimates of the kind we encounter early on in any course on real analysis. We give details below. This proof can be found in Wikipedia, and it deserves to be more widely known because it is simple and cuts straight to the heart of the issue. The necessary facts can be proved directly by elementary means, and would have been considered obvious before mathematicians started worrying about rigour in analysis around 1850. So Euler, Gauss, and other mathematicians of those periods could have discovered this proof.

We now state this property of the complex numbers formally, present the aforementioned proof, and explore some of its easier consequences.

**Theorem 2.4 (Fundamental Theorem of Algebra).** *If $p(z)$ is a non-constant polynomial over $\mathbb{C}$, then there exists $z_0 \in \mathbb{C}$ such that $p(z_0) = 0$.*

Such a number $z$ is called a *root* of the equation $p(t) = 0$, or a *zero* of the polynomial $p$. For example, i is a root of the equation $t^2 + 1 = 0$ and a zero of $t^2 + 1$. Polynomial equations may have more than one root; indeed, $t^2 + 1 = 0$ has at least one other root, −i.

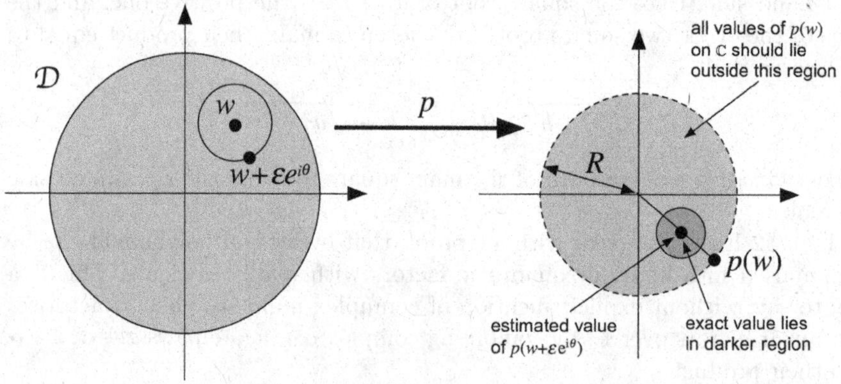


FIGURE 8: Idea of proof.

The idea behind the proof is illustrated in Figure 8 and can be summarised in a few lines. Assume for a contradiction that $p(z)$ is never zero. Then the real number $|p(z)|^2$ has a nonzero minimum value and attains that minimum at some point $w \in \mathbb{C}$. Consider points $v$ on a small circle centred at $w$, and use simple estimates to show that $|p(v)|^2$ must be less than $|p(w)|^2$ for some $v$. Contradiction.

Now for the details.

*Proof of Theorem* 2.4. Suppose for a contradiction that no such $z_0$ exists. For some $R > 0$ the set

$$\mathcal{D} = \{z : |p(z)|^2 \leq R\}$$

is non-empty. The map $\psi : \mathbb{C} \to \mathbb{R}^+$ defined by $\psi(z) = |p(z)|^2$ is continuous, so $\mathcal{D} = \psi^{-1}([0, R])$ is closed. If $|z|$ is sufficiently large, $|p(z)|^2 > R$ because $|p(z)| \to \infty$ as $|z| \to \infty$. Therefore $\mathcal{D}$ is bounded, hence compact. It follows that $|p(z)|^2$ attains its minimum value on $\mathcal{D}$. By the definition of $\mathcal{D}$ this is also its minimum value on $\mathbb{C}$. Assume that this minimum is attained at $w \in \mathbb{C}$. Then

$$|p(z)|^2 \geq |p(w)|^2$$

for all $z \in \mathbb{C}$, and by assumption $p(w) \neq 0$.

We now consider $|p(z)|^2$ as $z$ runs round a small circle centred at $w$, and derive a contradiction. Let $h \in \mathbb{C}$. Expand $p(w + h)$ in powers of $h$ to get

$$p(w + h) = p_0 + p_1 h + p_2 h^2 + \cdots + p_n h^n \tag{2.4}$$

where $n$ is the degree of $p$. Here the $p_j$ are specific complex numbers. They are in fact the Taylor series coefficients

$$p_j = p^{(j)}(w)/j!$$

but we do not actually need to use this, and (2.4) can be proved algebraically without difficulty.

Clearly $p_0 = p(w)$, and we are assuming this is nonzero, so $p_0 \neq 0$. If $p_1 = p_2 = \cdots = p_n = 0$, then $p(z) = p_0$ is constant, contrary to hypothesis. So some $p_j \neq 0$. Let $m$ be the smallest integer $\geq 1$ from which $p_m \neq 0$. In (2.4) let $h = \varepsilon e^{i\theta}$ for small $\varepsilon > 0$. Then

$$p(w + \varepsilon e^{i\theta}) = p_0 + p_m \varepsilon^m e^{mi\theta} + O(\varepsilon^{m+1})$$

where $O(\varepsilon^n)$ indicates terms of order $n$ or more in $\varepsilon$. (We use the same symbol $O(\varepsilon^n)$ for all such terms, even when they are different. For example $O(\varepsilon^n) + O(\varepsilon^n) = O(\varepsilon^n)$.) Therefore

$$\begin{aligned} |p(w + \varepsilon e^{i\theta})|^2 &= |p_0 + p_m \varepsilon^m e^{mi\theta}|^2 + O(\varepsilon^{m+1}) \\ &= p_0 \overline{p}_0 + \overline{p}_0 p_m \varepsilon^m e^{mi\theta} + p_0 \overline{p}_m \varepsilon^m e^{-mi\theta} + O(\varepsilon^{m+1}) \end{aligned}$$

Let $p_0 \overline{p}_m = r e^{i\phi}$ for $r \geq 0$. Since $p_0 \neq 0$ and $p_m \neq 0$, we have $r > 0$. Setting $h = 0$ we see that $p_0 \overline{p}_0 = |p(w)|^2$. Now

$$\begin{aligned} |p(w + \varepsilon e^{i\theta})|^2 &= p_0 \overline{p}_0 + r e^{i\phi} \varepsilon^m e^{mi\theta} + r e^{-i\phi} \varepsilon^m e^{-mi\theta} + O(\varepsilon^{m+1}) \\ &= |p(w)|^2 + 2\varepsilon^m r \cos(m\theta + \phi) + O(\varepsilon^{m+1}) \end{aligned}$$

Set $\theta = \frac{1}{m}(\pi - \phi)$, so that $\phi = \pi - m\theta$. Then $\cos(m\theta + \phi) = \cos(\pi) = -1$, and

$$|p(w + \varepsilon e^{i\theta})|^2 = |p(w)|^2 - 2\varepsilon^m r + O(\varepsilon^{m+1})$$

But $\varepsilon, r > 0$, so for sufficiently small $\varepsilon$, we have

$$|p(w + \varepsilon e^{i\theta})|^2 < |p(w)|^2$$

contradicting the definition of $w$. Therefore there exists $z_0 \in \mathbb{C}$ such that $p(z_0) = 0$. □

## 2.3 Implications

The Fundamental Theorem of Algebra has some useful implications. One is:

**Theorem 2.5 (Remainder Theorem).** *Let $p(t) \in \mathbb{C}[t]$ with $\partial p \geq 1$, and let $\alpha \in \mathbb{C}$.*

(1) *There exist $q(t) \in \mathbb{C}[t]$ and $r \in \mathbb{C}$ such that $p(t) = (t - \alpha)q(t) + r$.*

(2) *The constant $r$ satisfies $r = p(\alpha)$.*

*Proof.* Let $y = t - \alpha$ so that $t = y + \alpha$. Write $p(t) = p_n t^n + \cdots + p_0$ where $p_n \neq 0$ and $n \geq 1$. Then

$$p(t) = p_n(y + \alpha)^n + \cdots + p_0$$

Expand the powers of $y + \alpha$ by the binomial theorem, and collect terms to get

$$\begin{aligned} p(t) &= a_n y^n + \cdots + a_1 y + a_0 \qquad a_j \in \mathbb{C} \\ &= y(a_n y^{n-1} + \cdots + a_1) + a_0 \\ &= (t - \alpha)q(t) + r \end{aligned}$$

where

$$\begin{aligned} q(t) &= a_n(t - \alpha)^{n-1} + \cdots + a_2(t - \alpha) + a_1 0 \\ r &= a_0 \end{aligned}$$

Now substitute $t = \alpha$ in the identity $p(t) = (t - \alpha)q(t) + r$ to get

$$p(\alpha) = (\alpha - \alpha)q(\alpha) + r = 0.q(\alpha) + r = r$$

□

**Corollary 2.6.** *The complex number $\alpha$ is a zero of $p(t)$ if there exists $q(t)$ such that $p(t) = (t - \alpha)q(t)$ in $\mathbb{C}[t]$.* □

**Proposition 2.7.** *Let $p(t) \in \mathbb{C}[t]$ with $\partial p = n \geq 1$. Then there exist $\alpha_1, \ldots, \alpha_n \in \mathbb{C}$, and $0 \neq k \in \mathbb{C}$, such that*

$$p(t) = k(t - \alpha_1) \ldots (t - \alpha_n) \tag{2.5}$$

*Proof.* Use induction on $n$. The case $n = 1$ is obvious. If $n > 1$ we know, by the Fundamental Theorem of Algebra, that $p(t)$ has at least one zero in $\mathbb{C}$: call this zero $\alpha_n$. By the Remainder Theorem, there exists $q(t) \in \mathbb{C}[t]$ such that

$$p(t) = (t - \alpha_n)q(t) \tag{2.6}$$

(the remainder is $r = p(\alpha_n) = 0$). Then $\partial q = n - 1$, so by induction

$$q(t) = k(t - \alpha_1) \ldots (t - \alpha_{n-1}) \tag{2.7}$$

For suitable complex numbers $k, \alpha_1, \ldots, \alpha_{n-1}$. Substitute (2.7) in (2.6) and the induction step is complete. □

It follows immediately that the $\alpha_j$ are the *only* complex zeros of $p(t)$.

The zeros $\alpha_j$ need not be distinct. Collecting together those that are equal, we can rewrite (2.5) in the form

$$p(t) = k(t - \beta_1)^{m_1} \ldots (t - \beta_l)^{m_l}$$

where $k = a_n$, the $\beta_j$ are distinct, the $m_j$ are integers $\geq 1$, and $m_1 + \cdots + m_l = n$. We call $m_j$ the *multiplicity* of the zero $\beta_j$ of $p(t)$. If $m_j = 1$ we say that the zero $\beta_j$ is *simple*.

In particular, we have proved that every complex polynomial of degree $n$ has precisely $n$ complex zeros, counted according to multiplicity.

---

## EXERCISES

2.1 Let $p(t) \in \mathbb{Q}[t]$. Show that $p(t)$ has a unique expression in the form

$$p(t) = (t - \alpha_1) \ldots (t - \alpha_r) q(t)$$

(unique, that is, except for re-ordering the $\alpha_j$) where $\alpha_j \in \mathbb{Q}$ for $1 \leq j \leq r$ and $q(t)$ has no zeros in $\mathbb{Q}[t]$. Prove that here, the $\alpha_j$ are precisely the zeros of $p(t)$ in $\mathbb{Q}$.

2.2 A formal definition of $\mathbb{C}[t]$ runs as follows. Consider the set $S$ of all infinite sequences

$$(a_n)_{n \in \mathbb{N}} = (a_0, a_1, \ldots, a_n, \ldots)$$

where $a_n \in \mathbb{C}$ for all $n \in \mathbb{N}$, and such that $a_n = 0$ for all but a finite set of $n$. Define operations of addition and multiplication on $S$ by the rules

$$(a_n) + (b_n) = (u_n) \quad \text{where} \quad u_n = a_n + b_n$$
$$(a_n)(s_n) = (v_n) \quad \text{where} \quad v_n = a_n b_0 + a_{n-1} b_1 + \cdots + a_0 b_n$$

Prove that $\mathbb{C}[t]$, so defined, satisfies all of the usual laws of algebra for addition, subtraction, and multiplication. Define the map

$$\theta : \mathbb{C} \to S$$
$$\theta(k) = (k, 0, 0, 0, \ldots)$$

and prove that $\theta(\mathbb{C}) \subseteq S$ is isomorphic to $\mathbb{C}$.

Finally, prove that if we identify $a \in \mathbb{C}$ with $\theta(a) \in S$ and the 'indeterminate' $t$ with $(0, 1, 0, 0, 0, \ldots) \in S$, then $(a_n) = a_0 + \cdots + a_N t^N$, where $N$ is chosen so that $a_n = 0$ for $n > N$. Thus we can define polynomials as sequences of complex numbers corresponding to the coefficients.

2.3 Using (2.1, 2.2), prove that polynomials over $\mathbb{C}$ obey the following algebraic laws:
$f + g = g + f$, $f + (g + h) = (f + g) + h$, $fg = gf$, $f(gh) = (fg)h$, and $f(g + h) = fg + fh$.

2.4 Show that $\partial(f+g)$ can be less than $\max(\partial f, \partial g)$, and indeed that $\partial(f+g)$ can be less than $\min(\partial f, \partial g)$.

2.5* If $z_1, z_2, \ldots, z_n$ are distinct complex numbers, show that the determinant

$$D = \begin{vmatrix} 1 & 1 & \cdots & 1 \\ z_1 & z_2 & \cdots & z_n \\ z_1^2 & z_2^2 & \cdots & z_n^2 \\ \vdots & \vdots & \ddots & \vdots \\ z_1^{n-1} & z_2^{n-1} & \cdots & z_n^{n-1} \end{vmatrix}$$

is non-zero.

(*Hint:* Consider the $z_j$ as independent indeterminates over $\mathbb{C}$. Then $D$ is a polynomial in the $z_j$, of total degree $0+1+2+\cdots+(n-1) = \frac{1}{2}n(n-1)$. Moreover, $D$ vanishes whenever $z_j = z_k$, for $k \neq j$, since it then has two identical rows. Therefore $D$ is divisible by $z_j - z_k$ for all $j \neq k$, hence it is divisible by $\prod_{j<k}(z_j - z_k)$. Now compare degrees.)

The determinant $D$ is called a *Vandermonde determinant*, for obscure reasons (no such expression occurs in Vandermonde's published writings).

2.6 Use the Vandermonde determinant to prove that if a polynomial $f(t)$ vanishes for all $t \in \mathbb{C}$, then all coefficients of $f$ are zero. (*Hint.* Substitute $t = 1, 2, 3, \ldots$ and solve the resulting system of linear equations for the coefficients.)

2.7 Prove that the polynomial ring $\mathbb{C}[t]$ has no divisors of zero. That is, if $f(t)g(t) = \mathbf{0}$, then either $f(t) = \mathbf{0}$ or $g(t) = \mathbf{0}$. (*Hint*: Equivalently, prove that if $f(t) \neq \mathbf{0}$ and $g(t) \neq \mathbf{0}$, then $f(t)g(t) \neq \mathbf{0}$. To do so, consider the leading terms of $f(t), g(t)$, and $f(t)g(t)$, where the leading term of a polynomial $p(t) = a_n t^n + a_{n-1}t^{n-1} + \cdots + a_0$ is the term $a_m t^m$ where $m$ is the largest exponent such that $a_m \neq 0$.)

2.8 An alternative proof for the result of Exercise 2.6 runs like this:

Suppose that $f(t) = a_n t^n + \cdots + a_0 = 0$ for all $t$. We prove by induction on $n$ that all $a_j = 0$. This is clear when $n = 0$. When $n > 0$, set $t = 0$ to deduce that $a_0 = 0$. Therefore

$$t(a_n t^{n-1} + \cdots + a_1) = 0 \quad \forall t \in \mathbb{C}$$

Divide by $t$ to get $(a_n t^{n-1} + \cdots + a_1) = 0$ for all $t$. By induction, $a_n = \ldots = a_1 = 0$. Since $a_0 = 0$, all coefficients of $f$ are zero.

In this proof we first set $t = 0$; then later we divide through by $t$. So is this proof valid, or is it fallacious? (*Hint*: Consider Exercise 2.7.)

2.9 Prove, without using the Fundamental Theorem of Algebra, that every cubic polynomial over $\mathbb{R}$ can be expressed as a product of linear factors over $\mathbb{C}$.

2.10* Do the same for cubic polynomials over $\mathbb{C}$.

2.11 Mark the following true or false. Here $f, g$ are polynomials over $\mathbb{C}$.

(a) $\partial(f - g) \geq \min(\partial f, \partial g)$.

(b) $\partial(f - g) \leq \min(\partial f, \partial g)$.

(c) $\partial(f - g) \leq \max(\partial f, \partial g)$.

(d) $\partial(f - g) \geq \max(\partial f, \partial g)$.

(e) Every polynomial over $\mathbb{C}$ has at least one zero in $\mathbb{C}$.

(f) Every polynomial over $\mathbb{C}$ of degree $\geq 1$ has at least one zero in $\mathbb{R}$.

# Chapter 3

# Factorisation of Polynomials

Not only is there an algebra of polynomials: there is an arithmetic. That is, there are notions analogous to the integer-based concepts of divisibility, primes, prime factorisation, and greatest common divisors. These notions are essential for any serious understanding of polynomial equations, and we develop them in this chapter.

Mathematicians noticed early on that if $f$ is a product $gh$ of polynomials of smaller degree, then the solutions of $f(t) = 0$ are precisely those of $g(t) = 0$ together with those of $h(t) = 0$. For example, to solve the equation

$$t^3 - 6t^2 + 11t - 6 = 0$$

we can spot the factorisation $(t-1)(t-2)(t-3)$ and deduce that the roots are $t = 1, 2, 3$. From this simple idea emerged the arithmetic of polynomials—a systematic study of divisibility properties of polynomials with particular reference to analogies with the integers. In particular, there is an analogue for polynomials of the Euclidean Algorithm for finding the greatest common divisor of two integers.

In this chapter we define the relevant notions of divisibility and show that there are certain polynomials, the 'irreducible' ones, that play a similar role to prime numbers in the ring of integers. Every polynomial over a given subfield of $\mathbb{C}$ can be expressed as a product of irreducible polynomials over the same subfield, in an essentially unique way. We relate zeros of polynomials to the factorisation theory.

Throughout this chapter all polynomials are assumed to lie in $K[t]$, where $K$ is a subfield of the complex numbers, or in $R[t]$, where $R$ is a subring of the complex numbers. Some theorems are valid over $R$, while others are valid only over $K$: we will need both types.

## 3.1 The Euclidean Algorithm

In number theory, one of the key concepts is divisibility: an integer $a$ is divisible by an integer $b$ if there exists an integer $c$ such that $a = bc$. For instance, 60 is divisible by 3 since $60 = 3.20$, but 60 is not divisible by 7.

DOI: 10.1201/9781003213949-3

Divisibility properties of integers lead to such ideas as primes and factorisation. We wish to develop similar ideas for polynomials. As a start:

**Definition 3.1.** Let $f, g \in \mathbb{C}[t]$. Then $f$ *divides* $g$ if there exists $h(t) \in \mathbb{C}[t]$ such that $g(t) = f(t)h(t)$.

We also say that $f$ is a *factor* of $g$, and that $g$ is a *multiple* of $f$.

Many important results in the factorisation theory of polynomials derive from the observation that one polynomial may always be divided by another provided that a 'remainder' term is allowed. This is a generalisation of the Remainder Theorem, replacing $(t - \alpha)$ by any polynomial.

**Proposition 3.2 (Division Algorithm).** *Let $f$ and $g$ be polynomials over a subfield $K$ of $\mathbb{C}$, and suppose that $f$ is non-zero. Then there exist unique polynomials $q$ and $r$ over $K$ such that $g = fq + r$ and $r$ has strictly smaller degree than $f$.*

*Proof.* Use induction on the degree of $g$. If $\partial g = -\infty$ then $g = 0$ and we may take $q = r = 0$. If $\partial g = 0$ then $g = k$ is an element of $K$. If also $\partial f = 0$ then $f$ is an element of $K$, and we may take $q = k/f$ and $r = 0$. Otherwise $\partial f > 0$ and we may take $q = 0$ and $r = g$. This starts the induction.

Now assume the result whenever the degree of $g$ is less than $n$, and let $\partial g = n > 0$. If $\partial f > \partial g$, then we may as before take $q = 0$, $r = g$. Otherwise

$$f = a_m t^m + \cdots + a_0 \qquad g = b_n t^n + \cdots + b_0$$

where $a_m \neq 0 \neq b_n$ and $m \leq n$. Let

$$g_1 = b_n a_m^{-1} t^{n-m} f - g$$

Since the terms of highest degree cancel, we have $\partial g_1 < \partial g$. By induction there are polynomials $q_1$ and $r_1$ over $K$ such that $g_1 = fq_1 + r_1$ and $\partial r_1 < \partial f$. Let

$$q = b_n a_m^{-1} t^{n-m} - q_1 \qquad r = -r_1$$

Then

$$fq + r = b_n a_m^{-1} t^{n-m} f - q_1 f - r_1 = g + g_1 - g_1 = g$$

so $g = fq + r$; clearly $\partial r < \partial f$ as required.

Finally we prove uniqueness. Suppose that

$$g = fq_1 + r_1 = fq_2 + r_2 \quad \text{where } \partial r_1, \partial r_2 < \partial f$$

Then $f(q_1 - q_2) = r_2 - r_1$. By Proposition 2.2, the polynomial on the left has higher degree than that on the right, unless both are zero. Since $f \neq 0$ we must have $q_1 = q_2$ and $r_1 = r_2$. Thus $q$ and $r$ are unique. □

With the above notation, $q$ is the *quotient* and $r$ is the *remainder* on dividing $g$ by $f$. The inductive process employed to find $q$ and $r$ is the *Division Algorithm*.

**Example 3.3.** Divide $g(t) = t^4 - 7t^3 + 5t^2 + 4$ by $f = t^2 + 3$ and find the quotient and remainder.

Observe that

$$t^2(t^2 + 3) = t^4 + 3t^2$$

has the same leading coefficient as $g$. Subtracting,

$$g - t^2(t^2 + 3) = -7t^3 + 2t^2 + 4$$

which has the same leading coefficient as

$$-7t(t^2 + 3) = -7t^3 - 21t$$

Therefore

$$g - t^2(t^2 + 3) + 7t(t^2 + 3) = 2t^2 + 21t + 4$$

which has the same leading coefficient as

$$2(t^2 + 3) = 2t^2 + 6$$

Therefore

$$g - t^2(t^2 + 3) + 7t(t^2 + 3) - 2(t^2 + 3) = 21t - 2$$

So

$$g = (t^2 + 3)(t^2 - 7t + 2) + (21t - 2)$$

and the quotient $q(t) = t^2 - 7t + 2$, while the remainder $r(t) = 21t - 2$.

The next step is to introduce notions of divisibility for polynomials, and in particular the idea of 'greatest common divisor,' which is crucial to the arithmetic of polynomials. (Another term is 'highest common factor', but this has gone out of fashion.)

**Definition 3.4.** Let $f$ and $g$ be polynomials over a subfield $K$ of $\mathbb{C}$. We say that $f$ *divides* $g$ (or $f$ is a *factor* of $g$, or $g$ is a *multiple* of $f$) if there exists some polynomial $h$ over $K$ such that $g = fh$. The notation $f|g$ will mean that $f$ divides $g$, while $f \nmid g$ will mean that $f$ does not divide $g$.

**Definition 3.5.** A polynomial $d$ over a subfield $K$ of $\mathbb{C}$ is a *greatest common divisor* (gcd) of polynomials $f$ and $g$ over $K$ if $d|f$ and $d|g$ and further, whenever $e|f$ and $e|g$, we have $e|d$.

We have said *a* greatest common divisor rather than *the* greatest common divisor because gcd's need not be unique. However, the next lemma shows that they are unique apart from constant factors.

**Lemma 3.6.** *If $d$ is a gcd of the polynomials $f$ and $g$ over $K$, and if $0 \neq k \in K$, then $kd$ is also a gcd for $f$ and $g$.*

*If $d$ and $e$ are two gcd's for $f$ and $g$, then there exists a non-zero element $k \in K$ such that $e = kd$.*

*Proof.* Clearly $kd|f$ and $kd|g$. If $e|f$ and $e|g$, then $e|d$ so that $e|kd$. Hence $kd$ is a gcd.

If $d$ and $e$ are gcd's, then by definition $e|d$ and $d|e$. Thus $e = kd$ for some polynomial $k$. Since $e|d$ the degree of $e$ is less than or equal to the degree of $d$, so $k$ must have degree $\leq 0$. Therefore $k$ is a constant, and so belongs to $K$. Since $0 \neq e = kd$, we must have $k \neq 0$. □

To obtain a unique gcd we need a technical definition:

**Definition 3.7.** A polynomial $f(t) = a_n t^n + a_{n-1}t^{n-1} + \cdots + a_1 t + a_0$ over a subfield $K$ of $\mathbb{C}$ is *monic* if $a_n = 1$.

A unique choice of gcd can be obtained by requiring it to be monic. This determines a unique constant factor. We write $\gcd(f, g)$ for this choice of gcd.

We shall prove that any two non-zero polynomials have a gcd by providing a method to calculate one. This method is a generalisation of the technique used by Euclid (*Elements* Book 7 Proposition 2) around 300 BC for calculating gcd's of integers and is accordingly known as the *Euclidean Algorithm.*

**Algorithm 3.8 (Euclidean Algorithm).**

***Ingredients*** Two polynomials $f$ and $g$ over $K$, both non-zero.

***Recipe*** For notational convenience let $f = r_{-1}$, $g = r_0$. Use the Division Algorithm to find successively polynomials $q_j$ and $r_i$ such that

$$\begin{aligned}
r_{-1} &= q_1 r_0 + r_1 & \partial r_1 &< \partial r_0 \\
r_0 &= q_2 r_1 + r_2 & \partial r_2 &< \partial r_1 \\
r_1 &= q_3 r_2 + r_3 & \partial r_3 &< \partial r_2 \\
&\vdots \\
r_i &= q_{i+2} r_{i+1} + r_{i+2} & \partial r_{i+2} &< \partial r_{i+1} \\
&\vdots
\end{aligned} \tag{3.1}$$

Since the degrees of the $r_i$ decrease, we must eventually reach a point where the process stops; this can happen only if some $r_{s+2} = 0$. The last equation in the list then reads

$$r_s = q_{s+2} r_{s+1} \tag{3.2}$$

and it provides the answer we seek:

**Theorem 3.9.** *With the above notation, $r_{s+1}$ is a gcd for $f$ and $g$.*

*Proof.* First we show that $r_{s+1}$ divides both $f$ and $g$. We use descending induction to show that $r_{s+1}|r_i$ for all $i$. Clearly $r_{s+1}|r_{s+1}$. Equation (3.2) shows that $r_{s+1}|r_s$. Equation (3.1) implies that if $r_{s+1}|r_{i+2}$ and $r_{s+1}|r_{i+1}$ then $r_{s+1}|r_i$. Hence $r_{s+1}|r_i$ for all $i$; in particular $r_{s+1}|r_0 = g$ and $r_{s+1}|r_{-1} = f$.

Now suppose that $e|f$ and $e|g$. By (3.1) and induction, $e|r_i$ for all $i$. In particular, $e|r_{s+1}$. Therefore $r_{s+1}$ is a gcd for $f$ and $g$, as claimed. □

**Example 3.10.** Let $f = t^4 + 2t^3 + 2t^2 + 2t + 1$, $g = t^2 - 1$ over $\mathbb{Q}$. We compute a gcd as follows:

$$t^4 + 2t^3 + 2t^2 + 2t + 1 = (t^2 + 2t + 3)(t^2 - 1) + 4t + 4$$
$$t^2 - 1 = (4t + 4)(\tfrac{1}{4}t - \tfrac{1}{4})$$

Hence $4t + 4$ is a gcd. So is any rational multiple of it, in particular, $t + 1$.

We end this section by deducing from the Euclidean Algorithm an important property of the gcd of two polynomials.

**Theorem 3.11.** *Let $f$ and $g$ be non-zero polynomials over $K$, and let $d$ be a gcd for $f$ and $g$. Then there exist polynomials $a$ and $b$ over $K$ such that*

$$d = af + bg$$

*Proof.* Since gcd's are unique up to constant factors we may assume that $d = r_{s+1}$ where equations (3.1) and (3.2) hold. We claim as induction hypothesis that there exist polynomials $a_i$ and $b_i$ such that

$$d = a_i r_i + b_i r_{i+1}$$

This is clearly true when $i = s + 1$, for we may then take $a_i = 1$, $b_i = 0$. By (3.1)

$$r_{i+1} = r_{i-1} - q_{i+1} r_i$$

so by induction

$$d = a_i r_i + b_i (r_{i-1} - q_{i+1} r_i)$$

If we put

$$a_{i-1} = b_i \qquad b_{i-1} = a_i - b_i q_{i+1}$$

then

$$d = a_{i-1} r_{i-1} + b_{i-1} r_i$$

and by descending induction

$$d = a_{-1} r_{-1} + b_{-1} r_0 = af + bg$$

where $a = a_{-1}$, $b = b_{-1}$. This completes the proof. □

The induction step above affords a practical method of calculating $a$ and $b$ in any particular case.

## 3.2 Irreducibility

Now we investigate the analogue, for polynomials, of prime numbers. The concept required is 'irreducibility'. In particular, we prove that every polynomial over a subring of $\mathbb{C}$ can be expressed as a product of irreducibles in an 'essentially' unique way.

An integer is prime if it cannot be expressed as a product of smaller integers. The analogue for polynomials is similar: we interpret 'smaller' as 'smaller degree'. So the following definition yields the polynomial analogue of a prime number.

**Definition 3.12.** A non-constant polynomial over a subring $R$ of $\mathbb{C}$ is *reducible* if it is a product of two polynomials over $R$ of smaller degree. Otherwise it is *irreducible.*

**Examples 3.13.** (1) All polynomials of degree 1 are irreducible, since they certainly cannot be expressed as a product of polynomials of smaller degree.
(2) The polynomial $t^2 - 2$ is irreducible over $\mathbb{Q}$. To show this we suppose, for a contradiction, that it is reducible. Then

$$t^2 - 2 = (at + b)(ct + d)$$

where $a, b, c, d, \in \mathbb{Q}$. Dividing out if necessary we may assume $a = c = 1$. Then $b + d = 0$ and $bd = -2$, so that $b^2 = 2$. But no rational number has its square equal to 2 (Exercise 1.2).
(3) However, $t^2 - 2$ is reducible over the larger subfield $\mathbb{R}$, for now

$$t^2 - 2 = (t - \sqrt{2})(t + \sqrt{2})$$

This shows that an irreducible polynomial may become reducible over a larger subfield of $\mathbb{C}$.
(4) The polynomial $6t + 3$ is irreducible in $\mathbb{Z}[t]$. Although it has factors

$$6t + 3 = 3(2t + 1)$$

the degree of $2t + 1$ is the same as that of $6t + 6$. So this factorisation does not count.

Any reducible polynomial can be written as the product of two polynomials of smaller degree. If either of these is reducible it too can be split up into factors of smaller degree ... and so on. This process must terminate since the degrees cannot decrease indefinitely. This is the idea behind the proof of:

**Theorem 3.14.** *Any non-zero polynomial over a subring $R$ of $\mathbb{C}$ is a product of irreducible polynomials over $R$.*

*Proof.* Let $g$ be any non-zero polynomial over $R$. We proceed by induction on the degree of $g$. If $\partial g = 0$ or 1, then $g$ is automatically irreducible. If $\partial g > 1$, then either $g$ is irreducible or $g = hk$ where $\partial h, \partial k < \partial g$. By induction, $h$ and $k$ are products of irreducible polynomials, whence $g$ is such a product. The theorem follows by induction. □

**Example 3.15.** We can use Theorem 3.14 to prove irreducibility in some cases, especially for cubic polynomials over $\mathbb{Z}$. For instance, let $R = \mathbb{Z}$. The polynomial

$$f(t) = t^3 - 5t + 1$$

is irreducible. If not, then it must have a linear factor $t - \alpha$ over $\mathbb{Z}$, and then $\alpha \in \mathbb{Z}$ and $f(\alpha) = 0$. Moreover, there must exist $\beta, \gamma \in \mathbb{Z}$ such that

$$\begin{aligned} f(t) &= (t-\alpha)(t^2 + \beta t + \gamma) \\ &= t^3 + (\beta - \alpha)t^2 + (\gamma - \alpha\beta)t - \alpha\gamma \end{aligned}$$

so in particular $\alpha\gamma = -1$. Therefore $\alpha = \pm 1$. But $f(1) = -3 \neq 0$ and $f(-1) = 5 \neq 0$. Therefore no such factor exists.

Irreducible polynomials are analogous to prime numbers. The importance of prime numbers in $\mathbb{Z}$ stems in part from the possibility of factorising every integer into primes, but even more so from the *uniqueness* (up to order) of the prime factors. Likewise the importance of irreducible polynomials depends upon a uniqueness theorem. Uniqueness of factorisation is not obvious, see Stewart and Tall (2015b) Chapter 4. In certain cases it is possible to express every element as a product of irreducible elements, without this expression being in any way unique. We shall heed the warning and prove the uniqueness of factorisation for polynomials. To avoid technical issues concerning factors of ring elements, we restrict attention to polynomials over a sub*field* $K$ of $\mathbb{C}$. It is possible to prove more general theorems by introducing the idea of a 'unique factorisation domain', see Fraleigh (1989) Chapter 6.

For convenience we make the following:

**Definition 3.16.** If $f$ and $g$ are polynomials over a subfield $K$ of $\mathbb{C}$ with gcd equal to 1, we say that $f$ and $g$ are *coprime*, or $f$ *is prime to* $g$. (These are the original terms, see Hardy and Wright (1962). Instead of 'prime to', the phrase 'coprime to' is common, perhaps because it adds emphasis. However, the prefix 'co' and the 'to' say the same thing, so the 'co' is redundant. In any case, 'coprime *with*' is better grammatically.)

The key to unique factorisation is a statement analogous to an important property of primes in $\mathbb{Z}$ and is used in the same way:

**Lemma 3.17.** *Let $K$ be a subfield of $\mathbb{C}$, $f$ an irreducible polynomial over $K$, and $g, h$ polynomials over $K$. If $f$ divides $gh$, then either $f$ divides $g$ or $f$ divides $h$.*

*Proof.* Suppose that $f \nmid g$. We claim that $f$ and $g$ are coprime. For if $d$ is a gcd for $f$ and $g$, then since $f$ is irreducible and $d|f$, either $d = kf$ for some $k \in K$, or $d = k \in K$. In the first case $f|g$, contrary to hypothesis. In the second case, 1 is also a gcd for $f$ and $g$, so they are coprime. By Theorem 3.11, there exist polynomials $a$ and $b$ over $K$ such that

$$1 = af + bg$$

Then

$$h = haf + hbg$$

Now $f|haf$, and $f|hbg$ since $f|gh$. Hence $f|h$. This completes the proof. □

We may now prove the uniqueness theorem.

**Theorem 3.18.** *For any subfield $K$ of $\mathbb{C}$, factorisation of polynomials over $K$ into irreducible polynomials is unique up to constant factors and the order in which the factors are written.*

*Proof.* Suppose that $f = f_1 \dots f_r = g_1 \dots g_s$ where $f$ is a polynomial over $K$ and $f_1, \dots, f_r, g_1, \dots, g_s$ are irreducible polynomials over $K$. If all the $f_i$ are constant, then $f \in K$, so all the $g_j$ are constant. Otherwise we may assume that no $f_i$ is constant, by dividing out all of the constant terms. Then $f_1|g_1 \dots g_s$. By an obvious induction based on Lemma 3.17, $f_1|g_j$ for some $j$. We can choose notation so that $j = 1$, and then $f_1|g_1$. Since $f_1$ and $g_1$ are irreducible and $f_1$ is not a constant, we must have $f_1 = k_1 g_1$ for some constant $k_1$. Similarly $f_2 = k_2 g_2, \dots, f_r = k_r g_r$ where $k_2, \dots, k_r$ are constant. The remaining $g_l (l > r)$ must also be constant, or else the degree of the right-hand side would be too large. The theorem is proved. □

## 3.3 Gauss's Lemma

It is in general very difficult to decide—without using computer algebra, at any rate—whether a given polynomial is irreducible. As an example, think about

$$t^{16}+t^{15}+t^{14}+t^{13}+t^{12}+t^{11}+t^{10}+t^{9}+t^{8}+t^{7}+t^{6}+t^{5}+t^{4}+t^{3}+t^{2}+t+1 \quad (3.3)$$

This is not an idle example: we shall be considering precisely this polynomial in Chapter 20, in connection with the regular 17-gon, and its irreducibility (or not) will be crucial.

To test for irreducibility by trying all possible factors is usually futile. Indeed, at first sight there are infinitely many potential factors to try, although with suitable short cuts the possibilities can be reduced to a finite—usually unfeasibly large—number. In principle the resulting method can be applied to polynomials over $\mathbb{Q}$, for example: see van der Waerden (1953), Garling (1987). But the method is not really practicable.

Computer algebra packages generally have excellent methods for factorising polynomials over the integers. For hand calculation, and some theoretical purposes, there are a few useful tricks. In the next two sections we describe two of them: Eisenstein's Criterion and reduction modulo a prime. Both tricks apply in the first instance to polynomials over $\mathbb{Z}$. However, we now prove that irreducibility over $\mathbb{Z}$ is equivalent to irreducibility over $\mathbb{Q}$. This extremely useful result was proved by Gauss, and we use it repeatedly.

**Lemma 3.19 (Gauss's Lemma).** *Let $f$ be a polynomial over $\mathbb{Z}$ that is irreducible over $\mathbb{Z}$. Then $f$, considered as a polynomial over $\mathbb{Q}$, is also irreducible over $\mathbb{Q}$.*

*Proof.* The point of this lemma is that when we extend the subring of coefficients from $\mathbb{Z}$ to $\mathbb{Q}$, there are hosts of new polynomials which, perhaps, might be factors of $f$. We show that in fact they are not. For a contradiction, suppose that $f$ is irreducible over $\mathbb{Z}$ but reducible over $\mathbb{Q}$, so that $f = gh$ where $g$ and $h$ are polynomials over $\mathbb{Q}$, of smaller degree, and seek a contradiction. Multiplying through by the product of the denominators of the coefficients of $g$ and $h$, we can rewrite this equation in the form $nf = g'h'$, where $n \in \mathbb{Z}$ and $g'$, $h'$ are polynomials over $\mathbb{Z}$. We now show that we can cancel out the prime factors of $n$ one by one, without going outside $\mathbb{Z}[t]$.

Suppose that $p$ is a prime factor of $n$. We claim that if

$$g' = g_0 + g_1 t + \cdots + g_r t^r \qquad h' = h_0 + h_1 t + \cdots + h_s t^s$$

then either $p$ divides all the coefficients $g_i$, or else $p$ divides all the coefficients $h_j$. If not, there must be smallest values $i$ and $j$ such that $p \nmid g_i$ and $p \nmid h_j$. However, $p$ divides the coefficient of $t^{i+j}$ in $g'h'$, which is

$$h_0 g_{i+j} + h_1 g_{i+j-1} + \cdots + h_j g_i + \cdots + h_{i+j} g_0$$

and by the choice of $i$ and $j$, the prime $p$ divides every term of this expression except perhaps $h_j g_i$. But $p$ divides the whole expression, so $p | h_j g_i$. However, $p \nmid h_j$ and $p \nmid g_i$, a contradiction. This establishes the claim.

Without loss of generality, we may assume that $p$ divides every coefficient $g_i$. Then $g' = pg''$ where $g''$ is a polynomial over $\mathbb{Z}$ of the same degree as $g'$ (or $g$). Let $n = pn_1$. Then $pn_1 f = pg''h'$, so that $n_1 f = g''h'$. Proceeding in this way we can remove all the prime factors of $n$, arriving at an equation $f = \bar{g}\bar{h}$. Here $\bar{g}$ and $\bar{h}$ are polynomials over $\mathbb{Z}$, which are rational multiples of the original $g$ and $h$, so $\partial\bar{g} = \partial g$ and $\partial\bar{h} = \partial h$. But this contradicts the irreducibility of $f$ over $\mathbb{Z}$, so the lemma is proved. □

**Corollary 3.20.** *Let $f \in \mathbb{Z}[t]$ and suppose that over $\mathbb{Q}[t]$ there is a factorisation into irreducibles:*

$$f = g_1 \cdots g_s$$

*Then there exist* $a_i \in \mathbb{Q}$ *such that* $a_i g_i \in \mathbb{Z}[t]$ *and* $a_1 \dots a_s = 1$. *Furthermore,*

$$f = (a_1 g_1) \dots (a_s g_s)$$

*is a factorisation of* $f$ *into irreducibles in* $\mathbb{Z}[t]$.

*Proof.* Factorise $f$ into irreducibles over $\mathbb{Z}[t]$, obtaining $f = h_1 \dots h_r$. By Gauss's Lemma, each $h_j$ is irreducible over $\mathbb{Q}$. By uniqueness of factorisation in $\mathbb{Q}[t]$, we must have $r = s$ and $h_j = a_j g_j$ for $a_j \in \mathbb{Q}$. Clearly $a_1 \dots a_s = 1$. The Corollary is now proved. □

---

## 3.4 Eisenstein's Criterion

No, not 'Einstein'. Ferdinand Gotthold Eisenstein was a student of Gauss, and greatly impressed his tutor. We can apply the tutor's lemma to prove the student's criterion for irreducibility:

**Theorem 3.21 (Eisenstein's Criterion).** *Let*

$$f(t) = a_0 + a_1 t + \cdots + a_n t^n$$

*be a polynomial over* $\mathbb{Z}$. *Suppose that there is a prime* $q$ *such that*

(1) $q \nmid a_n$

(2) $q | a_i \ (i = 0, \dots, n-1)$

(3) $q^2 \nmid a_0$

*Then* $f$ *is irreducible over* $\mathbb{Q}$.

*Proof.* By Gauss's Lemma it is sufficient to show that $f$ is irreducible over $\mathbb{Z}$. Suppose for a contradiction that $f = gh$, where

$$g = b_0 + b_1 t + \cdots + b_r t^r \qquad h = c_0 + c_1 t + \cdots + c_s t^s$$

are polynomials of smaller degree over $\mathbb{Z}$. Then $r \geq 1, s \geq 1$, and $r + s = n$. Now $b_0 c_0 = a_0$ so by (2) $q | b_0$ or $q | c_0$. By (3) $q$ cannot divide both $b_0$ and $c_0$, so without loss of generality we can assume $q | b_0$, $q \nmid c_0$. If all $b_j$ are divisible by $q$, then $a_n$ is divisible by $q$, contrary to (1). Let $b_j$ be the first coefficient of $g$ not divisible by $q$. Then

$$a_j = b_j c_0 + \cdots + b_0 c_j$$

where $j < n$. This implies that $q$ divides $c_0$, since $q$ divides $a_j, b_0, \dots, b_{j-1}$, but not $b_j$. This is a contradiction. Hence $f$ is irreducible. □

**Example 3.22.** Consider

$$f(t) = \tfrac{2}{9}t^5 + \tfrac{5}{3}t^4 + t^3 + \tfrac{1}{3} \text{ over } \mathbb{Q}$$

This is irreducible over $\mathbb{Q}$ if and only if

$$9f(t) = 2t^5 + 15t^4 + 9t^3 + 3$$

is irreducible over $\mathbb{Q}$. Eisenstein's criterion now applies with $q = 3$, showing that $f$ is irreducible.

We now turn to the polynomial (3.3). This provides an instructive example that leads to a useful general result. In preparation, we prove a standard number-theoretic property of binomial coefficients:

**Lemma 3.23.** *If $p$ is prime, the binomial coefficient*

$$\binom{p}{r}$$

*is divisible by $p$ if $1 \leq r \leq p-1$.*

*Proof.* The binomial coefficient is an integer, and

$$\binom{p}{r} = \frac{p!}{r!(p-r)!}$$

The factor $p$ in the numerator cannot cancel with any factor in the denominator unless $r = 0$ or $r = p$. □

We then have:

**Lemma 3.24.** *If $p$ is a prime then the polynomial*

$$f(t) = t^{p-1} + t^{p-2} + \cdots + t + 1$$

*is irreducible over $\mathbb{Q}$.*

*Proof.* Note that $f(t) = (t^p - 1)/(t-1)$. Put $t = 1 + u$ where $u$ is a new indeterminate. Then $f(t)$ is irreducible over $\mathbb{Q}$ if and only if $f(1+u)$ is irreducible. But

$$f(1+u) = \frac{(1+u)^p - 1}{u} = u^{p-1} + ph(u)$$

where $h$ is a polynomial in $u$ over $\mathbb{Z}$ with constant term 1, by Lemma 3.23. By Eisenstein's Criterion, Theorem 3.21, $f(1+u)$ is irreducible over $\mathbb{Q}$. Hence $f(t)$ is irreducible over $\mathbb{Q}$. □

Setting $p = 17$ shows that the polynomial (3.3) is irreducible over $\mathbb{Q}$.

## 3.5 Reduction Modulo $p$

A second trick to prove irreducibility of polynomials in $\mathbb{Z}[t]$ involves 'reducing' the polynomial modulo a prime integer $p$.

Recall that if $n \in \mathbb{Z}$, then two integers $a, b$ are *congruent modulo* $n$ if $a - b$ is divisible by $n$. We write

$$a \equiv b \pmod{n}$$

The number $n$ is the *modulus*, and 'modulo' is Latin for 'to the modulus'. Congruence modulo $n$ is an equivalence relation, and the set of equivalence classes is denoted by $\mathbb{Z}_n$. Arithmetic in $\mathbb{Z}_n$ is just like arithmetic in $\mathbb{Z}$, except that $n \equiv 0$. (The notation $\mathbb{Z}_n$ is now slightly old-fashioned, mainly because $\mathbb{Z}_p$ tends to be used for $p$-adic integers, see Chapter 26. Instead, the symbols $\mathbb{Z}_{/n}$ or $\mathbb{Z}/n\mathbb{Z}$ are often used. Since we are not using $p$-adic numbers, the old-fashioned notation should cause no confusion and is neater.)

The test for irreducibility that we now wish to discuss is most easily explained by an example. The idea is this. There is a natural map $\mathbb{Z} \to \mathbb{Z}_n$ in which each $m \in \mathbb{Z}$ maps to its congruence class modulo $n$. The natural map extends in an obvious way to a map $\mathbb{Z}[t] \to \mathbb{Z}_n[t]$. Now a reducible polynomial over $\mathbb{Z}$ is a product $gh$ of polynomials of lower degree, and this factorisation is preserved by the map. Provided $n$ does not divide the highest coefficient of the given polynomial, the image is reducible over $\mathbb{Z}_n$. So if the image of a polynomial is irreducible over $\mathbb{Z}_n$, then the original polynomial must be irreducible over $\mathbb{Z}$. (The corresponding statement for reducible polynomials is in general false: consider $t^2 - 2 \in \mathbb{Z}[t]$ when $p = 2$.) Since $\mathbb{Z}_n$ is finite, there are only finitely many possibilities to check when deciding irreducibility.

In practice, the trick is to choose the right value for $n$.

**Example 3.25.** Consider

$$f(t) = t^4 + 15t^3 + 7 \text{ over } \mathbb{Z}$$

Over $\mathbb{Z}_5$ this becomes $t^4 + 2$. If this is reducible over $\mathbb{Z}_5$, then either it has a factor of degree 1, or it is a product of two factors of degree 2. The first possibility gives rise to an element $x \in \mathbb{Z}_5$ such that $x^4 + 2 = 0$. No such element exists (there are only five elements to check) so this case is ruled out. In the remaining case we have, without loss of generality,

$$t^4 + 2 = (t^2 + at + b)(t^2 + ct + d)$$

Therefore $a + c = 0, ac + b + d = 0, ad + bc = 0, bd = 2$. Combining $ad + bc = 0$ with $a + c = 0$ we get $a(b - d) = 0$. So either $a = 0$ or $b = d$.

If $a = 0$ then $c = 0$, so $b + d = 0, bd = 2$. That is, $b^2 = -2 = 3$ in $\mathbb{Z}_5$. But this is not possible.

If $b = d$ then $b^2 = 2$, also impossible in $\mathbb{Z}_5$.

Hence $t^4 + 2$ is irreducible over $\mathbb{Z}_5$, and therefore the original $f(t)$ is irreducible over $\mathbb{Z}$, hence over $\mathbb{Q}$.

If instead we work in $\mathbb{Z}_3$, then $f(t)$ becomes $t^4 + 1$, which equals $(t^2 + t - 1)(t^2 - t - 1)$ and so is reducible. Thus working (mod 3) fails to prove irreducibility.

---

## 3.6 Zeros of Polynomials

We have already studied the zeros of a polynomial over $\mathbb{C}$. It will be useful to employ similar terminology for polynomials over a subring $R$ of $\mathbb{C}$, because then we can keep track of where the zeros lie. We begin with a formal definition.

**Definition 3.26.** Let $R$ be a subring of $\mathbb{C}$, and let $f$ be a polynomial over $R$. An element $\alpha \in R$ such that $f(\alpha) = 0$ is a *zero of* $f$ *in* $R$.

To illustrate some basic phenomena associated with zeros, we consider polynomials over the real numbers. In this case, we can draw the graph $y = f(x)$ (in standard terminology, with $x \in \mathbb{R}$ in place of $t$). The graph might, for example, resemble Figure 9.

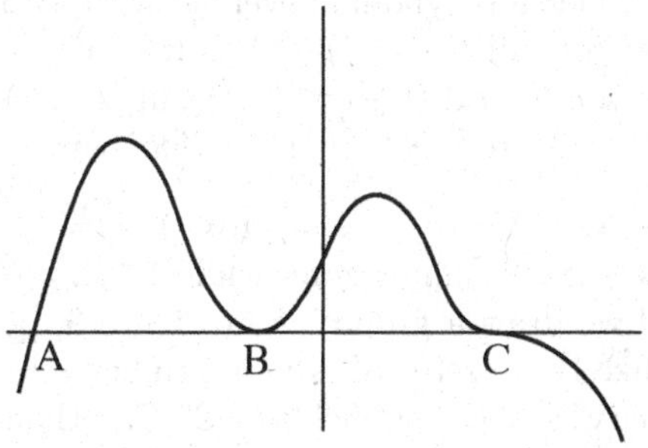


FIGURE 9: Multiple zeros of a (real) polynomial. The multiplicity is 1 at (A), 2 at (B), and 3 at (C).

The zeros of $f$ are the values of $x$ at which the curve crosses the $x$-axis. Consider the three zeros marked A, B, C in the diagram. At A the curve cuts straight through the axis; at B it 'bounces' off it; at C it 'slides' through horizontally. These phenomena are generally distinguished by saying that B and C are 'multiple zeros' of $f(t)$. The single zero B must be thought of as two equal zeros (or more) and C as three (or more).

But if they are equal, how can there be two of them? The answer is the concept of 'multiplicity' of a zero, introduced in Section 2.3. We now reformulate this concept *without* using the Fundamental Theorem of Algebra, which in this context is the proverbial nut-cracking sledgehammer. The key is to look at linear factors of $f$.

**Lemma 3.27.** *Let $f$ be a polynomial over the subfield $K$ of $\mathbb{C}$. An element $\alpha \in K$ is a zero of $f$ if and only if $(t-\alpha)|f(t)$ in $K[t]$.*

*Proof.* We know that $(t-\alpha)|f(t)$ in $\mathbb{C}[t]$ by Theorem 2.5, but we want slightly more. If $(t-\alpha)|f(t)$ in $K[t]$, then $f(t) = (t-\alpha)g(t)$ for some polynomial $g$ over $K$, so that $f(\alpha) = (\alpha - \alpha)g(\alpha) = 0$.

Conversely, suppose $f(\alpha) = 0$. By the Division Algorithm, there exist polynomials $q, r \in K[t]$ such that

$$f(t) = (t-\alpha)q(t) + r(t)$$

where $\partial r < 1$. Thus $r(t) = r \in K$. Substituting $\alpha$ for $t$,

$$0 = f(\alpha) = (\alpha - \alpha)q(\alpha) + r$$

so $r = 0$. Hence $(t-\alpha)|f(t) \in K[t]$ as required. □

We can now say what we mean by a multiple zero, without appealing to the Fundamental Theorem of Algebra.

**Definition 3.28.** Let $f$ be a polynomial over the subfield $K$ of $\mathbb{C}$. An element $\alpha \in K$ is a *simple zero* of $f$ if $(t-\alpha)|f(t)$ but $(t-\alpha)^2 \nmid f(t)$. The element $\alpha$ is a zero of $f$ of *multiplicity* $m$ if $(t-\alpha)^m|f(t)$ but $(t-\alpha)^{m+1} \nmid f(t)$. Zeros of multiplicity greater than 1 are *repeated* or *multiple zeros*.

For example, $t^3 - 3t + 2$ over $\mathbb{Q}$ has zeros at $\alpha = 1, -2$. It factorises as $(t-1)^2(t+2)$. Hence $-2$ is a simple zero, while 1 is a zero of multiplicity 2.

When $K = \mathbb{R}$ and we draw a graph, as in Figure 9, points like A are the simple zeros; points like B are zeros of even multiplicity; and points like C are zeros of odd multiplicity $> 1$. For subfields of $\mathbb{C}$ other than $\mathbb{R}$ (except perhaps $\mathbb{Q}$, or other subfields of $\mathbb{R}$) a graph has no evident meaning, but the simple geometric picture for $\mathbb{R}$ is often helpful.

**Lemma 3.29.** *Let $f$ be a non-zero polynomial over the subfield $K$ of $\mathbb{C}$, and let its distinct zeros be $\alpha_1, \ldots, \alpha_r$ with multiplicities $m_1, \ldots, m_r$ respectively. Then*

$$f(t) = (t-\alpha_1)^{m_1} \ldots (t-\alpha_r)^{m_r} g(t) \tag{3.4}$$

*where $g$ has no zeros in $K$.*

*Conversely, if* (3.4) *holds and $g$ has no zeros in $K$, then the zeros of $f$ in $K$ are $\alpha_1, \ldots, \alpha_r$, with multiplicities $m_1, \ldots, m_r$ respectively.*

*Proof.* For any $\alpha \in K$ the polynomial $t - \alpha$ is irreducible. Hence for distinct $\alpha, \beta \in K$ the polynomials $t - \alpha$ and $t - \beta$ are coprime in $K[t]$. By uniqueness of factorisation (Theorem 3.18) equation (3.4) must hold. Moreover, $g$ cannot have any zeros in $K$, or else $f$ would have extra zeros or zeros of larger multiplicity.

The converse follows easily from uniqueness of factorisation, Theorem 3.14 and Theorem 3.18. □

From this lemma we deduce a famous theorem:

**Theorem 3.30.** *The number of zeros of a nonzero polynomial over a subfield of* $\mathbb{C}$*, counted according to multiplicity, is less than or equal to its degree.*

*Proof.* In equation (3.4) we must have $m_1 + \cdots + m_r \leq \partial f$. □

---

## EXERCISES

3.1 For the following pairs of polynomials $f$ and $g$ over $\mathbb{Q}$, find the quotient and remainder on dividing $g$ by $f$.

(a) $g = t^7 - t^3 + 5$, $f = t^3 + 7$

(b) $g = t^2 + 1$, $f = t^2$

(c) $g = 4t^3 - 17t^2 + t - 3$, $f = 2t + 5$

(d) $g = t^4 - 1$, $f = t^2 + 1$

(e) $g = t^4 - 1$, $f = 3t^2 + 3t$

3.2 Find gcd's for these pairs of polynomials, and check that your results are common factors of $f$ and $g$.

3.3 Express these gcd's in the form $af + bg$.

3.4 Decide the irreducibility or otherwise of the following polynomials:

(a) $t^4 + 1$ over $\mathbb{R}$.

(b) $t^4 + 1$ over $\mathbb{Q}$.

(c) $t^7 + 11t^3 - 33t + 22$ over $\mathbb{Q}$.

(d) $t^4 + t^3 + t^2 + t + 1$ over $\mathbb{Q}$.

(e) $t^3 - 7t^2 + 3t + 3$ over $\mathbb{Q}$.

3.5 Decide the irreducibility or otherwise of the following polynomials:

(a) $t^4 + t^3 + t^2 + t + 1$ over $\mathbb{Q}$. (*Hint:* Substitute $t + 1$ in place of $t$ and appeal to Eisenstein's Criterion.)

(b) $t^5 + t^4 + t^3 + t^2 + t + 1$ over $\mathbb{Q}$.

(c) $t^6 + t^5 + t^4 + t^3 + t^2 + t + 1$ over $\mathbb{Q}$.

3.6 In each of the above cases, factorise the polynomial into irreducibles.

3.7 Say that a polynomial $f$ over a subfield $K$ of $\mathbb{C}$ is *prime* if whenever $f|gh$ either $f|g$ or $f|h$. Show that a polynomial $f \neq 0$ is prime if and only if it is irreducible.

3.8 Find the zeros of the following polynomials; first over $\mathbb{Q}$, then $\mathbb{R}$, then $\mathbb{C}$.

(a) $t^3 + 1$

(b) $t^3 - 6t^2 + 11t - 6$

(c) $t^5 + t + 1$

(d) $t^2 + 1$

(e) $t^4 + t^3 + t^2 + t + 1$

(f) $t^4 - 6t^2 + 11$

3.9 Use a computer algebra package to express the polynomials

(a) $t^{12} + 4t^9 - 19t^8 - t^7 - 65t^5 + 19t^3 + 44t^2 - 11$

(b) $t^{10} - 14t^8 - t^7 + 52t^6 + 56t^5 - 75t^4 - 325t^3 + 441t^2 - 189t + 27$

(c) $t^{12} - 10t^{11} + 15t^{10} + 51t^9 + 9t^8 + 75t^7 - 40t^6 - 278t^5 - 111t^4 - 135t^3 - 15t^2 + 375t + 225$

(d) $t^8 - 12t^7 + 58t^6 - 144t^5 + 195t^4 - 144t^3 + 58t^2 - 12t + 1$

as products of irreducible polynomials over $\mathbb{Q}$.

3.10* Use a computer algebra package to factorise the polynomial

$$p(t) = t^n + t^4 + t^2 + t + 1 \in \mathbb{Q}[t]$$

into irreducibles over $\mathbb{Q}$. These are irreducible for many values of $n$, but reducible for others. Try to spot patterns in the cases that are reducible and formulate a conjecture.

3.11 Either using a computer algebra package, or pencil and paper, find out which integers $n$ make the polynomial $t^4 + n$ reducible in $\mathbb{Z}[t]$. Prove your answer is correct.

3.12 Mark the following true or false. (Here 'polynomial' means 'polynomial over $\mathbb{C}$'.)

(a) Every polynomial of degree $n$ has $n$ distinct zeros.

(b) Every polynomial of degree $n$ has at most $n$ distinct zeros.

(c) Every polynomial of degree $n$ has at least $n$ distinct zeros.

(d) If $f, g$ are non-zero polynomials and $f$ divides $g$, then $\partial f < \partial g$.

(e) If $f, g$ are non-zero polynomials and $f$ divides $g$, then $\partial f \leq \partial g$.

(f) Every polynomial of degree 1 is irreducible.

(g) Every irreducible polynomial has prime degree.

(h) If a polynomial $f$ has integer coefficients and is irreducible over $\mathbb{Z}$, then it is irreducible over $\mathbb{Q}$.

(i) If a polynomial $f$ has integer coefficients and is irreducible over $\mathbb{Z}$, then it is irreducible over $\mathbb{R}$.

(j) If a polynomial $f$ has integer coefficients and is irreducible over $\mathbb{R}$, then it is irreducible over $\mathbb{Z}$.

# Chapter 4

# Field Extensions

Galois's original theory was couched in terms of polynomials over the complex field. The modern approach is a consequence of the methods used, starting around 1890 and flourishing in the 1920s and 1930s, to generalise the theory to arbitrary fields. From this viewpoint the central object of study ceases to be a polynomial, and becomes instead a 'field extension' related to a polynomial. Every polynomial $f$ over a field $K$ defines another field $L$ containing $K$ (or at any rate a subfield isomorphic to $K$). There are conceptual advantages in setting up the theory from this point of view. In this chapter we define field extensions and explain the link with polynomials. To keep the discussion concrete we work inside $\mathbb{C}$, but almost everything generalises to abstract fields and rings. The more general abstract approach is discussed in Chapters 16 onwards.

## 4.1 Field Extensions

Suppose that we wish to study the quartic polynomial

$$f(t) = t^4 - 4t^2 - 5$$

over $\mathbb{Q}$. Its irreducible factorisation over $\mathbb{Q}$ is

$$f(t) = (t^2 + 1)(t^2 - 5)$$

so the zeros of $f$ in $\mathbb{C}$ are $\pm\mathrm{i}$ and $\pm\sqrt{5}$. There is a natural subfield $L$ of $\mathbb{C}$ associated with these zeros; in fact, it is the unique smallest subfield that contains them. We claim that $L$ consists of all complex numbers of the form

$$p + q\mathrm{i} + r\sqrt{5} + s\mathrm{i}\sqrt{5} \qquad (p, q, r, s \in \mathbb{Q})$$

Clearly $L$ must contain every such element, and it is not hard to see that sums and products of such elements have the same form. It is harder to see that inverses of (non-zero) such elements also have the same form, but it is true: we postpone the proof to Example 4.8. Thus the study of a polynomial over $\mathbb{Q}$ leads us to consider a subfield $L$ of $\mathbb{C}$ that contains $\mathbb{Q}$. In the same

DOI: 10.1201/9781003213949-4

way the study of a polynomial over an arbitrary subfield $K$ of $\mathbb{C}$ will lead to a subfield $L$ of $\mathbb{C}$ that contains $K$. We shall call $L$ an 'extension' of $K$. For technical reasons this definition is too restrictive; we wish to allow cases where $L$ contains a subfield isomorphic to $K$, but not necessarily equal to it.

Recall the definition of 'monomorphism', Definition 1.3. In that terminology, we can state:

**Definition 4.1.** A *field extension* is a monomorphism $\iota : K \to L$, where $K$ and $L$ are subfields of $\mathbb{C}$. We say that $K$ is the *small* field and $L$ is the *large* field.

Notice that with a strict set-theoretic definition of function, the map $\iota$ determines both $K$ and $L$. See Definition 1.3 for the definition of 'monomorphism'. We often think of a field extension as being a pair of fields $(K, L)$, when it is clear which monomorphism is intended.

**Examples 4.2.** 1. The inclusion maps $\iota_1 : \mathbb{Q} \to \mathbb{R}$, $\iota_2 : \mathbb{R} \to \mathbb{C}$, and $\iota_3 : \mathbb{Q} \to \mathbb{C}$ are all field extensions.
2. Let $K$ be the set of all real numbers of the form $p + q\sqrt{2}$, where $p, q \in \mathbb{Q}$. Then $K$ is a subfield of $\mathbb{C}$ by Example 1.7. The inclusion map $\iota : \mathbb{Q} \to K$ is a field extension.
3. Let $K$ be the field of all $p+q\sqrt{2}$ for $p, q \in \mathbb{Q}$, let $L = \mathbb{C}$, and define $\iota : K \to L$ by

$$\iota(p + q\sqrt{2}) = p - q\sqrt{2}$$

Now $\iota$ is not the inclusion map, but it is easily shown to be a monomorphism.

If $\iota : K \to L$ is a field extension, then we can usually identify $K$ with its image $\iota(K)$, so that $\iota$ can be thought of as an inclusion map and $K$ can be thought of as a subfield of $L$. Under these circumstances we use the notation

$$L/K$$

for the extension, and say that $L$ is an *extension of* $K$. (The symbol / here does not denote a quotient, as it does for groups. It should be read as 'over'.) In future we identify $K$ and $\iota(K)$ whenever this is legitimate.

The next concept is one which pervades much of abstract algebra:

**Definition 4.3.** Let $X$ be a subset of $\mathbb{C}$. Then the subfield of $\mathbb{C}$ *generated by* $X$ is the intersection of all subfields of $\mathbb{C}$ that contain $X$.

An extension $L/K$ is *generated over* $K$ by $X$ if $L$ is generated by $K \cup X$. If $X$ is a finite set we say that $L/K$ is a *finitely generated extension*. We use the notation

$$L = K(X)$$

If $X = \{x_1, \ldots, x_n\}$ is finite we write this as

$$L = K(x_1, \ldots, x_n)$$

It is easy to see that this definition is equivalent to either of the following:

1. The (unique) smallest subfield of $\mathbb{C}$ that contains $X$.

2. The set of all elements of $\mathbb{C}$ that can be obtained from elements of $X$ by a finite sequence of field operations, provided $X \neq \{0\}$ or $\emptyset$.

**Proposition 4.4.** *Every subfield of* $\mathbb{C}$ *contains* $\mathbb{Q}$.

*Proof.* Let $K \subseteq \mathbb{C}$ be a subfield. Then $0, 1 \in K$ by definition, so inductively we find that $1 + \ldots + 1 = n$ lies in $K$ for every integer $n > 0$. Now $K$ is closed under additive inverses, so $-n$ also lies in $K$, proving that $\mathbb{Z} \subseteq K$. Finally, if $p, q \in \mathbb{Z}$ and $q \neq 0$, closure under products and multiplicative inverses shows that $pq^{-1} \in K$. Therefore $\mathbb{Q} \subseteq K$ as claimed. □

**Corollary 4.5.** *Let* $X$ *be a subset of* $\mathbb{C}$. *Then the subfield of* $\mathbb{C}$ *generated by* $X$ *contains* $\mathbb{Q}$.

Because of Corollary 4.5, we use the notation $\mathbb{Q}(X)$ for the subfield of $\mathbb{C}$ generated by $X$, and write this as $\mathbb{Q}(x_1, \ldots, x_n)$ when $X = \{x_1, \ldots, x_n\}$.

**Example 4.6.** We find the subfield $K$ of $\mathbb{C}$ generated by $X = \{1, \mathrm{i}\}$. By Proposition 4.4, $K$ must contain $\mathbb{Q}$. Since $K$ is closed under the arithmetical operations, it must contain all complex numbers of the form $p + q\mathrm{i}$, where $p, q \in \mathbb{Q}$. Let $M$ be the set of all such numbers. We claim that $M$ is a subfield of $\mathbb{C}$. Clearly $M$ is closed under sums, differences, and products. Further

$$(p + q\mathrm{i})^{-1} = \frac{p}{p^2 + q^2} - \frac{q}{p^2 + q^2}\mathrm{i}$$

so that every non-zero element of $M$ has a multiplicative inverse in $M$. Hence $M$ is a subfield, and contains $X$. Since $K$ is the smallest subfield containing $X$, we have $K \subseteq M$. But $M \subseteq K$ by definition. Hence $K = M$, and we have found a description of the subfield generated by $X$.

In the case of a field extension $L/K$ we are mainly interested in subfields lying between $K$ and $L$. This means that we can restrict attention to subsets $X$ that contain $K$; equivalently, to sets of the form $K \cup Y$ where $Y \subseteq L$.

**Definition 4.7.** If $L/K$ is a field extension and $Y$ is a subset of $L$, then the subfield of $\mathbb{C}$ generated by $K \cup Y$ is written $K(Y)$ and is said to be obtained from $K$ by *adjoining* $Y$.

Clearly $K(Y) \subseteq L$ since $L$ is a subfield of $\mathbb{C}$. Notice that $K(Y)$ is in general considerably larger than $K \cup Y$.

This notation is open to all sorts of useful abuses. If $Y$ has a single element $y$ we write $K(y)$ instead of $K(\{y\})$, and in the same spirit $K(y_1, \ldots, y_n)$ replaces $K(\{y_1, \ldots, y_n\})$.

**Example 4.8.** Let $K = \mathbb{Q}$ and let $Y = \{\mathrm{i}, \sqrt{5}\}$. Then $K(Y)$ must contain $K$ and $Y$. It also contains the product $\mathrm{i}\sqrt{5}$. Since $K \supseteq \mathbb{Q}$, the subfield $K(Y)$ must contain all elements

$$\alpha = p + q\mathrm{i} + r\sqrt{5} + s\mathrm{i}\sqrt{5} \qquad (p, q, r, s \in \mathbb{Q}).$$

Let $L \subseteq \mathbb{C}$ be the set of all such $\alpha$. If we prove that $L$ is a subfield of $\mathbb{C}$, then it follows that $K(Y) = L$. Moreover, it is easy to check that $L$ is a subring of $\mathbb{C}$, hence $L$ is a subfield of $\mathbb{C}$ if and only if for $\alpha \neq 0$ we can find an inverse $\alpha^{-1} \in L$. If fact, we shall prove that if $(p, q, r, s) \neq (0, 0, 0, 0)$ then $\alpha \neq 0$, and then

$$(p + q\mathrm{i} + r\sqrt{5} + s\mathrm{i}\sqrt{5})^{-1} \in L$$

First, suppose that $p + q\mathrm{i} + r\sqrt{5} + s\mathrm{i}\sqrt{5} = 0$. Then

$$p + r\sqrt{5} = -\mathrm{i}(q + s\sqrt{5})$$

Now both $p + r\sqrt{5}$ and $-(q + s\sqrt{5})$ are real, but i is imaginary. Therefore $p + r\sqrt{5} = 0$ and $q + s\sqrt{5} = 0$. If $r \neq 0$ then $\sqrt{5} = -p/r \in \mathbb{Q}$, but $\sqrt{5}$ is irrational. Therefore $r = 0$, whence $p = 0$. Similarly, $q = s = 0$.

Now we prove the existence of $\alpha^{-1}$ in two stages. Let $M$ be the subset of $L$ containing all $p + q\mathrm{i}$ $(p, q \in \mathbb{Q})$. Then we can write

$$\alpha = x + y\sqrt{5}$$

where $x = p + q\mathrm{i}$ and $y = r + s\mathrm{i} \in M$. Let

$$\beta = p + q\mathrm{i} - r\sqrt{5} - s\mathrm{i}\sqrt{5} = x - y\sqrt{5} \in L$$

Then

$$\alpha\beta = (x + y\sqrt{5})(x - y\sqrt{5}) = x^2 - 5y^2 = z$$

say, where $z \in M$. Since $\alpha \neq 0$ and $\beta \neq 0$, we have $z \neq 0$, so $\alpha^{-1} = \beta z^{-1}$. Now write $z = u + v\mathrm{i}$ $(u, v \in \mathbb{Q})$ and consider $w = u - v\mathrm{i}$. Since $zw = u^2 + v^2 \in \mathbb{Q}$, we have

$$z^{-1} = (u^2 + v^2)^{-1} w \in M$$

so $\alpha^{-1} = \beta z^{-1} \in L$.

Alternatively, we can obtain an explicit formula by working out the expression

$$(p + q\mathrm{i} + r\sqrt{5} + s\mathrm{i}\sqrt{5})(p - q\mathrm{i} + r\sqrt{5} - s\mathrm{i}\sqrt{5})$$
$$\times\ (p + q\mathrm{i} - r\sqrt{5} - s\mathrm{i}\sqrt{5})(p - q\mathrm{i} - r\sqrt{5} + s\mathrm{i}\sqrt{5})$$

and showing that it belongs to $\mathbb{Q}$, and then dividing out by

$$(p + q\mathrm{i} + r\sqrt{5} + s\mathrm{i}\sqrt{5})$$

See Exercise 4.6.

**Examples 4.9.** (1) The subfield $\mathbb{R}(i)$ of $\mathbb{C}$ must contain all elements $x + iy$ where $x, y \in \mathbb{R}$. But those elements comprise the whole of $\mathbb{C}$. Therefore $\mathbb{C} = \mathbb{R}(i)$.
(2) The subfield $P$ of $\mathbb{R}$ consisting of all numbers $p + q\sqrt{2}$ where $p, q \in \mathbb{Q}$ is easily seen to equal $\mathbb{Q}(\sqrt{2})$.
(3) It is not always true that a subfield of the form $K(\alpha)$ consists of all elements of the form $j + k\alpha$ where $j, k \in K$. It certainly contains all such elements, but they need not form a subfield.

For example, in $\mathbb{R}/\mathbb{Q}$ let $\alpha$ be the real cube root of 2, and consider $\mathbb{Q}(\alpha)$. As well as $\alpha$, the subfield $\mathbb{Q}(\alpha)$ must contain $\alpha^2$. We show that $\alpha^2 \neq j + k\alpha$ for $j, k \in \mathbb{Q}$. For a contradiction, suppose that $\alpha^2 = j + k\alpha$. Then $2 = \alpha^3 = j\alpha + k\alpha^2 = jk + (j + k^2)\alpha$. Therefore $(j + k^2)\alpha = 2 - jk$. Since $\alpha$ is irrational, $(j + k^2) = 0 = 2 - jk$. Eliminating $j$, we find that $k^3 = -2$, contrary to $k \in \mathbb{Q}$.

In fact, $\mathbb{Q}(\alpha)$ is precisely the set of all elements of $\mathbb{R}$ of the form $p + q\alpha + r\alpha^2$, where $p, q, r \in \mathbb{Q}$. To show this, we prove that the set of such elements is a subfield. The only (minor) difficulty is finding a multiplicative inverse: see Exercise 4.7.

## 4.2 Rational Expressions

We can perform the operations of addition, subtraction, and multiplication in the polynomial ring $\mathbb{C}[t]$, but (usually) not division. For example, $\mathbb{C}[t]$ does not contain an inverse $t^{-1}$ for $t$, see Exercise 4.8.

However, we can enlarge $\mathbb{C}[t]$ to provide inverses in a natural way. We have seen that we can think of polynomials $f(t) \in \mathbb{C}[t]$ as functions from $\mathbb{C}$ to itself. Similarly, we can think of fractions $p(t)/q(t) \in \mathbb{C}(t)$, where $q \neq \mathbf{0}$, as functions. These are called *rational functions* of the complex variable $t$, and their formal statements in terms of polynomials are *rational expressions* in the indeterminate $t$. However, there is now a technical difficulty. The domain of such a function is not the whole of $\mathbb{C}$: all of the zeros of $q(t)$ have to be removed, or else we are trying to divide by zero. Complex analysts often work in the Riemann sphere $\mathbb{C} \cup \{\infty\}$, and cheerfully let $1/\infty = 0$, but care must be exercised if this is done; the civilised way to proceed is to remove all the potential troublemakers. So we take the domain of $p(t)/q(t)$ to be

$$\{z \in \mathbb{C} : q(z) \neq 0\}$$

As we have seen, any complex polynomial $q$ has only finitely many zeros, so the domain here is 'almost all' of $\mathbb{C}$. We have to be careful, but we should not get into much trouble provided we are.

In the same manner we can also construct the set

$$\mathbb{C}(t_1, \ldots, t_n)$$

of all rational functions in $n$ variables (rational expressions in $n$ indeterminates). One use of such functions is to specify the subfield generated by a given set $X$. It is straightforward to prove that $\mathbb{Q}(X)$ consists of all rational expressions

$$\frac{p(\alpha_1,\ldots,\alpha_n)}{q(\beta_1,\ldots,\beta_n)}$$

for all $n$, where $p, q \in \mathbb{Q}[t_1,\ldots,t_n]$, the $\alpha_j$ and $\beta_j$ belong to $X$, and $q(\beta_1,\ldots,\beta_n) \neq 0$. See Exercise 4.9.

It is also possible to define such expressions without using functions. See 'field of fractions' in Chapter 16, immediately after Corollary 16.18. This approach is necessary in the more abstract development of the subject.

---

## 4.3 Simple Extensions

The basic building-blocks for field extensions are those obtained by adjoining one element:

**Definition 4.10.** A *simple extension* is a field extension $L/K$ such that $L = K(\alpha)$ for some $\alpha \in L$.

**Examples 4.11.** (1) As the notation shows, the extensions in Examples 4.9 are all simple.

(2) *Beware:* An extension may be simple without appearing to be. Consider $L = \mathbb{Q}(\mathrm{i}, -\mathrm{i}, \sqrt{5}, -\sqrt{5})$. As written, it appears to require the adjunction of four new elements. Clearly just two, i and $\sqrt{5}$, suffice. But we claim that in fact only one element is needed, because $L = L'$ where $L' = \mathbb{Q}(\mathrm{i} + \sqrt{5})$, which is obviously simple. To prove this, it is enough to show that $\mathrm{i} \in L'$ and $\sqrt{5} \in L'$, because these imply that $L \subseteq L'$ and $L' \subseteq L$, so $L = L'$. Now $L'$ contains

$$(\mathrm{i} + \sqrt{5})^2 = -1 + 2\mathrm{i}\sqrt{5} + 5 = 4 + 2\mathrm{i}\sqrt{5}$$

Thus it also contains

$$(\mathrm{i} + \sqrt{5})(4 + 2\mathrm{i}\sqrt{5}) = 14\mathrm{i} - 2\sqrt{5}$$

Therefore it contains

$$14\mathrm{i} - 2\sqrt{5} + 2(\mathrm{i} + \sqrt{5}) = 16\mathrm{i}$$

so it contains i. But then it also contains $(\mathrm{i} + \sqrt{5}) - \mathrm{i} = \sqrt{5}$. Therefore $L = L'$ as claimed, and the extension $\mathbb{Q}(\mathrm{i}, -\mathrm{i}, \sqrt{5}, -\sqrt{5})/\mathbb{Q}$ is in fact simple.

(3) The example in (2) is no accident. Below we prove the Primitive Element Theorem 6.13, which asserts that if $K, L$ are subfields of $\mathbb{C}$ and $L = K(\alpha_1,\ldots,\alpha_m)$ where the $\alpha_j$ are algebraic over $K$, then there exists $\theta \in L$ such that $L = K(\theta)$. That is, $L/K$ is simple.

(4) On the other hand, $\mathbb{R}/\mathbb{Q}$ is not a simple extension (Exercise 4.5).

Our aim in the next chapter will be to classify all possible simple extensions. We end this chapter by formulating the concept of isomorphism of extensions. In Chapter 5 we develop techniques for constructing all possible simple extensions up to isomorphism.

**Definition 4.12.** An *isomorphism* between two field extensions $\iota : K \to \hat{K}, \jmath : L \to \hat{L}$ is a pair $(\lambda, \mu)$ of field isomorphisms $\lambda : K \to L, \mu : \hat{K} \to \hat{L}$, such that for all $k \in K$

$$\jmath(\lambda(k)) = \mu(\iota(k))$$

Another, more pictorial, way of putting this is to say that the diagram

$$\begin{array}{ccc} K & \xrightarrow[\iota]{} & \hat{K} \\ \downarrow{\scriptstyle\lambda} & & \downarrow{\scriptstyle\mu} \\ L & \xrightarrow[\jmath]{} & \hat{L} \end{array}$$

*commutes*; that is, the two paths from $K$ to $\hat{L}$ compose to give the same map.

The reason for setting up the definition like this is that as well as the field structure being preserved by isomorphism, the embedding of the small field in the large one is also preserved.

Various identifications may be made. If we identify $K$ and $\iota(K)$, and $L$ and $\jmath(L)$, then $\iota$ and $\jmath$ are inclusions, and the commutativity condition now becomes

$$\mu|_K = \lambda$$

where $\mu|_K$ denotes the restriction of $\mu$ to $K$. If we further identify $K$ and $L$ then $\lambda$ becomes the identity, and so $\mu|_K$ is the identity. In what follows we shall attempt to use these 'identified' conditions wherever possible. But on a few occasions (notably Theorem 9.6) we shall need the full generality of the first definition.

---

## EXERCISES

4.1 Prove that isomorphism of field extensions is an equivalence relation.

4.2 Find the subfields of $\mathbb{C}$ generated by:

(a) $\{0, 1\}$
(b) $\{0\}$
(c) $\{0, 1, \mathrm{i}\}$
(d) $\{\mathrm{i}, \sqrt{2}\}$

(e) $\{\sqrt{2}, \sqrt{3}\}$

(f) $\mathbb{R}$

(g) $\mathbb{R} \cup \{\mathrm{i}\}$

4.3 Describe the subfields of $\mathbb{C}$ of the form

(a) $\mathbb{Q}(\sqrt{2})$

(b) $\mathbb{Q}(\mathrm{i})$

(c) $\mathbb{Q}(\alpha)$ where $\alpha$ is the real cube root of 2

(d) $\mathbb{Q}(\sqrt{5}, \sqrt{7})$

(e) $\mathbb{Q}(\mathrm{i}\sqrt{11})$

(f) $\mathbb{Q}(\mathrm{e}^2 + 1)$

(g) $\mathbb{Q}(\sqrt[3]{\pi})$

4.4 This exercise illustrates a technique that we will tacitly assume in several subsequent exercises and examples.

Prove that $1, \sqrt{2}, \sqrt{3}, \sqrt{6}$ are linearly independent over $\mathbb{Q}$.

(*Hint:* Suppose that $p + q\sqrt{2} + r\sqrt{3} + s\sqrt{6} = 0$ with $p, q, r, s \in \mathbb{Q}$. We may suppose that $r \neq 0$ or $s \neq 0$ (why?). If so, then we can write $\sqrt{3}$ in the form

$$\sqrt{3} = \frac{a + b\sqrt{2}}{c + d\sqrt{2}} = e + f\sqrt{2}$$

where $a, b, c, d, e, f \in \mathbb{Q}$. Square both sides and obtain a contradiction.)

4.5 Show that $\mathbb{R}$ is not a simple extension of $\mathbb{Q}$ as follows:

(a) $\mathbb{Q}$ is countable.

(b) Any simple extension of a countable field is countable.

(c) $\mathbb{R}$ is not countable.

4.6 Find a formula for the inverse of $p+q\mathrm{i}+r\sqrt{5}+s\mathrm{i}\sqrt{5}$, where $p, q, r, s \in \mathbb{Q}$.

4.7 Find a formula for the inverse of $p + q\alpha + r\alpha^2$, where $p, q, r \in \mathbb{Q}$ and $\alpha = \sqrt[3]{2}$.

4.8 Prove that $t$ has no multiplicative inverse in $\mathbb{C}[t]$.

4.9 Prove that $\mathbb{Q}(X)$ consists of all rational expressions

$$\frac{p(\alpha_1, \ldots, \alpha_n)}{q(\beta_1, \ldots, \beta_n)}$$

for all $n$, where $p, q \in \mathbb{Q}[t_1, \ldots, t_n]$, the $\alpha_j$ and $\beta_j$ belong to $X$, and $q(\beta_1, \ldots, \beta_n) \neq 0$.

4.10 Mark the following true or false.

(a) If $X$ is the empty set then $\mathbb{Q}(X) = \mathbb{Q}$.

(b) If $X$ is a subset of $\mathbb{Q}$ then $\mathbb{Q}(X) = \mathbb{Q}$.

(c) If $X$ contains an irrational number, then $\mathbb{Q}(X) \neq \mathbb{Q}$.

(d) $\mathbb{Q}(\sqrt{2}) = \mathbb{Q}$.

(e) $\mathbb{Q}(\sqrt{2}) = \mathbb{R}$.

(f) $\mathbb{R}(\sqrt{2}) = \mathbb{R}$.

(g) Every subfield of $\mathbb{C}$ contains $\mathbb{Q}$.

(h) Every subfield of $\mathbb{C}$ contains $\mathbb{R}$.

(i) If $\alpha \neq \beta$ and both are irrational, then $\mathbb{Q}(\alpha, \beta)$ is not a simple extension of $\mathbb{Q}$.

# Chapter 5

# Simple Extensions

The basic building block of field theory is the simple field extension. Here *one* new element $\alpha$ is adjoined to a given subfield $K$ of $\mathbb{C}$, along with all rational expressions in that element over $K$. Any finitely generated extension—one that is obtained by adjoining finitely many elements to $K$—can be obtained by a finite equence of simple extensions, so the structure of a simple extension provides vital information about all of the extensions that we shall encounter.

We first classify simple extensions into two very different kinds: transcendental and algebraic. If the new element $\alpha$ satisfies a polynomial equation over $K$, then the extension is algebraic; if not, it is transcendental. Up to isomorphism, $K$ has exactly one simple transcendental extension. For most fields $K$ there are many more possibilities for simple algebraic extensions; they are classified by the irreducible polynomials $m$ over $K$.

The structure of simple algebraic extensions can be described in terms of the polynomial ring $K[t]$, with operations being performed 'modulo $m$'. In Chapter 16 we generalise this construction using the notion of an ideal.

## 5.1 Algebraic and Transcendental Extensions

Recall that a simple extension of a subfield $K$ of $\mathbb{C}$ takes the form $K(\alpha)$ where in nontrivial cases $\alpha \notin K$. We classify the possible simple extensions for any $K$. There are two distinct types:

**Definition 5.1.** Let $K$ be a subfield of $\mathbb{C}$ and let $\alpha \in \mathbb{C}$. Then $\alpha$ is *algebraic* over $K$ if there exists a non-zero polynomial $p$ over $K$ such that $p(\alpha) = 0$. Otherwise, $\alpha$ is *transcendental* over $K$.

We shorten 'algebraic over $\mathbb{Q}$' to 'algebraic', and 'transcendental over $\mathbb{Q}$' to 'transcendental'.

**Examples 5.2.** (1) The number $\alpha = \sqrt{2}$ is algebraic, because $\alpha^2 - 2 = 0$.
(2) The number $\alpha = \sqrt[3]{2}$ is algebraic, because $\alpha^3 - 2 = 0$.
(3) The number $\pi = 3 \cdot 14159\ldots$ is transcendental. We postpone a proof to Chapter 24. In Chapter 7 we use the transcendence of $\pi$ to prove the impossibility of 'squaring the circle'.

DOI: 10.1201/9781003213949-5

(4) The number $\alpha = \sqrt{\pi}$ is algebraic over $\mathbb{Q}(\pi)$, because $\alpha^2 - \pi = 0$.
(5) However, $\alpha = \sqrt{\pi}$ is transcendental over $\mathbb{Q}$. To see why, suppose that $p(\sqrt{\pi}) = 0$ where $0 \neq p(t) \in \mathbb{Q}[t]$. Separating out terms of odd and even degree, we can write this as $a(\pi) + b(\pi)\sqrt{\pi} = 0$, so $a(\pi) = -b(\pi)\sqrt{\pi}$ and $a^2(\pi) = \pi b^2(\pi)$. Thus $f(\pi) = 0$, where

$$f(t) = a^2(t) - tb^2(t) \in \mathbb{Q}[t]$$

Now $\partial(a^2)$ is even, and $\partial(tb^2)$ is odd, so the difference $f(t)$ is not the zero polynomial. But this implies that $\pi$ is algebraic, a contradiction.

In the next few sections we classify all possible simple extensions and find ways to construct them. The transcendental case is very straightforward: if $K(t)$ is the set of rational expressions in the indeterminate $t$ over $K$, then $K(t)/K$ is the unique simple transcendental extension of $K$ up to isomorphism. If $K(\alpha)/K$ is algebraic, the possibilities are richer, but tractable. We show that there is a unique monic irreducible polynomial $m$ over $K$ such that $m(\alpha) = 0$, and that $m$ determines the extension uniquely up to isomorphism.

We begin by constructing a simple transcendental extension of any subfield.

**Theorem 5.3.** *The set of rational expressions $K(t)$ is a simple transcendental extension of the subfield $K$ of $\mathbb{C}$.*

*Proof.* Clearly $K(t)/K$ is a simple extension, generated by $t$. If $p$ is a polynomial over $K$ such that $p(t) = 0$ then $p = 0$ by definition of $K(t)$, so the extension is transcendental. □

---

## 5.2 The Minimal Polynomial

The construction of simple algebraic extensions is a much more delicate issue. It is controlled by a polynomial associated with the generator $\alpha$ of $K(\alpha)/K$, the 'minimal polynomial' of $\alpha$.

Recall the definition of a monic polynomial, Definition 3.7. Clearly every polynomial is a constant multiple of some monic polynomial, and for a nonzero polynomial this monic polynomial is unique. Further, the product of two monic polynomials is again monic. Now suppose that $K(\alpha)/K$ is a simple algebraic extension. There is a polynomial $p$ over $K$ such that $p(\alpha) = 0$. We may suppose that $p$ is monic. Therefore there exists at least one monic polynomial *of smallest degree* that has $\alpha$ as a zero. We claim that $p$ is unique. To see why, suppose that $p, q$ are two such. Clearly $\partial p = \partial q$. Then $p(\alpha) - q(\alpha) = 0$, so if $p \neq q$ then some constant multiple of $p - q$ is a monic polynomial with $\alpha$ as a zero, and $\partial(p - q) < \partial p$, contrary to the definition. Hence there is a unique monic polynomial $p$ of smallest degree such that $p(\alpha) = 0$. We give this a name:

**Definition 5.4.** Let $L/K$ be a field extension, and suppose that $\alpha \in L$ is algebraic over $K$. Then the *minimal polynomial* of $\alpha$ over $K$ is the unique monic polynomial $m$ over $K$ of smallest degree such that $m(\alpha) = 0$.

Another term sometimes used is *minimum polynomial.*

For example, $\mathrm{i} \in \mathbb{C}$ is algebraic over $\mathbb{R}$. If we let $m(t) = t^2 + 1$ then $m(\mathrm{i}) = 0$. Clearly $m$ is monic. The only monic polynomials over $\mathbb{R}$ of smaller degree are those of the form $t + r$, where $r \in \mathbb{R}$, or the constant polynomial 1. But i cannot be a zero of any of these, or else we would have $\mathrm{i} \in \mathbb{R}$. Hence the minimal polynomial of i over $\mathbb{R}$ is $t^2 + 1$.

It is natural to ask which polynomials can be minimal. The next lemma provides information on this question.

**Lemma 5.5.** *If $\alpha$ is an algebraic element over the subfield $K$ of $\mathbb{C}$, then the minimal polynomial of $\alpha$ over $K$ is irreducible over $K$. It divides every polynomial of which $\alpha$ is a zero.*

*Proof.* Suppose that the minimal polynomial $m$ of $\alpha$ over $K$ is reducible, so that $m = fg$ where $f$ and $g$ are of smaller degree. We may assume $f$ and $g$ are monic. Since $m(\alpha) = 0$ we have $f(\alpha)g(\alpha) = 0$, so either $f(\alpha) = 0$ or $g(\alpha) = 0$. But this contradicts the definition of $m$. Hence $m$ is irreducible over $K$.

Now suppose that $p$ is a polynomial over $K$ such that $p(\alpha) = 0$. By the Division Algorithm, there exist polynomials $q$ and $r$ over $K$ such that $p = mq + r$ and $\partial r < \partial m$. Then $0 = p(\alpha) = 0 + r(\alpha)$. If $r \neq 0$ then a suitable constant multiple of $r$ is monic, which contradicts the definition of $m$. Therefore $r = 0$, so $m$ divides $p$. □

Conversely, if $K$ is a subfield of $\mathbb{C}$, then it is easy to show that any irreducible polynomial over $K$ can be the minimal polynomiall of an algebraic element over $K$:

**Theorem 5.6.** *If $K$ is any subfield of $\mathbb{C}$ and $m$ is any irreducible monic polynomial over $K$, then there exists $\alpha \in \mathbb{C}$, algebraic over $K$, such that $\alpha$ has minimal polynomial $m$ over $K$.*

*Proof.* Let $\alpha$ be any zero of $m$ in $\mathbb{C}$. Then $m(\alpha) = 0$, so the minimal polynomial $f$ of $\alpha$ over $K$ divides $m$. But $m$ is irreducible over $K$ and both $f$ and $m$ are monic; therefore $f = m$. □

**Definition 5.7.** Let $\alpha \in \mathbb{C}$ be algebraic over a subfield $K \subseteq \mathbb{C}$, with minimal polynomial $m(t)$. Then the zeros of $m(t)$ are the *conjugates* of $\alpha$.

This use of 'conjugate' is not to be confused with its use in group theory (where $g^{-1}hg$ is conjugate to $g$) or with the complex conjugate $\bar{z}$ of $z \in \mathbb{C}$. As an example, when $K = \mathbb{Q}$ the conjugates of $\sqrt{5}$ are $\sqrt{5}$ and $-\sqrt{5}$.

## 5.3 Simple Algebraic Extensions

Next, we describe the structure of the field extension $K(\alpha)/K$ when $\alpha$ has minimal polynomial $m$ over $K$. We proceed by analogy with a basic concept of number theory. Recall from Section 3.5 that for any positive integer $n$ it is possible to perform arithmetic *modulo* $n$, and that integers $a, b$ are *congruent modulo* $n$, written

$$a \equiv b \pmod{n}$$

if $a-b$ is divisible by $n$. In the same way, given a polynomial $m \in K[t]$, we can calculate with polynomials *modulo* $m$. We say that polynomials $a, b \in K[t]$ are *congruent modulo* $m$, written

$$a \equiv b \pmod{m}$$

if $a(t) - b(t)$ is divisible by $m(t)$ in $K[t]$.

**Lemma 5.8.** *Suppose that* $a_1 \equiv a_2 \pmod{m}$ *and* $b_1 \equiv b_2 \pmod{m}$*. Then* $a_1 + b_1 \equiv a_2 + b_2 \pmod{m}$*, and* $a_1 b_1 \equiv a_2 b_2 \pmod{m}$*.*

*Proof.* We know that $a_1 - a_2 = am$ and $b_1 - b_2 = bm$ for polynomials $a, b \in K[t]$. Now

$$(a_1 + b_1) - (a_2 + b_2) = (a_1 - a_2) + (b_1 - b_2) = (a - b)m$$

which proves the first statement. For the product, we need a slightly more elaborate argument:

$$\begin{aligned} a_1 b_1 - a_2 b_2 &= a_1 b_1 - a_1 b_2 + a_1 b_2 - a_2 b_2 \\ &= a_1(b_1 - b_2) + b_2(a_1 - a_2) \\ &= (a_1 b + b_2 a)m \end{aligned}$$

□

**Lemma 5.9.** *Every polynomial* $a \in K[t]$ *is congruent modulo* $m$ *to a unique polynomial of degree* $< \partial m$*.*

*Proof.* Divide $a$ by $m$ with remainder $r$, so that $a = qm + r$ where $q, r \in K[t]$ and $\partial r < \partial m$. Then $a - r = qm$, so $a \equiv r \pmod{m}$. To prove uniqueness, suppose that $r \equiv s \pmod{m}$ where $\partial r, \partial s < \partial m$. Then $r - s$ is divisible by $m$ but has smaller degree than $m$. Therefore $r - s = 0$, so $r = s$, proving uniqueness. □

We call $r$ the *reduced form* of $a$ modulo $m$. Lemma 5.9 shows that we can calculate with polynomials modulo $m$ in terms of their reduced forms. Indeed, the reduced form of $a + b$ is the reduced form of $a$ plus the reduced form of

$b$, while the reduced form of $ab$ is the remainder, after dividing by $m$, of the product of the reduced form of $a$ and the reduced form of $b$.

The sum $a + b$ is easy to calculate using reduced forms, but this characterisation of the product $ab$ is tricky to work with. A more tractable, but also slightly more abstract, method is to work with equivalence classes. The relation $\equiv \pmod{m}$ is an equivalence relation on $K[t]$, so it partitions $K[t]$ into equivalence classes. We write $[a]$ for the equivalence class of $a \in K[t]$. Clearly

$$[a] = \{f \in K[t] : m|(a - f)\}$$

The sum and product of $[a]$ and $[b]$ can be defined as:

$$[a] + [b] = [a + b] \qquad [a][b] = [ab]$$

It is straightforward to show that these operations are well-defined; that is, they do not depend on the choice of elements from equivalence classes. Each equivalence class contains a unique polynomial of degree less than $\partial m$, namely, the reduced form of $a$. Therefore algebraic computations with equivalence classes are the same as computations with reduced forms, and both are the same as computations in $K[t]$ with the added convention that $m(t)$ is identified with 0. In particular, the classes $[0]$ and $[1]$ are additive and multiplicative identities respectively.

We write

$$K[t]/\langle m \rangle$$

for the set of equivalence classes of $K[t]$ modulo $m$. Readers who know about ideals in rings will see at once that $K[t]/\langle m \rangle$ is a thin disguise for the quotient ring of $K[t]$ by the ideal generated by $m$, and the equivalence classes are cosets of that ideal, but at this stage of the book these concepts are more abstract than we really need.

A key result is:

**Theorem 5.10.** *Every nonzero element of $K[t]/\langle m \rangle$ has a multiplicative inverse in $K[t]/\langle m \rangle$ if and only if $m$ is irreducible in $K[t]$.*

*Proof.* If $m$ is reducible then $m = ab$ where $\partial a, \partial b < \partial m$. Then $[a][b] = [ab] = [m] = [0]$. Suppose that $[a]$ has an inverse $[c]$, so that $[c][a] = [1]$. Then $[0] = [c][0] = [c][a][b] = [1][b] = [b]$, so $m$ divides $b$. Since $\partial b < \partial m$ we must have $b = 0$, so $m = 0$, contradiction.

If $m$ is irreducible, let $a \in K[t]$ with $[a] \neq [0]$; that is, $m \nmid a$. Therefore $a$ is prime to $m$, so their greatest common divisor is 1. By Theorem 3.11, there exist $h, k \in K[t]$ such that $ha + km = 1$. Then $[h][a] + [k][m] = [1]$, but $[m] = [0]$ so $[1] = [h][a] + [k][m] = [h][a] + [k][0] = [h][a] + [0] = [h][a]$. Thus $[h]$ is the required inverse. □

Again, in abstract terminology, what we have proved is that $K[t]/\langle m \rangle$ is a field if and only if $m$ is irreducible in $K[t]$. See Chapter 17 for a full explanation and generalisations.

## 5.4 Classifying Simple Extensions

We now demonstrate that the above methods suffice for the construction of all possible simple extensions (up to isomorphism). Again transcendental extensions are easily dealt with.

**Theorem 5.11.** *Every simple transcendental extension $K(\alpha)/K$ is isomorphic to the extension $K(t)/K$ of rational expressions in an indeterminate $t$ over $K$. The isomorphism $K(t) \to K(\alpha)$ can be chosen to map $t$ to $\alpha$, and to be the identity on $K$.*

*Proof.* Define a map $\phi : K(t) \to K(\alpha)$ by

$$\phi(f(t)/g(t)) = f(\alpha)/g(\alpha)$$

If $g \neq 0$ then $g(\alpha) \neq 0$ (since $\alpha$ is transcendental) so this definition makes sense. It is clearly a homomorphism, and a simple calculation shows that it is a monomorphism. It is clearly onto, and so is an isomorphism. Further, $\phi|_K$ is the identity, so that $\phi$ defines an isomorphism of extensions. Finally, $\phi(t) = \alpha$. □

The classification for simple algebraic extensions is just as straightforward, but more interesting:

**Theorem 5.12.** *Let $K(\alpha)/K$ be a simple algebraic extension, and let the minimal polynomial of $\alpha$ over $K$ be $m$. Then $K(\alpha)/K$ is isomorphic to $K[t]/\langle m\rangle/K$. The isomorphism $K[t]/\langle m\rangle \to K(\alpha)$ can be chosen to map $t$ to $\alpha$ and to be the identity on $K$.*

*Proof.* The isomorphism is defined by $[p(t)] \mapsto p(\alpha)$, where $[p(t)]$ is the equivalence class of $p(t)$ (mod $m$). This map is well-defined because $p(\alpha) = 0$ if and only if $m|p$. It is clearly a field monomorphism. It maps $t$ to $\alpha$, and its restriction to $K$ is the identity. □

**Corollary 5.13.** *Suppose $K(\alpha)/K$ and $K(\beta)/K$ are simple algebraic extensions such that $\alpha$ and $\beta$ have the same minimal polynomial $m$ over $K$. Then the two extensions are isomorphic, and the isomorphism of the large fields can be taken to map $\alpha$ to $\beta$ and to be the identity on $K$.*

*Proof.* Both extensions are isomorphic to $K[t]/\langle m\rangle$. The isomorphisms concerned map $t$ to $\alpha$ and $t$ to $\beta$ respectively. Call them $\iota, j$ respectively. Then $j\iota^{-1}$ is an isomorphism from $K(\alpha)$ to $K(\beta)$ that is the identity on $K$ and maps $\alpha$ to $\beta$. □

**Lemma 5.14.** *Let $K(\alpha)/K$ be a simple algebraic extension, let the minimal polynomial of $\alpha$ over $K$ be $m$, and let $\partial m = n$. Then $\{1, \alpha, \ldots, \alpha^{n-1}\}$ is a basis for $K(\alpha)$, considered as a vector space over $K$.*

*Proof.* The theorem is a restatement of Lemma 5.9. □

For certain later applications we need a slightly stronger version of Theorem 5.12, to cover extensions of isomorphic (rather than identical) fields. Before we can state the more general theorem we need the following:

**Definition 5.15.** Let $\iota : K \to L$ be a field monomorphism. Then there is a map $\hat{\iota} : K[t] \to L[t]$, defined by

$$\hat{\iota}(k_0 + k_1 t + \cdots + k_n t^n) = \iota(k_0) + \iota(k_1)t + \cdots + \iota(k_n)t^n$$

$(k_0, \ldots, k_n \in K)$. It is easy to prove that $\hat{\iota}$ is a monomorphism. If $\iota$ is an isomorphism, then so is $\hat{\iota}$.

The hat is unnecessary, once the statement is clear, and it may be dispensed with. So in future we use the same symbol $\iota$ for the map between subfields of $\mathbb{C}$ and for its extension to polynomial rings. This should not cause confusion since $\hat{\iota}(k) = \iota(k)$ for any $k \in K$.

**Theorem 5.16.** *Suppose that $K$ and $L$ are subfields of $\mathbb{C}$ and $\iota : K \to L$ is an isomorphism. Let $K(\alpha), L(\beta)$ be simple algebraic extensions of $K$ and $L$ respectively, such that $\alpha$ has minimal polynomial $m_\alpha(t)$ over $K$ and $\beta$ has minimal polynomial $m_\beta(t)$ over $L$. Suppose further that $m_\beta(t) = \iota(m_\alpha(t))$. Then there exists an isomorphism $j : K(\alpha) \to L(\beta)$ such that $j|_K = \iota$ and $j(\alpha) = \beta$.*

*Proof.* We can summarise the hypotheses in the diagram

$$\begin{array}{ccc} K & \longrightarrow & K(\alpha) \\ \downarrow{\scriptstyle \iota} & & \downarrow{\scriptstyle j} \\ L & \longrightarrow & L(\beta) \end{array}$$

where $j$ is yet to be determined. Using the reduced form, every element of $K(\alpha)$ is of the form $p(\alpha)$ for a polynomial $p$ over $K$ of degree $< \partial m_\alpha$. Define $j(p(\alpha)) = (\iota(p))(\beta)$ where $\iota(\rho)$ is defined as above. Everything else follows easily from Theorem 5.12. $\square$

The point of this theorem is that the given map $\iota$ can be extended to a map $j$ between the larger fields. Such *extension theorems*, saying that under suitable conditions maps between sub-objects can be extended to maps between objects, constitute important weapons in the mathematician's armoury. Using them we can extend our knowledge from small structures to large ones in a sequence of simple steps.

Theorem 5.16 implies that under the given hypotheses the extensions $K(\alpha)/K$ and $L(\beta)/L$ are isomorphic. This allows us to identify $K$ with $L$ and $K(\alpha)$ with $L(\beta)$, via the maps $\iota$ and $j$.

Theorems 5.6 and 5.12 together give a complete characterisation of simple algebraic extensions in terms of polynomials. To each extension corresponds an irreducible monic polynomial, and given the small field and this polynomial, we can reconstruct the extension.

---

## EXERCISES

5.1 Is the extension $\mathbb{Q}(\sqrt{5}, \sqrt{7})$ simple? If so, why? If not, why not?

5.2 Find the minimal polynomials over the small field of the following elements in the following extensions:

(a) $\mathrm{i}$ in $\mathbb{C}/\mathbb{Q}$

(b) $\mathrm{i}$ in $\mathbb{C}/\mathbb{R}$

(c) $\sqrt{2}$ in $\mathbb{R}/\mathbb{Q}$

(d) $(\sqrt{5}+1)/2$ in $\mathbb{C}/\mathbb{Q}$

(e) $(\mathrm{i}\sqrt{3}-1)/2$ in $\mathbb{C}/\mathbb{Q}$

5.3 Show that if $\alpha$ has minimal polynomial $t^2-2$ over $\mathbb{Q}$ and $\beta$ has minimal polynomial $t^2-4t+2$ over $\mathbb{Q}$, then the extensions $\mathbb{Q}(\alpha)/\mathbb{Q}$ and $\mathbb{Q}(\beta)/\mathbb{Q}$ are isomorphic.

5.4 For which of the following $m(t)$ and $K$ do there exist extensions $K(\alpha)$ of $K$ for which $\alpha$ has minimal polynomial $m(t)$?

(a) $m(t) = t^2 - 4, K = \mathbb{R}$

(b) $m(t) = t^2 - 3, K = \mathbb{R}$

(c) $m(t) = t^2 - 3, K = \mathbb{Q}$

(d) $m(t) = t^7 - 3t^6 + 4t^3 - t - 1, K = \mathbb{R}$

5.5 Let $K$ be any subfield of $\mathbb{C}$ and let $m(t)$ be a quadratic polynomial over $K$ ($\partial m = 2$). Show that all zeros of $m(t)$ lie in an extension $K(\alpha)$ of $K$ where $\alpha^2 = k \in K$. Thus allowing 'square roots' $\sqrt{k}$ enables us to solve all quadratic equations over $K$.

5.6 Construct extensions $\mathbb{Q}(\alpha)/\mathbb{Q}$ where $\alpha$ has the following minimal polynomial over $\mathbb{Q}$:

(a) $t^2 - 5$

(b) $t^4 + t^3 + t^2 + t + 1$

(c) $t^3 + 2$

5.7 Is $\mathbb{Q}(\sqrt{2}, \sqrt{3}, \sqrt{5})/\mathbb{Q}$ a simple extension?

5.8 Suppose that $m(t)$ is irreducible over $K$, and $\alpha$ has minimal polynomial $m(t)$ over $K$. Does $m(t)$ necessarily factorise over $K(\alpha)$ into linear (degree 1) polynomials? (*Hint:* Try $K = \mathbb{Q}, \alpha =$ the real cube root of 2.)

5.9 Mark the following true or false.

(a) Every field has non-trivial extensions.

(b) Every field has non-trivial algebraic extensions.

(c) Every simple extension is algebraic.

(d) Every extension is simple.

(e) All simple algebraic extensions of a given subfield of $\mathbb{C}$ are isomorphic.

(f) All simple transcendental extensions of a given subfield of $\mathbb{C}$ are isomorphic.

(g) Every minimal polynomial is monic.

(h) Monic polynomials are always irreducible.

(i) Every polynomial is a constant multiple of an irreducible polynomial.

# Chapter 6

# The Degree of an Extension

A technique which has become very useful in mathematics is that of associating with a given structure a different one, of a type better understood. In this chapter we exploit the technique by associating with any field extension a vector space. This places at our disposal the machinery of linear algebra—a very successful algebraic theory—and with its aid we can make considerable progress. The machinery is sufficiently powerful to solve three notorious problems which remained unanswered for over two thousand years. We shall discuss these problems in the next chapter, and devote the present chapter to developing the theory.

## 6.1 Definition of the Degree

It is not hard to define a vector space structure on a field extension. It already has one! More precisely:

**Theorem 6.1.** *If $L / K$ is a field extension, then the operations*

$$(\lambda, u) \mapsto \lambda u \qquad (\lambda \in K, u \in L)$$
$$(u, v) \mapsto u + v \qquad (u, v \in L)$$

*define on $L$ the structure of a vector space over $K$.*

*Proof.* The set $L$ is a vector space over $K$ if the two operations just defined satisfy the following axioms:

(1) $u + v = v + u$ for all $u, v \in L$.

(2) $(u + v) + w = u + (v + w)$ for all $u, v, w \in L$.

(3) There exists $0 \in L$ such that $0 + u = u$ for all $u \in L$.

(4) For any $u \in L$ there exists $-u \in L$ such that $u + (-u) = 0$.

(5) If $\lambda \in K, u, v \in L$, then $\lambda(u + v) = \lambda u + \lambda v$.

(6) If 1 is the multiplicative identity of $K$, then $1u = u$ for all $u \in L$.

DOI: 10.1201/9781003213949-6

(7) If $\lambda, \mu \in K, u \in L$, then $(\lambda + \mu)u = \lambda u + \mu u$.

(8) If $\lambda, \mu \in K$, then $\lambda(\mu u) = (\lambda\mu)u$ for all $u \in L$.

Each of these statements follows immediately because $K$ and $L$ are subfields of $\mathbb{C}$ and $K \subseteq L$. □

We know that a vector space $V$ over a subfield $K$ of $\mathbb{C}$ (indeed over *any* field, but we're not supposed to know about those yet) is uniquely determined, up to isomorphism, by its dimension. The dimension is the number of elements in a basis—a subset of vectors that spans $V$ and is linearly independent over $K$. The following definition is the traditional terminology in the context of field extensions:

**Definition 6.2.** The *degree* $[L : K]$ of a field extension $L/K$ is the dimension of $L$ considered as a vector space over $K$.

(For some reason the notation $[L/K]$ seems not to be used.)

**Examples 6.3.** (1) The complex numbers $\mathbb{C}$ are two-dimensional over the real numbers $\mathbb{R}$, because a basis is $\{1, \mathrm{i}\}$. Hence $[\mathbb{C}/\mathbb{R}] = 2$.
(2) The extension $\mathbb{Q}(\mathrm{i}, \sqrt{5})/\mathbb{Q}$ has degree 4. The elements $\{1, \sqrt{5}, \mathrm{i}, \mathrm{i}\sqrt{5}\}$ form a basis for $\mathbb{Q}(\mathrm{i}, \sqrt{5})$ over $\mathbb{Q}$, by Example 4.8.

Isomorphic field extensions obviously have the same degree.

---

## 6.2 The Tower Law

The next theorem lets us calculate the degree of a complicated extension if we know the degrees of certain simpler ones.

**Theorem 6.4 (Short Tower Law).** *If $K$, $L$, $M$ are subfields of $\mathbb{C}$ and $K \subseteq L \subseteq M$, then*

$$[M : K] = [M : L][L : K]$$

*Note:* For those who are happy with infinite cardinals this formula needs no extra explanation; the product on the right is just multiplication of cardinals. For those who are not, the formula needs interpretation if any of the degrees involved is infinite. This interpretation is the obvious one: if either $[M : L]$ or $[L : K] = \infty$ then $[M : K] = \infty$; and if $[M : K] = \infty$ then either $[M : L] = \infty$ or $[L : K] = \infty$.

*Proof.* Let $(x_i)_{i \in I}$ be a basis for $L$ as vector space over $K$ and let $(y_j)_{j \in J}$ be a basis for $M$ over $L$. For all $i \in I$ and $j \in J$, we have $x_i \in L$, $y_j \in M$. We shall show that $(x_i y_j)_{i \in I, j \in J}$ is a basis for $M$ over $K$ (where $x_i y_j$ is the product in the subfield $M$). Since dimensions are cardinalities of bases, the theorem follows.

First, we prove linear independence. Suppose that some finite linear combination of the putative basis elements is zero; that is,

$$\sum_{i,j} k_{ij}x_iy_j = 0 \qquad (k_{ij} \in K)$$

We can rearrange this as

$$\sum_j \left( \sum_i k_{ij}x_i \right) y_j = 0$$

Since the coefficients $\sum_i k_{ij}x_i$ lie in $L$ and the $y_j$ are linearly independent over $L$,

$$\sum_i k_{ij}x_i = 0$$

Repeating the argument inside $L$ we find that $k_{ij} = 0$ for all $i \in I$, $j \in J$. So the elements $x_iy_j$ are linearly independent over $K$.

Finally we show that the $x_iy_j$ span $M$ over $K$. Any element $x \in M$ can be written

$$x = \sum_j \lambda_j y_j$$

for suitable $\lambda_j \in L$, since the $y_j$ span $M$ over $L$. Similarly for any $j \in J$

$$\lambda_j = \sum_i \lambda_{ij}x_i$$

for $\lambda_{ij} \in K$. Putting the pieces together,

$$x = \sum_{i,j} \lambda_{ij}x_iy_j$$

as required. □

**Example 6.5.** Suppose we wish to find $[\mathbb{Q}(\sqrt{2}, \sqrt{3}) / \mathbb{Q}]$. It is easy to see that $\{1, \sqrt{2}\}$ is a basis for $\mathbb{Q}(\sqrt{2})$ over $\mathbb{Q}$. For let $\alpha \in \mathbb{Q}(\sqrt{2})$. Then $\alpha = p + q\sqrt{2}$ where $p, q \in \mathbb{Q}$, proving that $\{1, \sqrt{2}\}$ spans $\mathbb{Q}(\sqrt{2})$ over $\mathbb{Q}$ It remains to show that 1 and $\sqrt{2}$ are linearly independent over $\mathbb{Q}$. Suppose that $p + q\sqrt{2} = 0$, where $p, q \in \mathbb{Q}$. If $q \neq 0$ then $\sqrt{2} = p/q$, which is impossible since $\sqrt{2}$ is irrational. Therefore $q = 0$. But this implies $p = 0$.

In much the same way we can show that $\{1, \sqrt{3}\}$ is a basis for $\mathbb{Q}(\sqrt{2}, \sqrt{3})$ over $\mathbb{Q}(\sqrt{2})$. Every element of $\mathbb{Q}(\sqrt{2}, \sqrt{3})$ can be written as $p+q\sqrt{2}+r\sqrt{3}+s\sqrt{6}$ where $p, q, r, s \in \mathbb{Q}$. Rewriting this as

$$(p + q\sqrt{2}) + (r + s\sqrt{2})\sqrt{3}$$

we see that $\{1, \sqrt{3}\}$ spans $\mathbb{Q}(\sqrt{2}, \sqrt{3})$ over $\mathbb{Q}(\sqrt{2})$. To prove linear independence we argue much as above: if

$$(p + q\sqrt{2}) + (r + s\sqrt{2})\sqrt{3} = 0$$

then either $(r + s\sqrt{2}) = 0$, whence also $(p + q\sqrt{2}) = 0$, or else

$$\sqrt{3} = (p + q\sqrt{2})/(r + s\sqrt{2}) \in \mathbb{Q}(\sqrt{2})$$

Therefore $\sqrt{3} = a + b\sqrt{2}$ where $a, b \in \mathbb{Q}$. Squaring, we find that $ab\sqrt{2}$ is rational, which is possible only if either $a = 0$ or $b = 0$. But then $\sqrt{3} = a$ or $\sqrt{3} = b\sqrt{2}$, both of which are absurd. Then $(p + q\sqrt{2}) = (r + s\sqrt{2}) = 0$ and we have proved that $\{1, \sqrt{3}\}$ is a basis. Hence

$$\begin{aligned}[\mathbb{Q}(\sqrt{2}, \sqrt{3}) : \mathbb{Q}] &= [\mathbb{Q}(\sqrt{2}, \sqrt{3}) : \mathbb{Q}(\sqrt{2})][\mathbb{Q}(\sqrt{2}) : \mathbb{Q}] \\ &= 2 \times 2 = 4\end{aligned}$$

The theorem even furnishes a basis for $\mathbb{Q}(\sqrt{2}, \sqrt{3})$ over $\mathbb{Q}$: form all possible pairs of products from the two bases $\{1, \sqrt{2}\}$ and $\{1, \sqrt{3}\}$, to get the 'combined' basis $\{1, \sqrt{2}, \sqrt{3}, \sqrt{6}\}$.

By induction on $n$ we easily parlay the Short Tower Law into a useful generalisation:

**Corollary 6.6 (Tower Law).** *If $K_0 \subseteq K_1 \subseteq \cdots \subseteq K_n$ are subfields of $\mathbb{C}$, then*

$$[K_n : K_0] = [K_n : K_{n-1}][K_{n-1} : K_{n-2}] \cdots [K_1 : K_0]$$

□

In order to use the Tower Law we have to get started. The degree of a simple extension is fairly easy to find:

**Proposition 6.7.** *Let $K(\alpha)/K$ be a simple extension. If it is transcendental then $[K(\alpha) : K] = \infty$. If it is algebraic then $[K(\alpha) : K] = \partial m$, where $m$ is the minimal polynomial of $\alpha$ over $K$.*

*Proof.* For the transcendental case it suffices to note that the elements $1, \alpha, \alpha^2, \ldots$ are linearly independent over $K$. For the algebraic case, we appeal to Lemma 5.14. □

For example, we know that $\mathbb{C} = \mathbb{R}(\mathrm{i})$ where i has minimal polynomial $t^2 + 1$, of degree 2. Hence $[\mathbb{C} : \mathbb{R}] = 2$, which agrees with our previous remarks.

**Example 6.8.** We now illustrate a technique that we shall use, without explicit reference, whenever we discuss extensions of the form $\mathbb{Q}(\sqrt{\alpha_1}, \ldots, \sqrt{\alpha_n})/\mathbb{Q}$ with rational $\alpha_j$. The technique can be used to prove a general theorem about such extensions, see Exercise 6.14. The question we tackle is: find $[\mathbb{Q}(\sqrt{2}, \sqrt{3}, \sqrt{5}) : \mathbb{Q}]$.

By the Tower Law,

$$\begin{aligned}&[\mathbb{Q}(\sqrt{2}, \sqrt{3}, \sqrt{5}) : \mathbb{Q}] \\ &= [\mathbb{Q}(\sqrt{2}, \sqrt{3}, \sqrt{5}) : \mathbb{Q}(\sqrt{2}, \sqrt{3})][\mathbb{Q}(\sqrt{2}, \sqrt{3}) : \mathbb{Q}(\sqrt{2})][\mathbb{Q}(\sqrt{2}) : \mathbb{Q}]\end{aligned}$$

It is 'obvious' that each factor equals 2, but it takes some effort to prove it. As a cautionary remark: the degree $[\mathbb{Q}(\sqrt{6}, \sqrt{10}, \sqrt{15}) : \mathbb{Q}]$ is 4, not 8 (Exercise 6.14).

(a) Certainly $[\mathbb{Q}(\sqrt{2}) : \mathbb{Q}] = 2$.

(b) If $\sqrt{3} \notin \mathbb{Q}(\sqrt{2})$ then $[\mathbb{Q}(\sqrt{2}, \sqrt{3}) : \mathbb{Q}(\sqrt{2})] = 2$. So suppose $\sqrt{3} \in \mathbb{Q}(\sqrt{2})$, implying that

$$\sqrt{3} = p + q\sqrt{2} \qquad p, q \in \mathbb{Q}$$

We argue as in Example 6.5. Squaring,

$$3 = (p^2 + 2q^2) + 2pq\sqrt{2}$$

so

$$p^2 + 2q^2 = 3 \qquad pq = 0$$

If $p = 0$ then $2q^2 = 3$, which is impossible by Exercise 1.3. If $q = 0$ then $p^2 = 3$, which is impossible for the same reason. Therefore $\sqrt{3} \notin \mathbb{Q}(\sqrt{2})$, and $[\mathbb{Q}(\sqrt{2}, \sqrt{3}) : \mathbb{Q}(\sqrt{2})] = 2$.

(c) Finally, we claim that $\sqrt{5} \notin \mathbb{Q}(\sqrt{2}, \sqrt{3})$. Here we need a new idea. Suppose that

$$\sqrt{5} = p + q\sqrt{2} + r\sqrt{3} + s\sqrt{6} \qquad p, q, r, s \in \mathbb{Q}$$

Squaring:

$$5 = p^2 + 2q^2 + 3r^2 + 6s^2 + (2pq + 6rs)\sqrt{2} + (2pr + 4qs)\sqrt{3} + (2ps + 2qr)\sqrt{6}$$

whence

$$\begin{aligned} p^2 + 2q^2 + 3r^2 + 6s^2 &= 5 \\ pq + 3rs &= 0 \\ pr + 2qs &= 0 \\ ps + qr &= 0 \end{aligned} \tag{6.1}$$

The new idea is to observe that if $(p, q, r, s)$ satisfies (6.1), then so do $(p, q, -r, -s)$, $(p, -q, r, -s)$, and $(p, -q, -r, s)$. Therefore

$$\begin{aligned} p + q\sqrt{2} + r\sqrt{3} + s\sqrt{6} &= \sqrt{5} \\ p + q\sqrt{2} - r\sqrt{3} - s\sqrt{6} &= \pm\sqrt{5} \\ p - q\sqrt{2} + r\sqrt{3} - s\sqrt{6} &= \pm\sqrt{5} \\ p - q\sqrt{2} - r\sqrt{3} + s\sqrt{6} &= \pm\sqrt{5} \end{aligned}$$

Adding the first two equations, we get $p + q\sqrt{2} = 0$ or $p + q\sqrt{2} = \sqrt{5}$. The first implies that $p = q = 0$. The second implies that $p^2 + 2q^2 + 2pq\sqrt{2} = 5$, which is easily seen to be impossible. Adding the first and third, $r\sqrt{3} = 0$ or $r\sqrt{3} = \sqrt{5}$, so $r = 0$. Finally, $s = 0$ since $s\sqrt{6} = \sqrt{5}$ is impossible by Exercise 1.3.

Having proved the claim, we immediately deduce that

$$[\mathbb{Q}(\sqrt{2}, \sqrt{3}, \sqrt{5}) : \mathbb{Q}(\sqrt{2}, \sqrt{3})] = 2$$

which implies that $[\mathbb{Q}(\sqrt{2}, \sqrt{3}, \sqrt{5}) : \mathbb{Q}] = 8$.

Linear algebra is at its most powerful when dealing with finite-dimensional vector spaces. Accordingly we shall concentrate on field extensions that give rise to such vector spaces.

**Definition 6.9.** A *finite extension* is one whose degree is finite.

Proposition 6.7 implies that any simple algebraic extension is finite. The converse is not true, but certain partial results are: see Exercise 6.15. In order to state what is true we need:

**Definition 6.10.** An extension $L/K$ is *algebraic* if every element of $L$ is algebraic over $K$.

Algebraic extensions need not be finite, see Exercise 6.10, but every finite extension is algebraic. More generally:

**Lemma 6.11.** *An extension $L/K$ is finite if and only if $L = K(\alpha, \ldots, \alpha_r)$ where $r$ is finite and each $\alpha_i$ is algebraic over $K$.*

*Proof.* Induction using Theorem 6.4 and Proposition 6.7 shows that any extension of the form $K(\alpha_1, \ldots, \alpha_s)/K$ for algebraic $\alpha_j$ is finite.

Conversely, let $L/K$ be a finite extension. Then there is a basis $\{\alpha_1, \ldots, \alpha_s\}$ for $L$ over $K$, whence $L = K(\alpha_1, \ldots, \alpha_s)$. Each $\alpha_j$ is clearly algebraic. □

---

## 6.3 Primitive Element Theorem

Recall that a simple extension is one of the form $K(\theta)/K$. Many field extensions are presented in a manner that makes them appear not to be simple, but on further examination it turns out that they are. See Example 4.11(2), for instance.

We now give a classical criterion for a field extension in $\mathbb{C}$ to be simple. First, we need:

**Lemma 6.12.** *If $p(t) \in K[t]$ is irreducible, it has no multiple zeros.*

*Proof.* If $p(t)$ has a multiple zero $a$ then $p(t) = (t-a)^2 q(t)$ in $\mathbb{C}[t]$. The derivative

$$p'(t) = 2(t-a)q(t) + (t-a)q'(t)$$

is divisible by $(t-a)$ in $\mathbb{C}[t]$, and the usual formula for the derivative of $t^n$ implies that $p'(t) \in K[t]$. Therefore $p(t)$ and $p'(t)$ have a common factor in $K[t]$ of degree at least 1. But this is impossible since $\partial p' = \partial p - 1 < \partial p$ and $p(t)$ is irreducible. Therefore $p(t)$ has no multiple zeros. □

Using the degree, we can prove that every finite extension of a subfield of $\mathbb{C}$ is a simple extension, and conversely. A simpler and more general proof using Galois theory is given later in Theorem 17.29. The name of the theorem reflects classical terminology: if $L = K(\theta)$ then $\theta$ is a *primitive element* of $L$. The proof below comes from Brown (2010).

**Theorem 6.13 (Primitive Element Theorem).** *Let $K, L$ be subfields of $\mathbb{C}$ with $[L : K]$ finite. Then there exists $\theta \in L$ such that $L = K(\theta)$.*

*Proof.* By Lemma 6.11, $L = K(\alpha_1, \ldots, \alpha_n)$ where the $\alpha_j$ are algebraic over $K$. The main point is to prove the case $n = 2$ : if $\alpha_1, \alpha_2$ are algebraic over $K$ and $L = K(\alpha_1, \alpha_2)$, then there exists $\theta \in L$ such that $L = K(\theta)$. The theorem then follows by induction on $n$, since

$$L = K(\alpha_1, \ldots, \alpha_n) = K(\alpha_1, \alpha_2)(\alpha_3, \ldots, \alpha_n) = K(\theta, \alpha_3, \ldots, \alpha_n)$$

adjoining only $n - 1$ algebraic elements to $K$.

To prove the result for $n = 2$, write $\alpha_1 = \alpha, \alpha_2 = \beta$ to simplify notation. Consider an element

$$\gamma = \alpha + \lambda\beta \tag{6.2}$$

for $\lambda \in K$, and let $L = K(\alpha, \beta)$. We claim that $\gamma$ is a primitive element unless $\lambda$ is 'bad', that is, $\lambda$ belongs to a specific finite subset $S \subseteq K$, which we define during the proof; see (6.4).

To prove that $\gamma$ is primitive, it is enough to show that the simple extension $K(\gamma)$ contains $\beta$. This follows because it then also contains $\alpha = \gamma - \lambda\beta$, so $K(\alpha, \beta) \subseteq K(\gamma) \subseteq K(\alpha, \beta)$ and the two fields are equal.

This in turn follows if we can show that the minimal polynomial $p$ of $\beta$ over $K(\gamma)$ has degree 1, unless $\lambda \in S$. To do this, let $f, g$ be the minimal polynomials over $K$ of $\alpha, \beta$ respectively. Write $f, g$ as products of linear factors

$$\begin{aligned} f(t) &= (t - \rho_1) \ldots (t - \rho_m) \\ g(t) &= (t - \sigma_1) \ldots (t - \sigma_n) \end{aligned}$$

and set $\Sigma = K(\rho_1, \ldots, \rho_m, \sigma_1, \ldots, \sigma_n)$. Then $\Sigma/K$ is an extension in which both $f$ and $g$ are products of linear factors. Since $f(\gamma - \lambda\beta) = f(\alpha) = 0$, the element $\beta$ is a zero of the polynomial $h \in K(\gamma)[t]$ defined by $h(t) = f(\gamma - \lambda t)$. Therefore $p$ divides both $g$ and $h$, so it divides $\gcd(g, h)$.

It now suffices to prove that the degree of $\gcd(g, h)$ cannot be greater than 1. Suppose (for a contradiction) that the degree is greater than 1. By Lemma 6.12, $g$ and $h$ have a common zero $\beta' \in \Sigma$ with $\beta' \neq \beta$. Then $f(\gamma - \lambda\beta') = 0$; that is,

$$\gamma - \lambda\beta' = \alpha' \tag{6.3}$$

for some zero $\alpha'$ of $f$ in $\Sigma$. Since $\gamma = \alpha + \lambda\beta$, we can eliminate $\gamma$ by subtracting (6.3) from (6.2), and then solve for $\lambda$, obtaining

$$\lambda = \frac{\alpha' - \alpha}{\beta - \beta'} \tag{6.4}$$

(this makes sense since $\beta' \neq \beta$). The set $S$ of all such 'bad' choices of $\lambda$ is finite because the numbers of possible $\alpha'$ and $\beta'$ are finite. Since $K$ is an infinite field, we can choose $\lambda \in K \setminus S$, and then $K(\alpha, \beta) = K(\gamma)$. $\square$

---

## EXERCISES

6.1. Find the degrees of the following extensions:

(a) $\mathbb{C}/\mathbb{Q}$

(b) $\mathbb{R}(\sqrt{5})/\mathbb{R}$

(c) $\mathbb{Q}(\alpha)/\mathbb{Q}$ where $\alpha$ is the real cube root of 2

(d) $\mathbb{Q}(3, \sqrt{5}, \sqrt{11})/\mathbb{Q}$

(e) $\mathbb{Q}(\sqrt{6})/\mathbb{Q}$

(f) $\mathbb{Q}(\alpha)/\mathbb{Q}$ where $\alpha^7 = 3$

6.2. Show that every element of $\mathbb{Q}(\sqrt{5}, \sqrt{7})$ can be expressed uniquely in the form

$$p + q\sqrt{5} + r\sqrt{7} + s\sqrt{35}$$

where $p$, $q$, $r$, $s \in \mathbb{Q}$. Calculate explicitly the inverse of such an element.

6.3. If $[L : K]$ is a prime number show that the only fields $M$ such that $K \subseteq M \subseteq L$ are $K$ and $L$ themselves.

6.4. If $[L : K] = 1$ show that $K = L$.

6.5. Write out in detail the inductive proof of Corollary 6.6.

6.6. Let $L/K$ be an extension. Show that multiplication by a fixed element of $L$ is a linear transformation of $L$ considered as a vector space over $K$. When is this linear transformation nonsingular?

6.7. Let $L/K$ be a finite extension, and let $p$ be an irreducible polynomial over $K$. Show that if $\partial p$ does not divide $[L : K]$, then $p$ has no zeros in $L$.

6.8. If $L/K$ is algebraic and $M/L$ is algebraic, is $M/K$ algebraic? (You may not assume the extensions are finite.)

6.9. Prove that $\mathbb{Q}(\sqrt{3}, \sqrt{5}) = \mathbb{Q}(\sqrt{3} + \sqrt{5})$. Try to generalise your result.

6.10* Prove that the square roots of all prime numbers are linearly independent over $\mathbb{Q}$. Deduce that algebraic extensions need not be finite.

6.11 Find a basis for $\mathbb{Q}(\sqrt{1+\sqrt{3}})$ over $\mathbb{Q}$ and hence find the degree of $\mathbb{Q}(\sqrt{1+\sqrt{3}})/\mathbb{Q}$. (*Hint:* You will need to prove that $1+\sqrt{3}$ is not a square in $\mathbb{Q}(\sqrt{3})$.)

6.12 If $[L:K]$ is prime, show that $L$ is a simple extension of $K$.

6.13 Show that $[\mathbb{Q}(\sqrt{6},\sqrt{10},\sqrt{15}):\mathbb{Q}]=4$, not 8.

6.14* Let $K$ be a subfield of $\mathbb{C}$ and let $a_1,\ldots,a_n$ be elements of $K$ such that any product $a_{j_1}\cdots a_{j_k}$, with distinct indices $j_l$, is not a square in $K$. Let $\alpha_j=\sqrt{a_j}$ for $1\leq j\leq n$. Prove that $[K(\alpha_1,\ldots,\alpha_n):K]=2^n$.

If $K=\mathbb{Q}$, how can we verify the hypotheses on the $a_j$ by looking at their prime factorisations?

6.15* Let $L/K$ be an algebraic extension and suppose that $K$ is an infinite field. Prove that $L/K$ is simple if and only if there are only finitely many fields $M$ such that $K\subseteq M\subseteq L$, as follows.

(a) Assume only finitely many $M$ exist. Use Lemma 6.11 to show that $L/K$ is finite.

(b) Assume $L=K(\alpha_1,\alpha_2)$. For each $\beta\in K$ let $J_\beta=K(\alpha_1+\beta\alpha_2)$. Only finitely many distinct $J_\beta$ can occur: hence show that $L=J_\beta$ for some $\beta$.

(c) Use induction to prove the general case.

(d) For the converse, let $L=K(\alpha)$ be simple algebraic, with $K\subseteq M\subseteq L$. Let $m$ be the minimal polynomial of $\alpha$ over $K$, and let $m_M$ be the minimal polynomial of $\alpha$ over $M$. Show that $m_M|m$ in $L[t]$. Prove that $m_M$ determines $M$ uniquely, and that only finitely many $m_M$ can occur.

6.16 Mark the following true or false.

(a) Extensions of the same degree are isomorphic.

(b) Isomorphic extensions have the same degree.

(c) Every algebraic extension is finite.

(d) Every transcendental extension is not finite.

(e) Every element of $\mathbb{C}$ is algebraic over $\mathbb{R}$.

(f) Every extension of $\mathbb{R}$ that is a subfield of $\mathbb{C}$ is finite.

(g) Every algebraic extension of $\mathbb{Q}$ is finite.

# Chapter 7

## Ruler-and-Compass Constructions

Already we are in a position to see some payoff. The degree of a field extension is a surprisingly powerful tool. Even before we get into Galois theory proper, we can apply the degree to a warm-up problem—indeed, several. The problems come from classical Greek geometry, and we will do something much more interesting and difficult than solving them. We will prove that no solutions exist, subject to certain technical conditions on the permitted methods.

According to Plato the only 'perfect' geometric figures are the straight line and the circle. In the most widely known parts of ancient Greek geometry, this belief had the effect of restricting the (conceptual) instruments available for performing geometric constructions to two: the ruler and the compass. The ruler, furthermore, was a single unmarked straight edge.

Strictly, the term should be '*pair* of compasses', for the same reason we call a single cutting instrument a pair of scissors. However, 'compass' is shorter, and there is no serious danger of confusion with the navigational instrument that tells you which way is north. So 'compass' it is.

With these instruments alone it is possible to perform a wide range of constructions, as Euclid systematically set out in his *Elements* somewhere around 300 BC. This series of books opens with 23 definitions of basic objects ranging from points to parallels, five axioms (called 'postulates' in the translation by Sir Thomas Heath), and five 'common notions' about equality and inequality. The first three axioms state that certain constructions may be performed:

(1) To draw a straight line from any point to any (distinct) point.

(2) To produce a finite straight line continuously in a straight line.

(3) To describe a circle with any centre and any distance.

The first two model the use of a ruler (or straightedge); the third models the use of a compass.

**Definition 7.1.** A *ruler-and-compass construction* in the sense of Euclid is a finite sequence of operations of the above three types.

For readers who are unfamiliar with the classical construction methods in Euclid—which is not unusual nowadays—we give details in Sections 7.2 and 7.3. The website 'Euclid the game', listed in the References, offers a friendly interactive introduction to these constructions and other ideas in Euclid's

DOI: 10.1201/9781003213949-7

geometry. The restriction to finite constructions is important. Infinite constructions can sometimes make theoretical sense, and are more powerful: see Exercise 7.12. They provide arbitrarily good approximations if we stop after a finite number of steps.

Later Greek geometry introduced other 'drawing instruments', such as conic sections and a curve called the quadratrix. But long-standing tradition associates Euclid with geometric constructions carried out using an unmarked ruler and a compass. The *Elements* includes ruler-and-compass constructions to bisect a line or an angle, to divide a line into any specified number of equal parts, and to draw a regular pentagon.

However, there are many geometric problems that clearly 'should' have solutions, but for which the tools of ruler and compass are inadequate. In particular, there are three famous constructions that the Greeks could not perform using these tools: *duplicating the cube, trisecting the angle*, and *squaring the circle*. These ask respectively for a cube twice the volume of a given cube, an angle one-third the size of a given angle, and a square of area equal to a given circle.

It seems likely that Euclid would have included such constructions if he knew any, and it is a measure of his mathematical taste that he did not present fallacious constructions that are approximately correct but not exact. The Greeks were ingenious enough to find exact constructions if they existed, unless they had to be extraordinarily complicated. (The construction of a regular 17-gon is an example of a complicated construction that they missed: see Chapter 19.) We now know why they failed to find ruler-and-compass constructions for the three classical problems: they don't exist. But the Greeks lacked the algebraic techniques needed to prove that.

The impossibility of trisecting an arbitrary angle using ruler and compass was not proved until 1798 when Gauss was writing his *Disquisitiones Arithmeticae*, published in 1801. Discussing his construction of the regular 17-gon, he states without proof that such constructions do not exist for the 9-gon, 25-gon, and other numbers that are not a power of 2 times a product of distinct Fermat primes—those of the form $2^{2^n}+1$. He also writes that he can 'prove in all rigour that these higher-degree equations [involved in the construction] cannot be avoided in any way', but adds 'the limits of the present work exclude this demonstration here.' Constructing the regular 9-gon is clearly equivalent to trisecting $\frac{2\pi}{3}$, so Gauss's claim disposes of trisections. He did not publish a proof; the first person to do so was Pierre Wantzel in 1837.

This result does not imply that an angle one third the size of a given one does not exist, or that practical constructions with very small errors cannot be devised; it tells us that the specified instruments are inadequate to find it *exactly*. Wantzel also proved that it is impossible to duplicate the cube with ruler and compass. Squaring the circle had to wait even longer for an impossibility proof.

In this chapter we mention approximate constructions, which are entirely acceptable for practical work. We make some brief historical remarks to point out that the Greeks could solve the three classical problems using 'instruments' that went beyond just ruler and compass. We identify the Euclidean plane $\mathbb{R}^2$ with the complex plane $\mathbb{C}$, which lets us avoid considering the two coordinates of a point separately and greatly simplifies the discussion. We formalise the concept of ruler-and-compass construction by defining the notion of a constructible point in $\mathbb{C}$. We introduce a series of specific constructions that correspond to field operations $(+, -, \times, /)$ and square roots in $\mathbb{C}$. We characterise constructible points in terms of the 'Pythagorean closure' $\mathbb{Q}^{\text{py}}$ of $\mathbb{Q}$, and deduce a simple algebraic criterion for a point to be constructible. By applying this criterion, we prove that the three classical problems cannot be solved by ruler-and-compass construction. We also prove that there is no such construction for a regular heptagon (7-sided polygon).

## 7.1 Approximate Constructions and More General Instruments

For the technical drawing expert we emphasise that we are discussing *exact* constructions. There are many approximate constructions for trisecting the angle, for instance, but no exact methods. Dudley (1987) is a fascinating collection of approximate methods that were thought by their inventors to be exact. Figure 10 is a typical example. To trisect angle BOA, draw line BE parallel to OA. Mark off AC and CD equal to OA, draw arc DE with centre C and radius CD. Drop a perpendicular EF to OD and draw arc FT centre O radius OF to meet BE at T. Then angle TOA approximately trisects angle BOA. See Exercise 7.10.

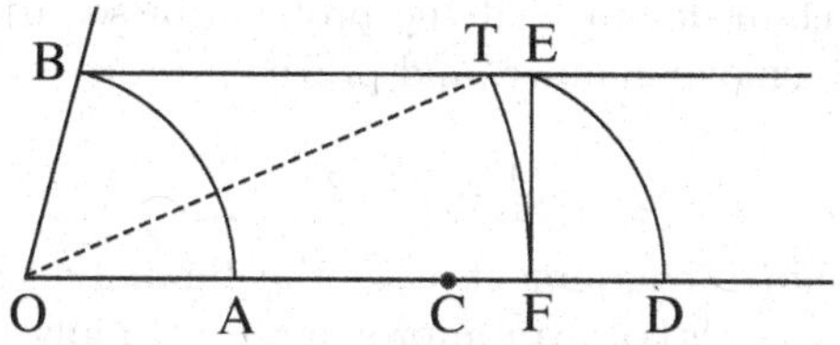


FIGURE 10: Close—but no banana.

The Greeks were well aware that by going outside the Platonic constraints, all three classical problems can be solved. Archimedes and others knew that angles can be trisected using a *marked* ruler, as in Figure 11. The ruler has marked on it two points distance $r$ apart. Given $\angle AOB = \theta$ draw a circle centre O with radius $r$, cutting OA at X, OB at Y. Place the ruler with its edge through X and one mark on the line OY at D; slide it until the other marked point lies on the circle at E. Then $\angle EDO = \theta/3$. For a proof, see Exercise 7.4. Exercise 7.14 shows how to duplicate the cube using a marked ruler.

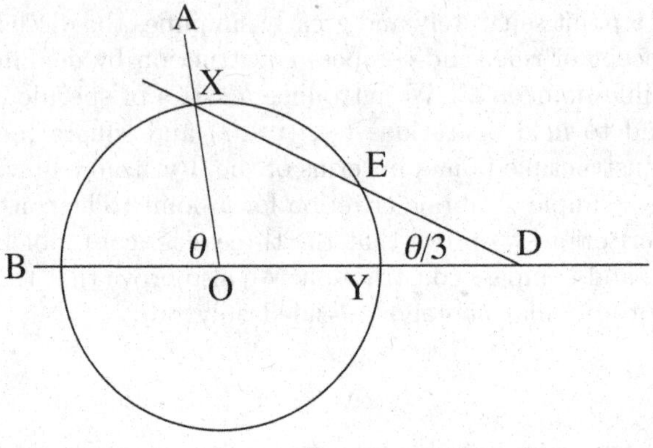


FIGURE 11: Trisecting an angle with a marked ruler.

Setting your compass up against the ruler so that the pivot point and the pencil effectively constitute such marks also provides a trisection, but again this goes beyond the precise concept of a 'ruler-and-compass construction'. Many other uses of 'exotic' instruments are catalogued in Dudley (1987), which examines the history of trisection attempts. Euclid may have limited himself to an unmarked ruler (plus compass) because it made his axiomatic treatment more convincing. It is not entirely clear what conditions should apply to a marked ruler—the distance between the marks causes difficulties. Presumably it ought to be constructible, for example.

The Greeks solved all three problems using conic sections, or more recondite curves such as the conchoid of Nichomedes or the quadratrix (Klein 1962, Coolidge 1963). Archimedes tackled the problem of squaring the circle in a characteristically ingenious manner, and proved a result which would now be written

$$3\tfrac{10}{71} < \pi < 3\tfrac{1}{7}$$

This was a remarkable achievement with the limited techniques available, and refinements of his method can approximate $\pi$ to any required degree of precision.

Such extensions of the apparatus solve the practical problem, but it is the theoretical one that holds the most interest. What, precisely, are the *limitations* on ruler-and-compass constructions? With the machinery now at our disposal it is relatively simple to characterise these limitations, and thereby give a complete answer to all three problems. We use coordinate geometry to express problems in algebraic terms, and apply the theory of field extensions to the algebraic questions that arise.

## 7.2 Constructions in $\mathbb{C}$

We begin by formalising the notion of a ruler-and-compass construction. Assume that initially we are given two distinct points in the plane. Equivalently, by Euclid's Axiom 1, we can begin with the line segment that joins them. These points let us choose an origin and set a scale. So we can identify the Euclidean plane $\mathbb{R}^2$ with $\mathbb{C}$, and assume that these two points are 0 and 1.

Euclid dealt with finite line segments (his condition (1) above) but could make them as long as he pleased by extending the line (condition (2)). We find it more convenient to work with infinitely long lines (modelling an infinitely long ruler), which in effect combines Euclid's conditions into just one: the possibility of drawing the (infinitely long) line that passes through two given distinct points. From now on, 'line' is always used in this sense.

If $z_1, z_2 \in \mathbb{C}$ and $0 \leq r \in \mathbb{R}$, define

$$\begin{aligned} L(z_1, z_2) &= \text{the line joining } z_1 \text{ to } z_2 \quad (z_1 \neq z_2) \\ C(z_1, r) &= \text{the circle centre } z_1 \text{ with radius } r > 0 \end{aligned}$$

We now define constructible points, lines, and circles recursively:

**Definition 7.2.** For each $n \in \mathbb{N}$ define sets $\mathcal{P}_n, \mathcal{L}_n$, and $\mathcal{C}_n$ of *n-constructible points, lines*, and *circles*, by:

$$\begin{aligned} \mathcal{P}_0 &= \{0, 1\} \\ \mathcal{L}_0 &= \emptyset \\ \mathcal{C}_0 &= \emptyset \\ \mathcal{L}_{n+1} &= \{L(z_1, z_2) : z_1, z_2 \in \mathcal{P}_n\} \\ \mathcal{C}_{n+1} &= \{C(z_1, |z_2 - z_3|) : z_1, z_2, z_3 \in \mathcal{P}_n\} \\ \mathcal{P}_{n+1} &= \{z \in \mathbb{C} : z \text{ lies on two distinct lines in } \mathcal{L}_{n+1}\} \cup \\ &\quad\ \{z \in \mathbb{C} : z \text{ lies on a line in } \mathcal{L}_{n+1} \text{ and a circle in } \mathcal{C}_{n+1}\} \cup \\ &\quad\ \{z \in \mathbb{C} : z \text{ lies on two distinct circles in } \mathcal{C}_{n+1}\} \end{aligned}$$

Figure 12 shows that

$$\mathcal{P}_1 = \{-1, 0, 1, 2, \frac{1 \pm i\sqrt{3}}{2}\}$$

**Lemma 7.3.** *For all* $n \in \mathbb{N}$,

$$\mathcal{P}_n \subseteq \mathcal{P}_{n+1} \qquad \mathcal{L}_n \subseteq \mathcal{L}_{n+1} \qquad \mathcal{C}_n \subseteq \mathcal{C}_{n+1}$$

*and each is a finite set.*

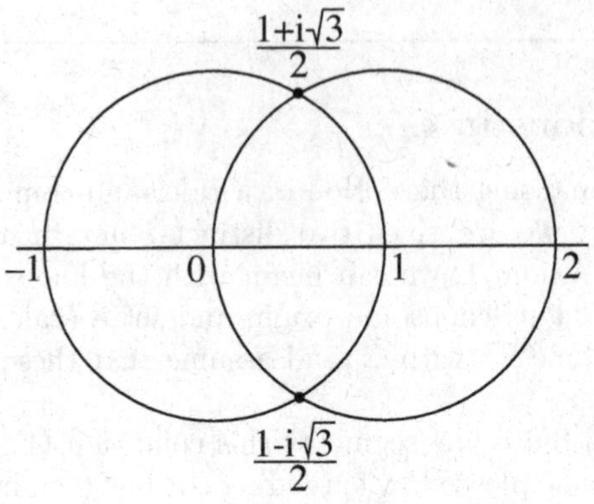


FIGURE 12: The set $\mathcal{P}_1$.

*Proof.* The inclusions are clear. Let $p_n$ be the number of points in $\mathcal{P}_n$, $l_n$ the number of lines in $\mathcal{L}_n$, and $c_n$ the number of circles in $\mathcal{C}_n$. Then

$$
\begin{aligned}
|\mathcal{L}_{n+1}| &\leq \tfrac{1}{2}p_n(p_n+1) \\
|\mathcal{C}_{n+1}| &\leq p_n \tfrac{1}{2}p_n(p_n+1) \\
|\mathcal{P}_{n+1}| &\leq \tfrac{1}{2}l_{n+1}(l_{n+1}+1) + 2l_n c_n + c_{n+1}(c_{n+1}+1)
\end{aligned}
$$

bearing in mind that a line or circle meets a distinct circle in $\leq 2$ points. By induction, all three sets are finite for all $n$. □

We formalise a Euclidean ruler-and-compass construction using these sets. The intuitive idea is that starting from 0 and 1, such a construction generates a finite sequence of points by drawing a line through two previously constructed points, or a circle whose centre is one previously constructed point and whose radius is the distance between two previously constructed points, and then defining a new point using intersections of these.

**Definition 7.4.** A point $z \in \mathbb{C}$ is *constructible* if there is a finite sequence of points

$$z_0 = 0, z_1 = 1, z_2, z_3, \ldots z_k = z \tag{7.1}$$

such that $z_{j+1}$ lies in at least one of:

$$
\begin{gathered}
L(z_{j_1}, z_{j_2}) \cap L(z_{j_3}, z_{j_4}) \\
L(z_{j_1}, z_{j_2}) \cap C(z_{j_3}, |z_{j_4} - z_{j_5}|) \\
C(z_{j_1}, |z_{j_2} - z_{j_3}|) \cap C(z_{j_4}, |z_{j_5} - z_{j_6}|)
\end{gathered}
$$

where all $j_i \leq j$ and the intersecting lines and circles are distinct.

In the first case, the lines must not be parallel in order to have non-empty intersection; in the other cases, the line must meet the circle and the two circles must meet. These technical conditions can be expressed as algebraic properties of the $z_j$.

We can now prove:

**Theorem 7.5.** *A point $z \in \mathbb{C}$ is constructible if and only if $z \in \mathcal{P}_n$ for some $n \in \mathbb{N}$.*

*Proof.* Let $z \in \mathbb{C}$ be constructible, using the sequence (7.1). Inductively, it is clear that $z = z_k \in \mathcal{P}_k$.

Conversely, let $z \in \mathcal{P}_k$. Then we can find a sequence $z_j \in \mathcal{P}_j$, where $0 \leq j \leq k$, satisfying (7.1). □

To characterise constructible points, we need:

**Definition 7.6.** The *Pythagorean closure* $\mathbb{Q}^{\text{py}}$ of $\mathbb{Q}$ is the union of all subfields $K$ such that there is a tower of field extensions

$$\mathbb{Q} = K_0 \subseteq K_1 \subseteq \ldots \subseteq K_n = K \tag{7.2}$$

such that

$$[K_{j+1} : K_j] \leq 2 \tag{7.3}$$

for $0 \leq j \leq n-1$.

**Remarks 7.7.** (1) The term 'Pythagorean closure' is not standard in the literature. We give Pythagoras credit because his theorem can be used to construct square roots of rational numbers. If $a^2 + b^2 = c^2$ then $c = \sqrt{a^2 + b^2}$.
(2) By definition, a complex number $z$ belongs to $\mathbb{Q}^{\text{py}}$ if and only if $z \in K_n$ for some tower (7.2) obeying condition (7.3).
(3) By removing duplicated subfields, we can replace (7.3) by $[K_{j+1} : K_j] = 2$, provided we allow the trivial tower $\mathbb{Q}$ of length $n = 0$.

**Proposition 7.8.** (1) *If $z \in \mathbb{Q}^{\text{py}}$ then its complex conjugate $\bar{z} \in \mathbb{Q}^{\text{py}}$.*

(2) *The field $\mathbb{Q}^{\text{py}}$ is the smallest subfield $P \subseteq \mathbb{C}$ with the property:*

$$z \in P \implies \pm\sqrt{z} \in P \tag{7.4}$$

*Proof.* (1) If $X \subseteq \mathbb{C}$ define $\overline{X} = \{\bar{x} : x \in X\}$. Suppose that $z \in \mathbb{Q}^{\text{py}}$. By Remark 7.7 (2), $z \in K$ for some tower (7.2) obeying condition (7.3). Then

$$\mathbb{Q} = K_0 \subseteq K_1 \subseteq \ldots \subseteq K_n = K \ni z$$

with $[K_{j+1} : K_j] \leq 2$. Take complex conjugates:

$$\mathbb{Q} = K_0 \subseteq \overline{K_1} \subseteq \ldots \subseteq \overline{K_n} = \overline{K} \ni \bar{z}$$

($\mathbb{Q}$ is its own complex conjugate since $\mathbb{Q} \subseteq \mathbb{R}$) and

$$[\overline{K_{j+1}} : \overline{K_j}] = [K_{j+1} : K_j] \leq 2$$

so $\bar{z} \in \mathbb{Q}^{\text{py}}$.

(2) We show that:

(a) Any subfield $P$ with property (7.4) contains $\mathbb{Q}^{\text{py}}$, and

(b) $\mathbb{Q}^{\text{py}}$ satisfies (7.4).

To prove (a), we show by induction on $n$ that every field $K$ at the top of a tower (7.2) is contained in $P$. When $n = 0$ we have $K_0 = \mathbb{Q} \subseteq P$ since $P$ is a subfield of $\mathbb{C}$. Assume that $K_{n-1} \subseteq P$. Then either $K_n = K_{n-1}$ or $K_n = K_{n-1}(\alpha)$ where $\alpha^2 \in K_{n-1}$. By (7.4), $K_n \subseteq P$. Taking the union over all such $K_n$, we obtain $\mathbb{Q}^{\text{py}} \subseteq P$.

To prove (b) let $z \in \mathbb{Q}^{\text{py}}$. By Remark 7.7 (2), $z \in K$ for some tower (7.2) obeying condition (7.3). Then $\sqrt{z} \in K_{n+1} = K_n(\sqrt{z})$ and $[K_{n+1} : K_n] \leq 2$. Therefore $\sqrt{z}$ belongs to a tower of length $n+1$ satisfying (7.3), so $\sqrt{z} \in Q^{\text{py}}$, and $Q^{\text{py}}$ satisfies (7.4). □

The main theorem of this section is:

**Theorem 7.9.** *A point $z \in \mathbb{C}$ is constructible if and only if $z \in \mathbb{Q}^{\text{py}}$. Equivalently,*

$$\bigcup_{n=0}^{\infty} \mathcal{P}_n = \mathbb{Q}^{\text{py}} \tag{7.5}$$

*Pre-proof Discussion.*

We can summarise the main idea succinctly. Coordinate geometry in $\mathbb{C}$ shows that each step in a ruler-and-compass construction leads to points that can be expressed using rational functions of the previously constructed points together with the square root of a rational function of those points. Conversely, all rational functions of given points can be constructed, and so can square roots of given points. Therefore anything that can be constructed lies in $\mathbb{Q}^{\text{py}}$, and anything in $\mathbb{Q}^{\text{py}}$ can be constructed.

The details require some algebraic computations in $\mathbb{C}$ and some basic Euclidean geometry. We prove Theorem 7.9 in two stages. In this section we show that

(A) $\mathcal{P}_n \subseteq \mathbb{Q}^{\text{py}}$ for all $n \in \mathbb{N}$.

In the next section, after describing some basic constructions for arithmetical operations and square roots, we complete the proof by establishing

(B) If $z \in \mathbb{Q}^{\text{py}}$ then $z \in \mathcal{P}_n$ for some $n \in \mathbb{N}$.

Theorem 7.9 is an immediate consequence of (A) and (B).

The proof uses:

**Lemma 7.10.** *Suppose that $z \in \mathcal{P}_n$. Then its complex conjugate $\bar{z} \in \mathcal{P}_{n+1}$.*

*Proof.* Let $z \in \mathcal{P}_n$. Then we can construct $\bar{z}$ from the points $0, 1, z$ as follows: If $z = 1$ then $\bar{z} = 1$ and we are done. If not, Draw $C(z, |z-1|)$ meeting $L(0,1)$ at 1. Either this is the only such intersection point, in which case $L(z, 1)$ is perpendicular to $L(0, 1)$, or there is a second intersection point $s$. In the first case, the circle $C(1, |z-1|)$ meets the line $L(z, 1$ in two points: one is $z$ and the other is $\bar{z}$. In the second case, $C(z, |z-1|)$ meets $L(0,1)$ at 1 and at $s \neq 1$. Now the circles $C(1, |z-1|)$ and $C(s, |z-1|)$ meet at two points: one is $z$ and the other is $\bar{z}$.

All points involved lie in $\mathcal{P}_n$, so $\bar{z} \in \mathcal{P}_{n+1}$. □

With a little more effort, it can be proved that if $z \in \mathcal{P}_n$ then $\bar{z} \in \mathcal{P}_n$. See Exercise 7.1.

*Proof of Part* (A). Part (A) follows by coordinate geometry in $\mathbb{C} \equiv \mathbb{R}^2$. The details are tedious, but we give them for completeness. Use induction on $n$. Since $\mathcal{P}_0 = \{0,1\} \subseteq \mathbb{Q}$, we have $\mathcal{P}_0 \subseteq \mathbb{Q}^{\text{py}}$. Suppose inductively that $\mathcal{P}_n \subseteq \mathbb{Q}^{\text{py}}$, and let $z \in \mathcal{P}_{n+1}$. We have to prove that $z \in \mathbb{Q}^{\text{py}}$.

There are three cases: line meets line, line meets circle, circle meets circle.

*Case 1*: Line meets line. Here $\{z\} = L(z_1, z_2) \cap L(z_3, z_4)$ where the $z_j \in \mathcal{P}_n \subseteq \mathbb{Q}^{\text{py}}$ (induction hypothesis) and the lines are distinct. Therefore there exist real $\alpha, \beta$ such that

$$\begin{aligned} z &= \alpha z_1 + (1-\alpha) z_2 \\ z &= \beta z_3 + (1-\beta) z_4 \end{aligned}$$

Therefore

$$\alpha = \frac{\beta(z_3 - z_4) + z_4 - z_2}{z_1 - z_2}$$

Since $\alpha, \beta \in \mathbb{R}$, we also have

$$\alpha = \frac{\beta(\bar{z}_3 - \bar{z}_4) + \bar{z}_4 - \bar{z}_2}{\bar{z}_1 - \bar{z}_2}$$

where the bar is complex conjugate. (Recall from Proposition 7.8(1) that if $z \in \mathbb{Q}^{\text{py}}$ then $\bar{z} \in \mathbb{Q}^{\text{py}}$.) The above two equations have a unique solution for $\alpha, \beta$ because we are assuming that the lines meet at a unique point $z$, and the solution is:

$$\begin{aligned} \alpha &= \frac{z_2(\bar{z}_4 - \bar{z}_3) + \bar{z}_2(z_3 - z_4) - z_3\bar{z}_4 + z_4\bar{z}_3}{(z_1 - z_2)(\bar{z}_3 - \bar{z}_4) + (z_4 - z_3)(\bar{z}_1 - \bar{z}_2)} \\ \beta &= \frac{z_3(\bar{z}_1 - \bar{z}_2) + \bar{z}_3(z_2 - z_1) - z_2\bar{z}_1 + z_1\bar{z}_2}{(z_4 - z_3)(\bar{z}_2 - \bar{z}_1) + (z_1 - z_2)(\bar{z}_4 - \bar{z}_3)} \end{aligned}$$

so $\alpha, \beta \in \mathbb{Q}^{\text{py}}$. Then $z = \alpha z_1 + (1-\alpha) z_2 \in \mathbb{Q}^{\text{py}}$.

*Case 2*: Line meets circle. Here $z \in L(z_1, z_2) \cap C(z_3, |z_4 - z_5|)$ where the $z_j \in \mathcal{P}_n \subseteq \mathbb{Q}^{\text{py}}$ (induction hypothesis). Let $r = |z_4 - z_5|$. There exist $\alpha, \theta \in \mathbb{R}$ such that

$$\begin{aligned} z &= \alpha z_1 + (1 - \alpha z_2) \\ z &= z_3 + r e^{\mathrm{i}\theta} \end{aligned}$$

Therefore

$$\alpha(z_1 - z_2) + z_2 = z_3 + re^{i\theta}$$
$$\alpha(\overline{z}_1 - \overline{z}_2) + \overline{z}_2 = \overline{z}_3 + re^{-i\theta}$$

where we take the complex conjugate to get the second equation. We can eliminate $\theta$ to get

$$(\alpha(z_1 - z_2) + z_2 - z_3)(\alpha(\overline{z}_1 - \overline{z}_2) + \overline{z}_2 - \overline{z}_3) = re^{i\theta}.re^{-i\theta} = r^2 = (z_4 - z_5)(\overline{z}_4 - \overline{z}_5)$$

which is a quadratic equation for $\alpha$ with coefficients in $\mathbb{Q}^{\mathrm{py}}$. Since the quadratic formula involves only rational functions of the coefficients and a square root, $\alpha \in \mathbb{Q}^{\mathrm{py}}$. Therefore $z \in \mathbb{Q}^{\mathrm{py}}$.

*Case 3*: Circle meets circle. Here $z \in C(z_1, |z_2 - z_3|) \cap C(z_4, |z_5 - z_6|)$ where the $z_j \in \mathcal{P}_n \subseteq \mathbb{Q}^{\mathrm{py}}$ (induction hypothesis). Let $r = |z_2 - z_3|, s = |z_5 - z_6|$. There exist $\theta, \phi \in \mathbb{R}$ such that

$$z = z_1 + re^{i\theta}$$
$$z = z_4 + se^{i\phi}$$

Take conjugates and eliminate $\theta, \phi$ as above to get

$$(z - z_1)(\overline{z} - \overline{z}_1) = r^2$$
$$(z - z_4)(\overline{z} - \overline{z}_4) = s^2$$

Solving for $z$ and $\overline{z}$ (left as an exercise) we find that $z$ satisifies a quadratic equation with coefficients in $\mathbb{Q}^{\mathrm{py}}$. Therefore $z \in \mathbb{Q}^{\mathrm{py}}$, and we have proved (A). □

---

## 7.3 Specific Constructions

It remains to prove the converse (B). We first discuss constructions that implement algebraic operations and square roots in $\mathbb{C}$. The next lemma begins the process of assembling useful constructions and bounding the number of steps they require.

**Lemma 7.11.** (1) *A line can be bisected using a 2-step construction.*

(2) *An angle can be bisected using a 2-step construction.*

(3) *An angle can be copied (so that its vertex is a given point and one leg lies along a given line through that point) using a 3-step construction.*

(4) *A perpendicular to a given line at a given point can be constructed using a 2-step construction.*

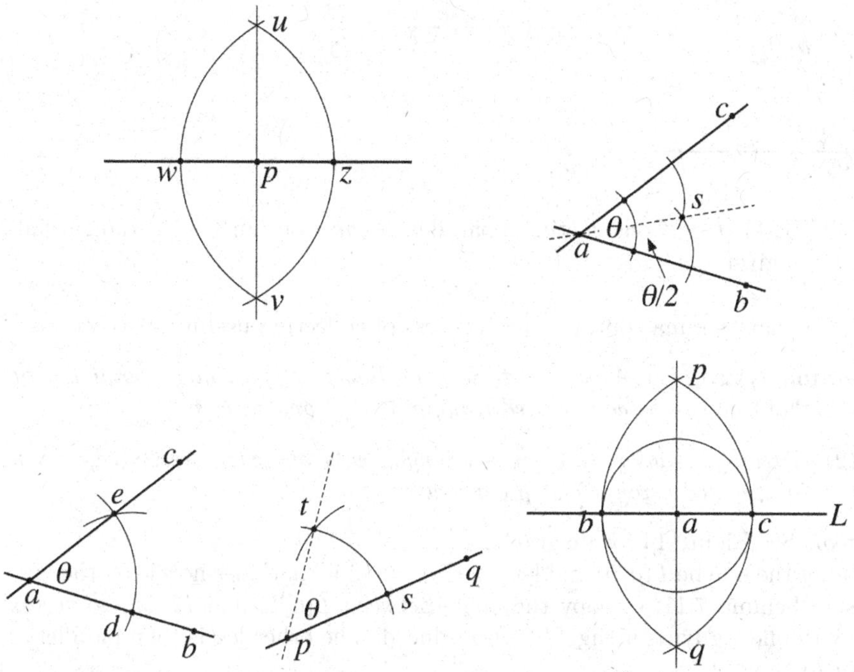


FIGURE 13: Four basic constructions. Top left: Bisecting a line segment. Top right: Bisecting an angle. Bottom left: Copying an angle. Bottom right: Constructing a perpendicular.

*Proof.* See Figure 13 for diagrams.

(1) Let the line be $L(z, w)$.

Draw circles $C(z, |z-w|)$ and $C(w, |z-w|)$. These meet at two points $u, v$. The midpoint $p$ of $L(z, w)$ is its intersection with $L(u, v)$.

(2) Let $\theta$ be the angle between $L(a, b)$ and $L(a, c)$.

Draw $C(a, 1)$ meeting $L(a, b)$ at $p$ and $L(a, c)$ at $q$.

Draw $C(p, 1)$ and $C(q, 1)$ meeting at $s, t$. Then $L(a, s)$ (or $L(a, t)$) bisects $\theta$.

(3) Let $\theta$ be the angle between $L(a, b)$ and $L(a, c)$.

Suppose $p, q \in \mathbb{C}$ are given, and we wish to construct angle $\theta$ at $p$ with one side $L(p, q)$.

Let $C(a, 1)$ meet $L(a, b)$ at $d$ and $L(a, c)$ at $e$.

Let $C(p, 1)$ meet $L(p, q)$ at $s$.

Let $C(s, |d-e|)$ meet $C(p, 1)$ at $t$ as shown. Then the angle between $L(p, t)$ and $L(p, q)$ is $\theta$ for the appropriate choice of $t$.

(4) Let $a$ lie on a line $L$. Let the circle $C(a, 1)$ meet $L$ at $b, c$.

Let $C(b, |b-c|)$ meet $C(c, |b-c|)$ at $p, q$.

Then $L(p, q)$ is the required perpendicular. □

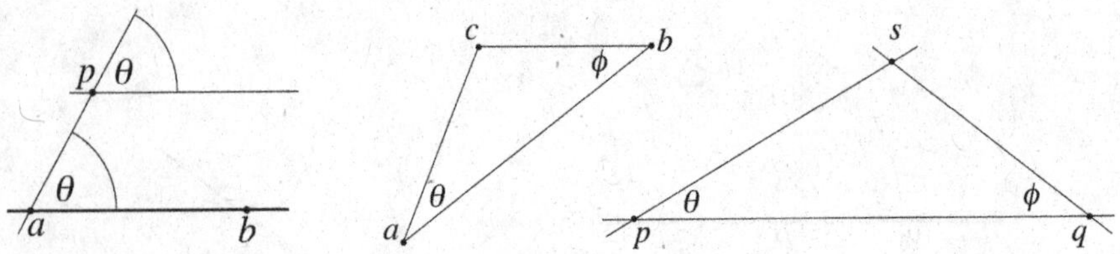


FIGURE 14: *Left*: Constructing a parallel. *Right*: Constructing a triangle similar to a given one.

The next lemma continues the process of collecting useful constructions.

**Lemma 7.12.** (1) *A parallel to a given line through a given point not on that line can be constructed using a 3-step construction.*

(2) *A triangle similar to a given triangle, with one edge prescribed, can be constructed using a 7-step construction.*

*Proof.* See Figure 14 for diagrams.
(1) Let the line be $L(a, b)$ and let $p \in \mathbb{C}$ be a point that does not lie on the line. Using Lemma 7.11(3), copy the angle between $L(a, b)$ and $L(a, p)$ to vertex $p$, with one leg lying along $L(a, p)$ produced. The other leg is then parallel to $L(a, b)$.
(2) Let the vertices of the first triangle be $a, b, c$. Suppose two vertices $p, q$ of the required similar triangle are given, such that the similarity maps $a$ to $p$ and $b$ to $q$.

Using Lemma 7.11(3), copy angles $\theta, \phi$ at $a, b$ to locations $p, q$, with one leg of each lying along $L(p, q)$. Then the other legs meet at $s$, which is the third vertex of the similar triangle required. □

We can now prove the existence of constructions that produce useful algebraic results:

**Theorem 7.13.** *Let $z, w \in \mathbb{C}$. Then, assuming $z$ and $w$ are already constructed:*

(1) *$z + w$ can be constructed using a 7-step construction.*

(2) *$-z$ can be constructed using a 1-step construction.*

(3) *$zw$ can be constructed using a 7-step construction.*

(4) *$1/z$ can be constructed using an 8-step construction.*

(5) *$\pm\sqrt{z}$ can be constructed using an 8-step construction.*

*Proof.* See Figure 15 for diagrams.
(1) If $z, w$ are not collinear with 0, complete the parallelogram with vertices $0, z, w$. The remaining vertex is $z + w$. If $z, w$ are collinear with 0, circle $C(z, |w|)$ meets $L(0, z)$ in two points, $z + w$ and $z - w$.

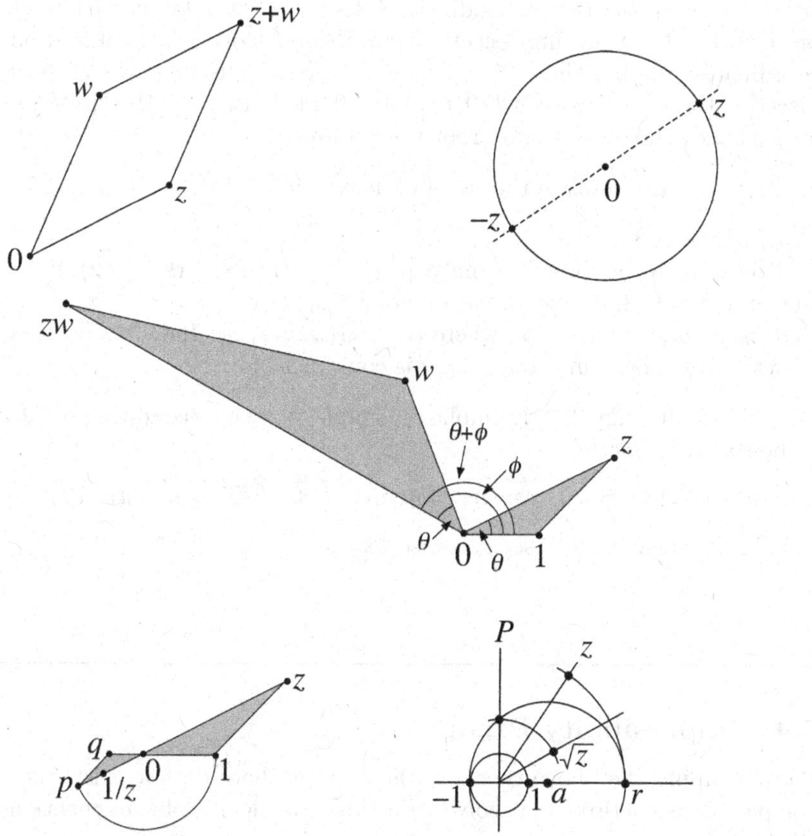


FIGURE 15: Constructions for five operations. Top left: $z+w$. Top right: $-z$. Middle $zw$. Bottom left: $1/z$. Bottom right: $\pm\sqrt{z}$.

(2) The circle $C(0,|z|)$ meets the line $L(0,z))$ at $z$ and at $-z$.
(3) Consider the triangle $T$ with vertices $0,1,z$. Construct point $p$ so that the triangle with vertices $0,w,p$ is similar to $T$. We claim that $p = zw$. By similarity $|p|/|w| = |z|/1$, so $|p| = |z||w|$. Further, $\arg(p) = \arg z + \arg w$, where arg denotes the argument. Therefore $p = zw$.
(4) Let $z = re^{i\theta}$ with $r > 0$. Then $1/z = (1/r)e^{-i\theta}$ Let $C(0,1)$ meet $L(0,z)$ at $p$ (with 0 lying between $z$ and $p$). Then $|p| = 1$. Construct a triangle with vertices $0,p,q$ similar to $0,z,1$. Then $|q|/1 = |p|/|z| = 1/|z|$, so $|q| = 1/|z|$. Let $C(0,|q|)$ meet $L(p,z)$ at $s$, on the same side of the origin as $p$. Then $s = (1/r)e^{-i\theta}$. Drop a perpendicular $P$ from $s$ to meet the real axis at $u$ and extend it. Then $C(u,|s-u|)$ meets $P$ in two points; the one not equal to $s$ is $\bar{s} = (1/r)e^{-i\theta} = 1/z$.
(5) Let $z = e^{i\theta}$. Then $\sqrt{z} = e^{i\theta/2}$ or $e^{i(\pi+\theta/2)}$, so we have to bisect $\theta$ and construct $\sqrt{r} \in \mathbb{R}^+$. Use $C(0,1)$ to construct $-1$. Bisect $L(1,r)$ to get $a =$

$(r-1)/2$. Construct the perpendicular $P$ to $L(0,1)$ at 0. Let circle $C(a,|r-a|)$ meet $P$ at $s$. Then the intersecting chords theorem (or a short calculation with coordinates) implies that $s.s = 1.r$, so $s = \sqrt{r}$. Construct line $L$ through 0 bisecting the angle between $L(0,r)$ and $L(0,z)$. This meets the circle $C(0,|s|)$ at $\pm\sqrt{z}$. For the other square root use (2) above. □

Now we can complete the proof of Theorem 7.9:

*Proof of Part* (B).

To complete the proof, we must prove (B). By Remark 7.7 (2), if $z \in \mathbb{Q}^{\text{py}}$ then there is a finite sequence of points $z_0 = 0, z_1 = 1, \ldots z_k = z$ such that $z_{l+1} \in \mathbb{Q}(z_0, \ldots, z_l, \alpha)$ where $\alpha^2 \in \mathbb{Q}(z_0, \ldots, z_l)$. Inductively, $z_l$ is constructible by Theorem 7.13, so $z_{l+1}$ is constructible. □

This result immediately implies a simple *necessary* condition for a point to be constructible:

**Theorem 7.14.** *If $\alpha$ is constructible then $[\mathbb{Q}(\alpha):\mathbb{Q}]$ is a power of* 2. □

The converse is false; see Exercise 22.9.

---

## 7.4 Impossibility Proofs

We now apply the above theory to prove that there do not exist ruler-and-compass constructions that solve the three classical problems mentioned in the introduction to this chapter.

We first prove the impossibility of Duplicating the Cube, where the method is especially straightforward.

**Theorem 7.15.** *The cube cannot be duplicated by ruler-and-compass construction.*

*Proof.* Duplicating the cube is equivalent to constructing $\alpha = \sqrt[3]{2}$. Suppose for a contradiction that $\alpha \in \mathbb{Q}^{\text{py}}$, and let $m$ be its minimal polynomial over $\mathbb{Q}$. By Theorem 7.14, $\partial m = 2^k$ for some $k$.

However, since $\alpha^3 = 2$, the minimal polynomial of $\alpha$ divides $x^3 - 2$. But this is irreducible over $\mathbb{Q}$. If not, it would have a linear factor $x - a$ with $a \in \mathbb{Q}$, and then $a^3 = 2$, so $a = \alpha$. But $\alpha$ is irrational. Therefore $\partial m = 3$, which is not a power of 2, contradicting Theorem 7.14. □

Some angles can be trisected, for example $\pi/2$. However, the required construction should work for any angle, so to prove impossibility it is enough to exhibit one specific angle that cannot be trisected. We prove:

**Theorem 7.16.** *The angle $\frac{2\pi}{3}$ cannot be trisected by ruler-and-compass construction.*

*Proof.* We know that $\omega = e^{2\pi i/3} \in \mathbb{Q}^{\mathrm{py}}$, since $\omega = \frac{-1+i\sqrt{3}}{2}$. Suppose for a contradiction that such a construction exists. Then $\zeta = e^{2\pi i/9} \in \mathbb{Q}^{\mathrm{py}}$. Therefore $\alpha = \zeta + \zeta^{-1} \in \mathbb{Q}^{\mathrm{py}}$, so its minimal polynomial $m$ over $\mathbb{Q}$ has degree $\partial m = 2^k$ for some $k$. Now $\zeta^3 = \omega$ and $\omega^2 + \omega + 1 = 0$, so $\zeta^6 + \zeta^3 + 1 = 0$. Therefore $\zeta^6 + \zeta^3 = -1$. But

$$\begin{aligned}\alpha^3 &= (\zeta + \zeta^{-1})^3 \\ &= \zeta^3 + 3\zeta + 3\zeta^{-1} + \zeta^{-3} \\ &= \zeta^3 + 3\zeta + 3\zeta^{-1} + \zeta^6 \\ &= 3\alpha - 1\end{aligned}$$

Therefore $m$ divides $x^3 - 3x + 1$. But this is irreducible over $\mathbb{Q}$ by Gauss's lemma, so $m = x^3 - 3x + 1$ and $\partial m = 3$, contradicting Theorem 7.14. □

This is the place for a word of warning to would-be trisectors, who are often aware of Wantzel's impossibility proof but somehow imagine that they can succeed despite it, arguing things like 'Yes, I know that you can prove it's impossible using algebra, but my construction uses geometry'. For specific instances, see (Dudley 1987). If you claim a trisection of a general angle using ruler and compass according to our standing conventions (such as 'unmarked ruler') then you are in particular claiming a trisection of $\pi/3$ using those instruments. The above proof shows that you are therefore claiming that 3 is a power of 2; in particular, since $3 \neq 1$, you are claiming that 3 is an even number.

Do you *really* want to go down in history as believing you have proved this?

The final problem of antiquity is more difficult:

**Theorem 7.17.** *The circle cannot be squared by ruler-and-compass construction.*

*Proof.* Such a construction is equivalent to constructing the point $\sqrt{\pi} \in \mathbb{C}$ from the initial set of points $P_0 = \{0, 1\}$. From this we can easily construct $\pi$. So if such a construction exists, then $[\mathbb{Q}(\pi) : \mathbb{Q}]$ is a power of 2, and in particular $\pi$ is algebraic over $\mathbb{Q}$. On the other hand, a famous theorem of Ferdinand Lindemann asserts that $\pi$ is *not* algebraic over $\mathbb{Q}$. The theorem follows. □

We prove Lindemann's theorem in Chapter 24. We could give the proof now, but it involves ideas off the main track of the book. If you are willing to take the result on trust, you can skip the proof.

As a bonus, and to set the scene for Chapter 19 on regular polygons, we dispose of another construction that the ancients might well have wondered about. They knew constructions for regular polygons with 3, 4, 5, sides, and it is easy to double these to get 6, 8, 10, 12, 16, 20, and so on. The impossibility of trisecting $2\pi/3$ also proves that a regular 9-gon (enneagon) cannot be constructed with ruler and compass. The first 'missing' case is the regular 7-gon (heptagon). Our methods easily prove this impossible, too:

**Theorem 7.18.** *The regular 7-gon (heptagon) cannot be constructed with ruler and compass.*

*Proof.* Constructing the regular heptagon is equivalent to proving that

$$\zeta = \mathrm{e}^{2\pi i/7} \in \mathbb{Q}^{\mathrm{py}}$$

and this complex 7th root of unity satisfies the polynomial equation

$$\zeta^6 + \zeta^5 + \zeta^4 + \zeta^3 + \zeta^2 + \zeta + 1 = 0$$

because $\zeta^7 - 1 = 0$ and the polynomial $t^7 - 1$ factorises as

$$t^7 - 1 = (t-1)(t^6 + t^5 + t^4 + t^3 + t^2 + t + 1)$$

Since 7 is prime, Lemma 3.24, implies that $t^6 + t^5 + t^4 + t^3 + t^2 + t + 1$ is irreducible. Its degree is 6, which is not a power of 2, so the regular 7-gon is not constructible.

There is an alternative approach in this case, which does not appeal to Eisenstein's Criterion. Rewrite the above equation as

$$\zeta^3 + \zeta^2 + \zeta + 1 + \zeta^{-1} + \zeta^{-2} + \zeta^{-3} = 0$$

Now $\zeta \in \mathbb{Q}^{\mathrm{py}}$ if and only if $\alpha = \zeta + \zeta^{-1} \in \mathbb{Q}^{\mathrm{py}}$, as above. Observe that

$$\begin{aligned} \alpha^3 &= \zeta^3 + 3\zeta + 3\zeta^{-1} + \zeta^{-3} \\ \alpha^2 &= \zeta^2 + 2 + \zeta^{-2} \end{aligned}$$

so

$$\alpha^3 + \alpha^2 - 3\alpha - 1 = 0$$

The polynomial $x^3 + x^2 - 3x - 1$ is irreducible by Gauss's Lemma, Lemma 3.19, so the degree of the minimal polynomial of $\alpha$ over $\mathbb{Q}$ is 3. Therefore $\alpha \notin \mathbb{Q}^{\mathrm{py}}$. □

---

## 7.5 Construction from a Given Set of Points

There is a 'relative' version of the theory of this chapter, in which we start not with $\{0, 1\}$ but some finite subset $P \subseteq \mathbb{C}$, satisfying some simple technical conditions. This set-up is more appropriate for discussing constructions such as '*given* an angle, bisect it', without assuming that the original angle is itself constructible. In this context, Definition 7.4 is modified to:

**19.** Let $P$ be a finite subset of $\mathbb{C}$ containing at least two distinct $0, 1 \in P$ (to identify the plane with $\mathbb{C}$). For each $n \in \mathbb{N}$ define ıd $\mathcal{C}_n$ of *points, lines,* and *circles* that are $n$*-constructible from*

$P$

$\emptyset$

$\emptyset$

$\{L(z_1, z_2) : z_1, z_2 \in \mathcal{P}_n\}$

$\{C(z_1, |z_2 - z_3|) : z_1, z_2, z_3 \in \mathcal{P}_n\}$

$\{z \in \mathbb{C} : z \text{ lies on two distinct lines in } \mathcal{L}_{n+1}\} \cup$

$\{z \in \mathbb{C} : z \text{ lies on a line in } \mathcal{L}_{n+1} \text{ and a circle in } \mathcal{C}_{n+1}\} \cup$

$\{z \in \mathbb{C} : z \text{ lies on two distinct circles in } \mathcal{L}_{n+1}\}$

*structible from* $P$ if it is $n$*-constructible from* $P$ for some $n$.

theory then goes through, with essentially the same proofs, ground field $\mathbb{Q}$ must be replaced by $\mathbb{Q}(P)$ throughout. Bear- at the complex conjugate $\bar{P}$ is easily constructed from $P$, the oints are precisely those in $\mathbb{Q}(P)^{\text{py}}$, defined by:

**20.** A complex number $\alpha$ is an element of $\mathbb{Q}(P)^{\text{py}}$ if and only ver of field extensions

$$\mathbb{Q}(P) = K_0 \subseteq K_1 \subseteq \ldots \subseteq K_n$$

$K_n$ and

$$[K_{j+1} : K_j] = 2$$

- 1.

---

# ES

Lemma 7.10 to: If $z \in \mathcal{P}_n$ then $\bar{z} \in \mathcal{P}_n$. (*Hint*: Show inductively s closed under complex conjugation, by observing that any con- can be reflected in the real axis to give another construction.)

n the language of this chapter methods of constructing, by ruler pass:

perpendicular bisector of a line.

The points trisecting a line.

Division of a line into $n$ equal parts.

The tangent to a circle at a given point.

Common tangents to two circles.

mate the degrees of the field extensions corresponding to the con-
ctions in Exercise 7.2, by giving reasonably good upper bounds.

e using Euclidean geometry that the 'marked ruler' construction of
re 11 does indeed trisect the given angle AOB.

the angle $2\pi/5$ be trisected using ruler and compass?

v that it is impossible to construct a regular 9-gon using ruler and
pass.

onsidering a formula for $\cos 5\theta$ find a ruler-and-compass construc-
for the regular pentagon.

e that the angle $\theta$ can be trisected by ruler and compass if and only
e polynomial

$$4t^3 - 3t - \cos\theta$$

ducible over $\mathbb{Q}(\cos\theta)$.

fy the following approximate construction for $\pi$ due to Ramanujan
2) page 35, see Figure 16. Let AB be the diameter of a circle centre
Bisect AO at M, trisect OB at T. Draw TP perpendicular to AB
ting the circle at P. Draw BQ = PT, and join AQ. Draw OS, TR
llel to BQ. Draw AD = AS, and AC = RS tangential to the circle at
oin BC, BD, CD. Make BE = BM. Draw EX parallel to CD. Then
square on BX has approximately the same area as the circle.

1 will need to know that $\pi$ is approximately $\frac{355}{113}$. This approximation
rst found in the works of the Chinese astronomer Zu Chongzhi in
1t AD 450.)

e that the construction in Figure 10 is correct if and only if the
tity

$$\sin\frac{\theta}{3} = \frac{\sin\theta}{2+\cos\theta}$$

s. Disprove the identity and estimate the error in the construction.

v that the 'compass' operation can be replaced by 'draw a circle cen-
$P_0$ and passing through some point other than $P_0$' without altering
set of constructible points.

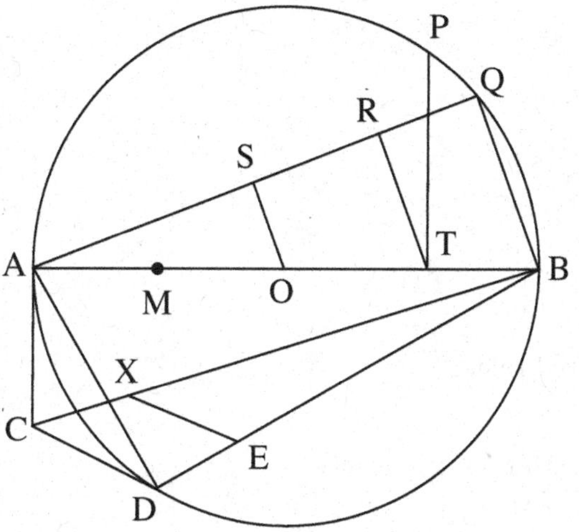


: Srinivasa Ramanujan's approximate squaring of the circle.

onstruction with infinitely many steps that trisects any given
in the sense that the angle $\phi_n$ obtained by stopping the con-
after $n$ steps converges to $\phi = \theta/3$ when $n$ tends to infinity.
nsider the infinite series

$$\frac{1}{4} + \frac{1}{16} + \frac{1}{64} + \cdots$$

nverges to $\frac{1}{3}$.)

alien creatures living in $n$-dimensional hyperspace $\mathbb{R}^n$ wishes to
the hypercube by ruler-and-compass construction. For which
y succeed?

7 shows a regular hexagon of side AB = 1 and some related
XY = 1, show that YB = $\sqrt[3]{2}$. Deduce that the cube can be
d using a marked ruler.

angles $\frac{\theta}{3}, \frac{\theta}{3} + \frac{2\pi}{3}, \frac{\theta}{3} + \frac{4\pi}{3}$ are all distinct, but equal $\theta$ when
d by 3, it can be argued that every angle has three distinct
s. Show that Archimedes's construction with a marked ruler
1) can find them all.

at the regular 11-gon cannot be constructed with ruler and
(*Hint*: Let $\zeta = e^{2\pi i/11}$ and mimic the proof for a heptagon.)

at the regular 13-gon cannot be constructed with ruler and
(*Hint*: Let $\zeta = e^{2\pi i/13}$ and mimic the proof for a heptagon.)

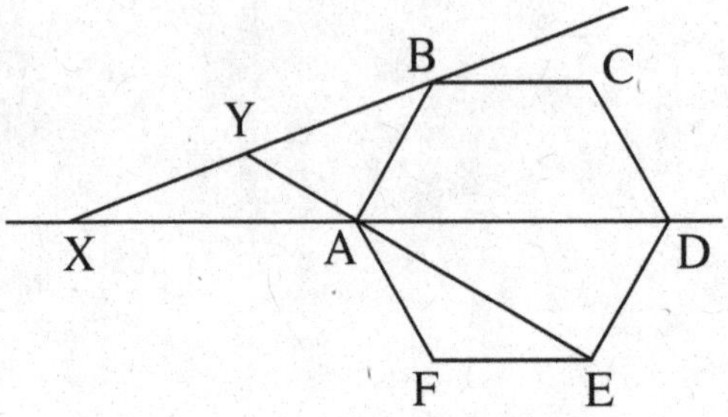


FIGURE 17: Duplicating the cube using a marked ruler.

regular 15-gon and 16-gon can be constructed with ruler and com-
. So the next regular polygon to consider is the 17-gon.

y does the method used in the previous questions *fail* for the 17-gon?

e that an angle (which you must specify and which must itself be tructible) cannot be divided into five equal pieces with ruler and pass. (*Hint*: Do not start with $2\pi/3$ or $\pi/2$, both of which *can* be ded into five equal pieces with ruler and compass (why?).)

$\in \mathbb{Q}$, prove that the angle $\theta$ such that $\tan\theta = \alpha$ is constructible.

$\theta$ be such that $\tan\theta = a/b$ where $a, b \in \mathbb{Z}$ are coprime and $b \neq 0$. e the following:

If $a + b$ is odd, then $\theta$ can be trisected using ruler and compass if only if $a^2 + b^2$ is a perfect cube.

If $a + b$ is even, then $\theta$ can be trisected using ruler and compass if only if $(a^2 + b^2)/2$ is a perfect cube.

The angles $\tan^{-1}\frac{2}{11}$ and $\tan^{-1}\frac{9}{13}$ can be trisected using ruler and pass.

t: Use the fact that the ring of Gaussian integers $\mathbb{Z}[i] = \{p + iq : \in \mathbb{Z}\}$ has the property of unique prime factorisation, together with standard formula for $\tan 3\theta$ in terms of $\tan\theta$.)

Exercise is based on Chang and Gordon (2014).

k the following true or false.

There exist ruler-and-compass constructions that trisect the angle to an arbitrary degree of approximation.

Such constructions are sufficient for practical purposes but insufficient for mathematical ones.

A point is constructible if it lies in a subfield of $\mathbb{C}$ whose degree over $\mathbb{Q}$ is a power of 2.

angle $\pi$ cannot be trisected using ruler and compass.

ne of length $\pi$ cannot be constructed using ruler and compass.

impossible to triplicate the cube (that is, construct one with e times the volume of a given cube) by ruler and compass.

real number $\pi$ is transcendental over $\mathbb{Q}$.

real number $\pi$ is transcendental over $\mathbb{R}$.

cannot be constructed by ruler and compass, then $\alpha$ is tran-dental over $\mathbb{Q}$.

# Chapter 8

# The Idea behind Galois Theory

Having satisfied ourselves that field extensions are good for something, we can focus on the main theme of this book: the elusive quintic, and Galois's deep insights into the solubility of equations by radicals. We start by outlining the main theorem that we wish to prove, and the steps required to prove it. We also explain where it came from.

We have already associated a vector space to each field extension. For some problems this is too coarse an instrument; it measures the size of the extension, but not its shape, so to speak. Galois went deeper into the structure. To any polynomial $p \in \mathbb{C}[t]$, he associated a group of permutations, now called the *Galois group* of $p$ in his honour. Complicated questions about the polynomial can sometimes be reduced to much simpler questions about the group, especially when it comes to solution by radicals. What makes his work so astonishing is that in his day the group concept existed only in rudimentary form. Others had investigated ideas that we now interpret as early examples of groups, but Galois was arguably the first to recognise the concept in sufficient generality, and to understand its importance.

We introduce the main ideas in a very simple context: a quartic polynomial equation whose roots are obvious. We show that the reason for the roots being obvious can be stated in terms of the *symmetries* of the polynomial—in an appropriate sense—and that any polynomial equation with those symmetries will also have 'obvious' roots.

With a little extra effort, we then subvert the entire reason for the existence of this book, by proving that the 'general' polynomial equation of the $n$th degree cannot be solved by radicals—of a particular, special kind—when $n \geq 5$. This is a spectacular application of the Galois group, but in a very limited context: it corresponds roughly to what Ruffini proved (or came close to proving) in 1813. By stealing one further idea from Abel, we can even remove Ruffini's assumption, and prove that there is no general radical expression in the coefficients of a quintic, or any polynomial of degree $\geq 5$, that determines a zero.

We could stop there. But Galois went much further: his methods are not only more elegant, they give much stronger results. The material in this chapter provides a springboard, from which we can launch into the full beauty of the theory.

DOI: 10.1201/9781003213949-8

## 8.1 A First Look at Galois Theory

Galois theory is a fascinating mixture of classical and modern mathematics, and it takes a certain amount of effort to get used to its thought patterns. This section is intended to give a quick survey of the basic principles of the subject, and explain how the abstract treatment has developed from Galois's original ideas.

The aim of Galois theory is to study the solutions of polynomial equations

$$f(t) = t^n + a_{n-1}t^{n-1} + \cdots + a_0 = 0$$

and, in particular, to distinguish those that can be solved by a 'formula' from those that cannot. By a formula we mean a *radical expression*: anything that can be built up from the coefficients $a_j$ by the operations of addition, subtraction, multiplication, and division, and also—the essential ingredient—by $n$th roots, $n = 2, 3, 4, \ldots$.

In Chapter 1 we saw that polynomial equations over $\mathbb{C}$ of degree 1, 2, 3, or 4 can be solved by radicals. The central objective of this book is a proof that the quintic equation is different. It cannot, in general, be solved by radicals. Along the way we come to appreciate the deep, general reason *why* quadratics, cubics, and quartics *can* be solved using radicals.

In modern terms, Galois's main idea is to look at the symmetries of the polynomial $f(t)$. These form a group, its *Galois group*, and the solution of the polynomial equation is reflected in various properties of the Galois group.

## 8.2 Galois Groups According to Galois

Galois had to invent the concept of a group, quite aside from sorting out how it relates to the solution of equations. Not surprisingly, his approach was relatively concrete by today's standards, but by those of his time it was highly abstract. Indeed Galois is one of the founders of modern abstract algebra. So to understand the modern approach, it helps to take a look at something rather closer to what Galois had in mind.

As an example, consider the polynomial equation

$$f(t) = t^4 - 4t^2 - 5 = 0$$

which we encountered in Chapter 4. As we saw, this factorises as

$$(t^2 + 1)(t^2 - 5) = 0$$

so there are four roots $t = \mathrm{i}, -\mathrm{i}, \sqrt{5}, -\sqrt{5}$. These form two natural pairs: i and $-$i go together, and so do $\sqrt{5}$ and $-\sqrt{5}$. Indeed, it is impossible to distinguish i from $-$i, or $\sqrt{5}$ from $-\sqrt{5}$, by algebraic means, in the following sense. Write down any polynomial equation, with rational coefficients, that is satisfied by some selection from the four roots. If we let

$$\alpha = \mathrm{i} \qquad \beta = -\mathrm{i} \qquad \gamma = \sqrt{5} \qquad \delta = -\sqrt{5}$$

then such equations include

$$\alpha^2 + 1 = 0 \qquad \alpha + \beta = 0 \qquad \delta^2 - 5 = 0 \qquad \gamma + \delta = 0 \qquad \alpha\gamma - \beta\delta = 0$$

and so on. There are infinitely many valid equations of this kind. On the other hand, infinitely many other algebraic equations, such as $\alpha + \gamma = 0$, are manifestly false.

Experiment suggests that if we take any valid equation connecting $\alpha$, $\beta$, $\gamma$, and $\delta$, and interchange $\alpha$ and $\beta$, we again get a valid equation. The same is true if we interchange $\gamma$ and $\delta$. For example, the above equations lead by this process to

$$\beta^2 + 1 = 0 \qquad \beta + \alpha = 0 \qquad \gamma^2 - 5 = 0 \qquad \delta + \gamma = 0$$
$$\beta\gamma - \alpha\delta = 0 \qquad \alpha\delta - \beta\gamma = 0 \qquad \beta\delta - \alpha\gamma = 0$$

and all of these are valid. In contrast, if we interchange $\alpha$ and $\gamma$, we obtain equations such as

$$\gamma^2 + 1 = 0 \qquad \gamma + \beta = 0 \qquad \alpha + \delta = 0$$

which are false. Exercise 8.1 outlines a simple proof that these operations preserve all valid equations connecting $\alpha$, $\beta$, $\gamma$, and $\delta$.

The operations that we are using here are *permutations* of the zeros $\alpha$, $\beta$, $\gamma$, $\delta$. In fact, in the usual permutation notation, the interchange of $\alpha$ and $\beta$ is

$$R = \begin{pmatrix} \alpha & \beta & \gamma & \delta \\ \beta & \alpha & \gamma & \delta \end{pmatrix} \tag{8.1}$$

and that of $\gamma$ and $\delta$ is

$$S = \begin{pmatrix} \alpha & \beta & \gamma & \delta \\ \alpha & \beta & \delta & \gamma \end{pmatrix} \tag{8.2}$$

These are elements of the symmetric group $\mathbb{S}_4$ on four symbols, which includes all 24 possible permutations of $\alpha$, $\beta$, $\gamma$, $\delta$.

If these two permutations turn valid equations into valid equations, then so must the permutation obtained by performing them both in turn, which is

$$T = \begin{pmatrix} \alpha & \beta & \gamma & \delta \\ \beta & \alpha & \delta & \gamma \end{pmatrix}$$

Are there any other permutations that preserve all the valid equations? Yes, of course, the identity

$$I = \begin{pmatrix} \alpha & \beta & \gamma & \delta \\ \alpha & \beta & \gamma & \delta \end{pmatrix}$$

It can be checked that only these four permutations preserve valid equations: the other 20 all turn some valid equation into a false one. For example, if $\alpha, \delta$ are fixed and $\beta, \gamma$ are swapped, the value equation $\alpha + \beta = 0$ becomes the invalid equation $\alpha + \gamma = 0$.

It is a general fact, and an easy one to prove, that the invertible transformations of a mathematical object that preserve some feature of its structure always form a group under the operation of composition. We call this the *symmetry group* of the object. This terminology is especially common when the object is a geometrical figure and the transformations are rigid motions, but the same idea applies more widely. And indeed these four permutations do form a group, which we denote by $G$.

What Galois realised is that the structure of this group to some extent controls how we should set about solving the equation.

He did not use today's notation for permutations, and this led to potential confusion. To him, a *permutation* of, say, $\{1, 2, 3, 4\}$, was an ordered list, such as 2413. Given a second list, say 3214, he then considered the *substitution* that changes 2413 to 3214; that is, the map $2 \mapsto 3, 4 \mapsto 2, 1 \mapsto 1, 3 \mapsto 4$. Nowadays we would write this as

$$\begin{pmatrix} 2\ 4\ 1\ 3 \\ 3\ 2\ 1\ 4 \end{pmatrix}$$

or, reordering the top row,

$$\begin{pmatrix} 1\ 2\ 3\ 4 \\ 1\ 3\ 4\ 2 \end{pmatrix}$$

but Galois did not even have the $\mapsto$ notation or associated concepts, so he had to write the substitution as 1342. His use of similar notation for both permutations and substitutions takes some getting used to, and probably did not make life easier for the people asked to referee his papers. Today's definition of 'function' or 'map' dates from about 1950; it certainly helps to clarify the ideas.

To see why permutations/substitutions of the roots matter, consider the subgroup $H = \{I, R\}$ of $G$. Certain expressions in $\alpha$, $\beta$, $\gamma$, $\delta$ are fixed by the permutations in this group. For example, if we apply $R$ to $\alpha^2 + \beta^2 - 5\gamma\delta^2$, then we obtain $\beta^2 + \alpha^2 - 5\gamma\delta^2$, which is clearly the same. In fact an expression is fixed by $R$ if and only if it is symmetric in $\alpha$ and $\beta$.

It is not hard to show that any polynomial in $\alpha, \beta, \gamma, \delta$ that is symmetric in $\alpha$ and $\beta$ can be rewritten as a polynomial in $\alpha + \beta, \alpha\beta, \gamma$, and $\delta$. For example, the above expression can be written as $(\alpha + \beta)^2 - 2\alpha\beta - 5\gamma\delta^2$. But we know that $\alpha = \mathrm{i}, \beta = -\mathrm{i}$, so that $\alpha + \beta = 0$ and $\alpha\beta = 1$. Hence the expression reduces to $-2 - 5\gamma\delta^2$. Now $\alpha$ and $\beta$ have been eliminated altogether.

## 8.3 How to Use the Galois Group

Pretend for a moment that we do not know the explicit zeros $\mathrm{i}, -\mathrm{i}, \sqrt{5}, -\sqrt{5}$, but we do know the Galois group $G$. In fact, consider any quartic polynomial $g(t)$ with the same Galois group as our example $f(t)$ above; that way we cannot possibly know the zeros explicitly. Let them be $\alpha, \beta, \gamma, \delta$. Consider three subfields of $\mathbb{C}$ related to $\alpha, \beta, \gamma, \delta$, namely

$$\mathbb{Q} \subseteq \mathbb{Q}(\gamma, \delta) \subseteq \mathbb{Q}(\alpha, \beta, \gamma, \delta)$$

Let $H = \{I, R\} \subseteq G$. Assume that we also know the following two facts:

(1) The numbers fixed by $H$ are precisely those in $\mathbb{Q}(\gamma, \delta)$.

(2) The numbers fixed by $G$ are precisely those in $\mathbb{Q}$.

Then we can work out how to solve the quartic equation $g(t) = 0$, as follows.

The numbers $\alpha + \beta$ and $\alpha\beta$ are obviously both fixed by $H$. By fact (1) they lie in $\mathbb{Q}(\gamma, \delta)$. But since

$$(t - \alpha)(t - \beta) = t^2 - (\alpha + \beta)t + \alpha\beta$$

this means that $\alpha$ and $\beta$ satisfy a quadratic equation whose coefficients are in $\mathbb{Q}(\gamma, \delta)$. That is, we can use the formula for solving a quadratic to express $\alpha, \beta$ in terms of rational functions of $\gamma$ and $\delta$, together with nothing worse than square roots. Thus we obtain $\alpha$ and $\beta$ as radical expressions in $\gamma$ and $\delta$.

We can then repeat the trick to find $\gamma$ and $\delta$. The numbers $\gamma + \delta$ and $\gamma\delta$ are fixed by the whole of $G$: they are clearly fixed by $R$, and also by $S$, and these generate $G$. Therefore $\gamma + \delta$ and $\gamma\delta$ belong to $\mathbb{Q}$ by fact (2) above. Therefore $\gamma$ and $\delta$ satisfy a quadratic equation over $\mathbb{Q}$, so they are given by radical expressions in rational numbers. Plugging these into the formulas for $\alpha$ and $\gamma$ we find that all four zeros are radical expressions in rational numbers.

We have not found the formulas explicitly. But we have shown that certain information about the Galois group necessarily implies that they exist. Given more information, we can finish the job.

This example illustrates that the subgroup structure of the Galois group $G$ is closely related to the possibility of solving the equation $g(t) = 0$. Galois discovered that this relationship is very deep and detailed. For example, the proof that an equation of the fifth degree cannot be solved by a formula boils down to this: *the quintic has the wrong sort of Galois group.* Galois's surviving papers do not make this proof explicit, probably because he considered the insolubility of the quintic to be a known theorem proved by Abel, but it is an easy deduction from results that he does state: see Chapter 25.

We present a simplified version of this argument, in a restricted setting, in Section 8.7. In Section 8.8 we remove this technical restriction using Abel's classical methods.

## 8.4 The Abstract Setting

The modern approach follows Galois closely in principle, but differs in several respects in practice. The permutations of $\alpha, \beta, \gamma, \delta$ that preserve all algebraic relations between them turns out to be the symmetry group of the subfield $\mathbb{Q}(\alpha, \beta, \gamma, \delta)$ of $\mathbb{C}$ generated by the zeros of $g$, or more precisely its *automorphism group*, which is a fancy name for the same thing.

Moreover, we wish to consider polynomials not just with integer or rational coefficients, but coefficients that lie in a subfield $K$ of $\mathbb{C}$ (or, later, any field). The zeros of a polynomial $f(t)$ with coefficients in $K$ determine another field $L$ which contains $K$, but may well be larger. Thus the primary object of consideration is a pair of fields $K \subset L$, or in a slight generalisation, a field extension $L/K$. Thus when Galois talks of polynomials, the modern approach talks of field extensions. And the Galois group of the polynomial becomes the group of $K$-automorphisms of $L$, that is, of bijections $\theta : L \to L$ such that for all $x, y \in L$ and $k \in K$

$$\begin{aligned} \theta(x+y) &= \theta(x) + \theta(y) \\ \theta(xy) &= \theta(x)\theta(y) \\ \theta(k) &= k \end{aligned}$$

Thus the bulk of the theory is described in terms of field extensions and their groups of $K$-automorphisms. This point of view was introduced in 1894 by Dedekind, who also gave axiomatic definitions of subrings and subfields of $\mathbb{C}$.

The method used above to solve $g(t) = 0$ relies crucially on knowing the conditions (1) and (2) at the start of Section 8.3. But can we lay hands on that kind of information if we do not already know the zeros of $g$? The answer is that we can—though not easily—provided we make a general study of the automorphism groups of field extensions, their subgroups, and the subfields fixed by those subgroups. This study leads to the *Galois correspondence* between subgroups of the Galois group and subfields $M$ of $L$ that contain $K$. Chapters 9–11 set up the Galois correspondence and prove its key properties, and the main theorem is stated and proved in Chapter 12. Chapter 13 studies one example in detail to drive the ideas home. Chapters 15 and 18 derive the spectacular consequences for the quintic. Then, starting in Chapter 16, we generalise the Galois correspondence to arbitrary fields, and develop the resulting theory in several directions.

## 8.5 Polynomials and Extensions

In this section we define the Galois group of a field extension $L/K$. We begin by defining a special kind of automorphism.

**Definition 8.1.** Let $L/K$ be a field extension, so that $K$ is a subfield of the subfield $L$ of $\mathbb{C}$. A *$K$-automorphism* of $L$ is an automorphism $\alpha$ of $L$ such that

$$\alpha(k) = k \quad \text{for all } k \in K \tag{8.3}$$

We say that $\alpha$ *fixes* $k \in K$ if (8.3) holds.

Effectively condition (8.3) makes $\alpha$ an automorphism of the *extension* $L/K$, rather than an automorphism of the large field $L$ alone. The idea of considering automorphisms of a mathematical object relative to a sub-object is a useful general method; it falls within the scope of the famous 1872 'Erlangen Programme' of Felix Klein. Klein's idea was to consider every 'geometry' as the theory of invariants of an associated transformation group. Thus Euclidean geometry is the study of invariants of the group of distance-preserving transformations of the plane; projective geometry arises if we allow projective transformations; topology comes from the group of all continuous maps possessing continuous inverses (called 'homeomorphisms' or 'topological transformations'). According to this interpretation any field extension is a geometry, and we are simply studying the geometrical figures.

The pivot upon which the whole theory turns is a result, which is not in itself hard to prove. As Lewis Carroll said in *The Hunting of the Snark,* it is a 'maxim tremendous but trite':

**Theorem 8.2.** *If $L/K$ is a field extension, then the set of all $K$-automorphisms of L forms a group under composition of maps.*

*Proof.* Suppose that $\alpha$ and $\beta$ are $K$-automorphisms of $L$. Then $\alpha\beta$ is clearly an automorphism; further if $k \in K$ then $\alpha\beta(k) = \alpha(k) = k$, so that $\alpha\beta$ is a $K$-automorphism. The identity map on $L$ is obviously a $K$-automorphism. Finally, $\alpha^{-1}$ is an automorphism of $L$, and for any $k \in K$, we have

$$k = \alpha^{-1}\alpha(k) = \alpha^{-1}(k)$$

so that $\alpha^{-1}$ is a $K$-automorphism. Composition of maps is associative, so the set of all $K$-automorphisms of $L$ is a group. □

**Definition 8.3.** The *Galois group* $\Gamma(L/K)$ of a field extension $L/K$ is the group of all $K$-automorphisms of $L$ under the operation of composition of maps.

**Examples 8.4.** (1) The extension $\mathbb{C}/\mathbb{R}$. Suppose that $\alpha$ is an $\mathbb{R}$-automorphism of $\mathbb{C}$. Let $j = \alpha(\mathrm{i})$. Then

$$j^2 = (\alpha(\mathrm{i}))^2 = \alpha(\mathrm{i}^2) = \alpha(-1) = -1$$

since $\alpha(r) = r$ for all $r \in \mathbb{R}$. Hence either $j = \mathrm{i}$ or $j = -\mathrm{i}$. Now for any $x$, $y \in \mathbb{R}$

$$\alpha(x + \mathrm{i}y) = \alpha(x) + \alpha(\mathrm{i})\alpha(y) = x + jy$$

Thus we have two candidates for $\mathbb{R}$-automorphisms:

$$\alpha_1 : x + \mathrm{i}y \mapsto x + \mathrm{i}y$$
$$\alpha_2 : x + \mathrm{i}y \mapsto x - \mathrm{i}y$$

Obviously $\alpha_1$ is the identity, and thus is an $\mathbb{R}$-automorphism of $\mathbb{C}$. The map $\alpha_2$ is complex conjugation and is an automorphism by Example 1.7(1). Moreover,

$$\alpha_2(x + 0\mathrm{i}) = x - 0\mathrm{i} = x$$

so $\alpha_2$ is an $\mathbb{R}$-automorphism. Obviously $\alpha_2^2 = \alpha_1$, so the Galois group $\Gamma(\mathbb{C}/\mathbb{R})$ is a cyclic group of order 2.
(2) Let $c$ be the real cube root of 2, and consider $\mathbb{Q}(c)/\mathbb{Q}$. If $\alpha$ is a $\mathbb{Q}$-automorphism of $\mathbb{Q}(c)$, then

$$(\alpha(c))^3 = \alpha(c^3) = \alpha(2) = 2$$

Since $\mathbb{Q}(c) \subseteq \mathbb{R}$ we must have $\alpha(c) = c$. Hence $\alpha$ is the identity map, and $\Gamma(\mathbb{Q}(c)/\mathbb{Q})$ has order 1.
(3) Let the field extension be $\mathbb{Q}(\sqrt{2}, \sqrt{3}, \sqrt{5})/\mathbb{Q}$, as in Example 6.8. The analysis presented in that example shows that $t^2 - 5$ is irreducible over $\mathbb{Q}(\sqrt{2}, \sqrt{3})$. Similarly, $t^2 - 2$ is irreducible over $\mathbb{Q}(\sqrt{3}, \sqrt{5})$ and $t^2 - 3$ is irreducible over $\mathbb{Q}(\sqrt{2}, \sqrt{5})$. Thus there are three $\mathbb{Q}$-automorphisms of $\mathbb{Q}(\sqrt{2}, \sqrt{3}, \sqrt{5})$, defined by

$$\begin{array}{llll} \rho_2 : & \sqrt{2} \mapsto -\sqrt{2} & \sqrt{3} \mapsto \sqrt{3} & \sqrt{5} \mapsto \sqrt{5} \\ \rho_3 : & \sqrt{2} \mapsto \sqrt{2} & \sqrt{3} \mapsto -\sqrt{3} & \sqrt{5} \mapsto \sqrt{5} \\ \rho_5 : & \sqrt{2} \mapsto \sqrt{2} & \sqrt{3} \mapsto \sqrt{3} & \sqrt{5} \mapsto -\sqrt{5} \end{array}$$

Its is easy to see that these maps commute, and hence generate the group $\mathbb{Z}_2 \times \mathbb{Z}_2 \times \mathbb{Z}_2$ Moreover, any $\mathbb{Q}$-automorphism of $\mathbb{Q}(\sqrt{2}, \sqrt{3}, \sqrt{5})$ must map $\sqrt{2} \mapsto \pm\sqrt{2}$, $\sqrt{3} \mapsto \pm\sqrt{3}$, and $\sqrt{5} \mapsto \pm\sqrt{5}$ by considering minimal polynomials. All combinations of signs occur in the group $\mathbb{Z}_2 \times \mathbb{Z}_2 \times \mathbb{Z}_2$, so this must be the Galois group.

## 8.6 The Galois Correspondence

It is easy to prove that the set of all $K$-automorphisms of $L/K$ forms a group, but to be of any use, the Galois group must reflect aspects of the structure of $L/K$. Galois made the discovery (which he expressed in terms of polynomials) that, under certain extra hypotheses, there is a one-to-one correspondence between:

(1) Subgroups of the Galois group of $L/K$.

(2) Subfields $M$ of $L$ such that $K \subseteq M$.

As it happens, this correspondence *reverses* inclusion relations: larger subfields correspond to smaller groups. First, we explain how the correspondence is set up.

If $L/K$ is a field extension, we call any field $M$ such that $K \subseteq M \subseteq L$ an *intermediate* field. To each intermediate field $M$ we associate the group $M^* = \Gamma(L/M)$ of all $M$-automorphisms of $L$. Thus $K^*$ is the whole Galois group, and $L^* = 1$ (the group consisting of just the identity map on $L$). Clearly if $M \subseteq N$ then $M^* \supseteq N^*$, because any automorphism of $L$ that fixes the elements of $N$ certainly fixes the elements of $M$. This is what we mean by 'reverses inclusions'.

Conversely, to each subgroup $H$ of $\Gamma(L/K)$ we associate the set $H^\dagger$ of all elements $x \in L$ that are *fixed* by $H$, that is, satisfy the condition $\alpha(x) = x$ for all $\alpha \in H$. In fact, this set is an intermediate field:

**Lemma 8.5.** *If $H$ is a subgroup of $\Gamma(L/K)$, then $H^\dagger$ is a subfield of $L$ containing $K$.*

*Proof.* Let $x, y \in H^\dagger$, and $\alpha \in H$. Then

$$\alpha(x+y) = \alpha(x) + \alpha(y) = x + y$$

so $x + y \in H^\dagger$. Similarly $H^\dagger$ is closed under subtraction, multiplication, and division (by nonzero elements), so $H^\dagger$ is a subfield of $L$. Since $\alpha \in \Gamma(L/K)$, we have $\alpha(k) = k$ for all $k \in K$, so $K \subseteq H^\dagger$. □

**Definition 8.6.** With the above notation, $H^\dagger$ is the *fixed field* of $H$.

It is easy to see that like $*$, the map $\dagger$ reverses inclusions: if $H \subseteq G$ then $H^\dagger \supseteq G^\dagger$. It is also easy to verify that if $M$ is an intermediate field and $H$ is a subgroup of the Galois group, then

$$M \subseteq M^{*\dagger} \qquad H \subseteq H^{\dagger *} \tag{8.4}$$

where for brevity we write

$$M^{*\dagger} = (M^*)^\dagger \qquad H^{\dagger *} = (H^\dagger)^*$$

Indeed, every element of $M$ is fixed by every automorphism that fixes all of $M$, and every element of $H$ fixes those elements that are fixed by all of $H$. Example 8.4(2) shows that these inclusions are not always equalities, for there

$$\mathbb{Q}^{*\dagger} = \mathbb{Q}(c) \neq \mathbb{Q}$$

If we let $\mathcal{F}$ denote the set of intermediate fields, and $\mathcal{G}$ the set of subgroups of the Galois group, then we have defined two maps

$$^* : \mathcal{F} \to \mathcal{G} \qquad ^\dagger : \mathcal{G} \to \mathcal{F}$$

which reverse inclusions and satisfy equation (8.4). These two maps constitute the *Galois correspondence* between $\mathcal{F}$ and $\mathcal{G}$. Galois's results can be interpreted as giving conditions under which $^*$ and $^\dagger$ are mutual inverses, setting up a bijection between $\mathcal{F}$ and $\mathcal{G}$. The extra conditions needed are called *separability* (which is automatic over $\mathbb{C}$) and *normality*. We discuss them in Chapter 9.

**Example 8.7.** The polynomial equation

$$f(t) = t^4 - 4t^2 - 5 = 0$$

was discussed in Section 8.2. Its roots are $\alpha = \mathrm{i}$, $\beta = -\mathrm{i}$, $\gamma = \sqrt{5}$, $\delta = -\sqrt{5}$. The associated field extension is $L/\mathbb{Q}$ where $L = \mathbb{Q}(\mathrm{i}, \sqrt{5})$, which we discussed in Example 4.8. There are four $\mathbb{Q}$-automorphisms of $L$, namely $I, R, S, T$ where $I$ is the identity, and in cycle notation $R = (\alpha\,\beta), S = (\gamma\,\delta)$, and $T = (\alpha\,\beta)(\gamma\,\delta)$. Recall that a *cycle* $(a_1 \dots a_k) \in \mathbb{S}_n$ is the permutation $\sigma$ such that $\sigma(a_j) = a_{j+1}$ when $1 \leq j \leq k-1$, $\sigma(a_k) = a_1$, and $\sigma(a) = a$ when $a \notin \{a_1, \dots, a_k\}$. Every element of $\mathbb{S}_n$ is a product of disjoint cycles, which commute, and this expression is unique except for the order in which the cycles are composed.

In fact $I, R, S, T$ are all possible $\mathbb{Q}$-automorphisms of $L$, because any $\mathbb{Q}$-automorphism must send i to $\pm\mathrm{i}$ and $\sqrt{5}$ to $\pm\sqrt{5}$. Therefore the Galois group is

$$G = \{I, R, S, T\}$$

The proper subgroups of $G$ are

$$1 \qquad \{I, R\} \qquad \{I, S\} \qquad \{I, T\}$$

where $1 = \{I\}$. It is easy to check that the corresponding fixed fields are respectively

$$L \qquad \mathbb{Q}(\sqrt{5}) \qquad \mathbb{Q}(\mathrm{i}) \qquad \mathbb{Q}(\mathrm{i}\sqrt{5})$$

Extensive but routine calculations (Exercise 8.2) show that these, together with $K$, are the only subfields of $L$. So in this case the Galois correspondence is bijective.

## 8.7 Diet Galois

To provide further motivation, we now pursue a modernised version of Lagrange's train of thought in his memoir of 1770–1771, which paved the way for Galois. Indeed we will follow a line of argument that is very close to the work of Ruffini and Abel, and prove that the general quintic is not soluble by radicals. Why, then, does the rest of this book exist? Because 'general' has a paradoxically special meaning in this context, and we have to place a very strong restriction on the kind of radical that is permitted. A major feature of Galois theory is that it does not assume this restriction. However, quadratics, cubics, and quartics *are* soluble by these restricted types of radical, so the discussion here does have some intrinsic merit. It could profitably be included as an application in a first course of group theory, or a digression in a course on rings and fields.

In passing, we mention that an elegant alternative treatment was found by Vladimir Arnold in the early 2000s. He devised a beautiful proof of the insolubility of the quintic using simple notions from complex analysis, which was written up by his student Alekseev (2004). It could have been found at the time of Cauchy, and it would have saved both Ruffini and Abel a lot of work. To be fair, the point of view involved is slightly sophisticated and became natural only in the 1900s. Examples and expanded discussions of many details can be found in Katz (2013) and Ramond (2020).

We have already encountered the symmetric group $\mathbb{S}_n$, which comprises all permutations of the set $\{1, 2, \ldots, n\}$. Its order is $n!$. When $n \geq 2$ the group $\mathbb{S}_n$ has a subgroup of index 2 (that is, of order $n!/2$), namely the *alternating group* $\mathbb{A}_n$, which consists of all products of an even number of transpositions $(a\,b)$. The elements of $\mathbb{A}_n$ are the *even permutations*. The group $\mathbb{A}_n$ is a normal subgroup of $\mathbb{S}_n$. It is well known that $\mathbb{A}_n$ is generated by all 3-cycles $(a\,b\,c)$: see Exercise 8.7. The group $\mathbb{A}_5$ holds the secret of the quintic, as we now explain.

Introduce the polynomial ring $\mathbb{C}[t_1, \ldots, t_n]$ in $n$ indeterminates. Let its field of fractions be $\mathbb{C}(t_1, \ldots, t_n)$, consisting of rational expressions in the $t_j$. Consider the polynomial

$$F(t) = (t - t_1) \ldots (t - t_n)$$

over $\mathbb{C}(t_1, \ldots, t_n)$, whose zeros are $t_1, \ldots, t_n$. Expanding and using induction, we see that

$$F(t) = t^n - s_1 t^{n-1} + s_2 t^{n-2} + \cdots + (-1)^n s_n \tag{8.5}$$

where the $s_j$ are the *elementary symmetric polynomials*

$$\begin{aligned} s_1 &= t_1 + \cdots + t_n \\ s_2 &= t_1 t_2 + t_1 t_3 + \cdots + t_{n-1} t_n \\ &\ldots \\ s_n &= t_1 \ldots t_n \end{aligned}$$

Here $s_r$ is the sum of all products of $r$ distinct $t_j$.

The symmetric group $\mathbb{S}_n$ acts as symmetries of $\mathbb{C}(t_1, \ldots, t_n)$:

$$\sigma f(t_1, \ldots, t_n) = f(t_{\sigma(1)}, \ldots, t_{\sigma(n)})$$

for $f \in \mathbb{C}(t_1, \ldots, t_n)$. The fixed field $K$ of $\mathbb{S}_n$ consists, by definition, of all symmetric rational functions in the $t_j$, which is known to be generated over $\mathbb{C}$ by the $n$ elementary symmetric polynomials in the $t_j$. That is, $K = \mathbb{C}(s_1, \ldots, s_n)$. Moreover, the $s_j$ satisfy no nontrivial polynomial relation: they are independent. There is a classical proof of these facts based on induction, using 'symmetrised monomials'

$$t_1^{a_1} t_2^{a_2} \cdots t_n^{a_n} + \text{ all permutations thereof}$$

and the so-called 'lexicographic ordering' of the list of exponents $a_1, \ldots, a_n$. See Exercise 8.5. A more modern but less constructive proof is given in Chapter 18.

Assuming that the $s_j$ generate the fixed field, consider the extension

$$\mathbb{C}(t_1, \ldots, t_n)/\mathbb{C}(s_1, \ldots, s_n)$$

We know that in $\mathbb{C}(t_1, \ldots, t_n)$ the polynomial $F(t)$ in (8.5) factorises completely as

$$F(t) = (t - t_1) \ldots (t - t_n)$$

Since the $s_j$ are independent indeterminates, $F(t)$ is traditionally called the *general* polynomial of degree $n$. The reason for this name is that this polynomial has a universal property. If we can solve $F(t) = 0$ by radicals, then we can solve any *specific* complex polynomial equation of degree $n$ by radicals. Just substitute specific numbers for the coefficients $s_j$. The converse, however, is not obvious. We might be able to solve every specific complex polynomial equation of degree $n$ by radicals, but using a different formula each time. Then we would not be able to deduce a radical expression to solve $F(t) = 0$. So the adjective 'general' is somewhat misleading; 'generic' would be better and is sometimes used.

The next definition is not standard, but its name is justified because it reflects the assumptions made by Ruffini in his attempted proof that the quintic is insoluble.

**Definition 8.8.** The general polynomial equation $F(t) = 0$ is *soluble by Ruffini radicals* if there exists a finite tower of subfields

$$\mathbb{C}(s_1, \ldots, s_n) = K_0 \subseteq K_1 \subseteq \cdots \subseteq K_r = \mathbb{C}(t_1, \ldots, t_n) \qquad (8.6)$$

such that for $j = 1, \ldots, r$,

$$K_j = K_{j-1}(\alpha_j) \qquad \text{and} \qquad \alpha_j^{n_j} \in K_{j-1} \qquad \text{for} \qquad n_j \geq 2,\ n_j \in \mathbb{N}$$

A *Ruffini radical* is any element of such a field $K_r$.

The aim of this definition is to exclude possibilities like the $\sqrt{-121}$ in Cardano's solution (1.10) of the quartic equation $t^4 - 15t - 4 = 0$, which does not lie in the field generated by the roots but is used to express them by radicals.

Ruffini tacitly assumed that if $F(t) = 0$ is soluble by radicals, then those radicals are all expressible as rational functions of the roots $t_1, \ldots, t_n$. Indeed, this was the situation studied by his predecessor Lagrange in his deep but inconclusive researches on the quintic. So Lagrange and Ruffini considered only solubility by Ruffini radicals. However, this is a strong assumption. It is conceivable that a solution by radicals might exist, for which some of the $\alpha_j$ constructed along the way do *not* lie in $\mathbb{C}(t_1, \ldots, t_n)$, but in some extension of $\mathbb{C}(t_1, \ldots, t_n)$. For example, $\sqrt[5]{s_1}$ might be useful. (It *is* useful to solve $t^5 - s_1 = 0$, for instance, but the solutions of this equation do not belong to $\mathbb{C}(t_1, \ldots, t_n)$.) However, the more we think about this possibility, the less likely it seems. Abel thought about it very hard, and *proved* that if $F(t) = 0$ is soluble by radicals, then those radicals are all expressible in terms of rational functions of the roots—they are Ruffini radicals after all. This step, historically called 'Abel's Theorem', is more commonly referred to as the 'Theorem on Natural Irrationalities'. From today's perspective, it is the main difficulty in the impossibility proof. So, following Lagrange and Ruffini, we start by defining the main difficulty away. In compensation, we gain excellent motivation for the remainder of this book. For completeness, we prove the Theorem on Natural Irrationalities in Section 8.8, using classical (pre-Galois) methods.

As preparation for all of the above, we need:

**Proposition 8.9.** *If there is a finite tower of subfields* (8.6), *then it can be refined (if necessary increasing its length) to make all $n_j$ prime.*

*Proof.* For fixed $j$ write $n_j = p_1 \ldots p_k$ where the $p_l$ are prime. Let $\beta_l = \alpha_j^{p_{l+1} \ldots p_k}$, for $0 \leq l \leq k$. Then $\beta_0 \in K_j$ and $\beta_l^{p_l} \in K_j(\beta_{l-1})$, and the rest is easy. □

For the remainder of this chapter we assume that this refinement has been performed, and write $p_j$ for $n_j$ as a reminder. With this preliminary step completed, we will prove:

**Theorem 8.10.** *The general polynomial equation $F(t) = 0$ is insoluble by Ruffini radicals if $n \geq 5$.*

All we need is a simple group-theoretic lemma.

**Lemma 8.11.** (1) *The symmetric group $\mathbb{S}_n$ has a cyclic quotient group of prime order $p$ if and only if $p = 2$ and $n \geq 2$, in which case the kernel is the alternating group $\mathbb{A}_n$.*
(2) *The alternating group $\mathbb{A}_n$ has a cyclic quotient group of prime order $p$ if and only if $p = 3$ and $n = 3, 4$.*

*Proof.* (1) We may assume $n \geq 3$ since there is nothing to prove when $n = 1, 2$. Suppose that $N$ is a normal subgroup of $\mathbb{S}_n$ and $\mathbb{S}_n/N \cong \mathbb{Z}_p$. Then $\mathbb{S}_n/N$ is abelian, so $N$ contains every *commutator* $ghg^{-1}h^{-1}$ for $g, h \in \mathbb{S}_n$. To see why, let $\bar{g}$ denote the image of $g \in \mathbb{S}_n$ in the quotient group $\mathbb{S}_n/N$. Since $\mathbb{S}_n/N$ is abelian, $\bar{g}\bar{h}\bar{g}^{-1}\bar{h}^{-1} = \bar{1}$ in $\mathbb{S}_n/N$; that is, $ghg^{-1}h^{-1} \in N$.

Let $g, h$ be 2-cycles of the form $g = (a\,b), h = (a\,c)$ where $a, b, c$ are distinct. Then

$$ghg^{-1}h^{-1} = (b\,c\,a)$$

is a 3-cycle, and all possible 3-cycles can be obtained in this way. Therefore $N$ contains all 3-cycles. But the 3-cycles generate $\mathbb{A}_n$, so $N \supseteq \mathbb{A}_n$. Therefore $p = 2$ since $|\mathbb{S}_n/\mathbb{A}_n| = 2$.

(2) Suppose that $N$ is a normal subgroup of $\mathbb{A}_n$ and $\mathbb{A}_n/N \cong \mathbb{Z}_p$. Again, $N$ contains every commutator. If $n = 2$ then $\mathbb{A}_n$ is trivial. When $n = 3$ we know that $\mathbb{A}_n \cong \mathbb{Z}_3$.

Suppose first that $n = 4$. Consider the commutator $ghg^{-1}h^{-1}$ where $g = (a\,b\,c), h = (abd)$ for $a, b, c, d$ distinct. Computation shows that

$$ghg^{-1}h^{-1} = (a\,b)(c\,d)$$

so $N$ must contain $(1\,2)(3\,4), (1\,3)(2\,4)$, and $(1\,4)(2\,3)$. It also contains the identity. But these four elements form a group $\mathbb{V}$. Thus $\mathbb{V} \subseteq N$. Since $\mathbb{V}$ is a normal subgroup of $\mathbb{A}_4$ and $\mathbb{A}_4/\mathbb{V} \cong \mathbb{Z}_3$, we are done.

The symbol $\mathbb{V}$ comes from Klein's term *Vierergruppe*, or 'fours-group'. Nowadays it is usually called the *Klein four-group*.

Finally, assume that $n \geq 5$. The same argument shows that $N$ contains all permutations of the form $(a\,b)(c\,d)$. If $a, b, c, d, e$ are all distinct (which is why the case $n = 4$ is special) then

$$(a\,b)(c\,d) \cdot (a\,b)(c\,e) = (c\,e\,d)$$

so $N$ contains all 3-cycles. But the 3-cycles generate $\mathbb{A}_n$, so this case cannot occur. □

As our final preparatory step, we recall the expression (1.13)

$$\delta = \prod_{j<k}^{n} (t_j - t_k)$$

It is not a symmetric polynomial in the $t_j$, but its square $\Delta = \delta^2$ is, because

$$\Delta = (-1)^{n(n-1)/2} \prod_{j \neq k}^{n} (t_j - t_k) \tag{8.7}$$

The expression $\Delta$, mentioned in passing in Section 1.4, is called the *discriminant* of $F(t)$. If $\sigma \in \mathbb{S}_n$, then the action of $\sigma$ sends $\delta$ to $\pm\delta$. The even permutations (those in $\mathbb{A}_n$) fix $\delta$, and the odd ones map $\delta$ to $-\delta$. Indeed, this is a standard way to define odd and even permutations.

We are now ready for the:

*Proof of Theorem* 8.10.

Assume that $F(t) = 0$ is soluble by Ruffini radicals, with a tower (8.6) of subfields $K_j$ in which all $n_j = p_j$ are prime. Let $K = \mathbb{C}(s_1, \ldots, s_n)$ and $L = \mathbb{C}(t_1, \ldots, t_n)$. Consider the first step in the tower,

$$K \subseteq K_1 \subseteq L$$

where $K_1 = K(\alpha_1), \alpha_1^p \in K, \alpha_1 \notin K$, and $p = p_1$ is prime.

Since $\alpha_1 \in L$ we can act on it by $\mathbb{S}_n$, and since every $\sigma \in \mathbb{S}_n$ fixes $K$, we have

$$(\sigma(\alpha_1))^p = \alpha_1^p$$

Therefore $\sigma(\alpha_1) = \zeta^{j(\sigma)}\alpha_1$, for $\zeta$ a primitive $p$th root of unity and $j(\sigma)$ an integer between 0 and $p-1$. The set of all $p$th roots of unity in $\mathbb{C}$ is a group under multiplication, and this group is cyclic, isomorphic to $\mathbb{Z}_p$. Indeed $\zeta^a\zeta^b = \zeta^{a+b}$ where $a + b$ is taken modulo $p$.

Clearly the map $j : \mathbb{S}_n \to \mathbb{Z}_p$ is a group homomorphism. Since $\alpha_1 \notin K$, some $\sigma(\alpha_1) \neq \alpha_1$, so $j$ is nontrivial. Since $\mathbb{Z}_p$ has prime order, hence no nontrivial proper subgroups, $j$ must be onto. Therefore $\mathbb{S}_n$ has a homomorphic image that is cyclic of order $p$. By Lemma 8.11, $p = 2$ and the kernel is $\mathbb{A}_n$. Therefore $\alpha_1$ is fixed by $\mathbb{A}_n$.

We claim that this implies that $\alpha_1 \in K(\delta)$. Since $p = 2$, the relation $\alpha_1^p \in K$ becomes $\alpha_1^2 \in K$, so $\alpha_1$ is a zero of $t^2 - \alpha_1^2 \in K[t]$. The images of $\alpha_1$ under $\mathbb{S}_n$ must all be zeros of this, namely $\pm\alpha_1$. Now $\alpha_1$ is fixed by $\mathbb{A}_n$ but not by $\mathbb{S}_n$, so some permutation $\sigma \in \mathbb{S}_n \setminus \mathbb{A}_n$ satisifes $\sigma(\alpha_1) = -\alpha_1$. Then $\delta\alpha_1$ is fixed by both $\mathbb{A}_n$ and $\sigma$, hence by $\mathbb{S}_n$. So $\delta\alpha_1 \in K$ and $\alpha_1 \in K(\delta)$.

If $r = 1$ we are finished. Otherwise consider the extension

$$K(\delta) \subseteq K_2 = K(\delta)(\alpha_2)$$

By a similar argument, $\alpha_2$ defines a group homomorphism $j : \mathbb{A}_n \to \mathbb{Z}_p$, which again must be onto. By Lemma 8.11, $p = 3$ and $n = 3, 4$. In particular, no tower of Ruffini radicals exists when $n \geq 5$. □

It is plausible that any tower of radicals that leads from $\mathbb{C}(s_1, \ldots, s_n)$ to a subfield containing $\mathbb{C}(t_1, \ldots, t_n)$ must give rise to a tower of Ruffini radicals. However, it is not at all clear how to prove this, and in fact, this is where the main difficulty of the problem really lies, once the role of permutations is understood. Ruffini appeared not to notice that this needed proof. Abel tackled the obstacle head on.

Galois worked round it by way of the Galois group—an extremely elegant solution. The actual details of his work differ considerably from the modern presentation, see Neumann (2011), both notationally and strategically. However, the underlying idea of studying what we now interpret as the symmetry group of the polynomial, and deriving properties related to solubility by radicals, is central to Galois's approach. His method also went much further: it applies not just to the general polynomial $F(t)$, but to any polynomial whatsoever. And it provides necessary *and sufficient* conditions for solutions by radicals to exist.

Exercises 8.9–8.11 provide enough hints for you to show that when $n = 2, 3, 4$ the equation $F(t) = 0$, where $F$ is defined by (8.5), can be solved by Ruffini radicals. Therefore, despite the special nature of Ruffini radicals, we see that the quintic equation differs (radically) from the quadratic, cubic, and quartic equations. We also appreciate the significant role of group theory and symmetries of the roots of a polynomial for the existence—or not—of a solution by radicals. This will serve us in good stead when the going gets tougher.

---

## 8.8 Natural Irrationalities

With a little more effort we can go the whole hog. Abel's proof contains one further idea, which lets us delete the word 'Ruffini' from Theorem 8.10. This section is an optional extra, and nothing later depends on it. We continue to work with the general polynomial, so throughout this section $L = \mathbb{C}(t_1, \ldots, t_n)$ and $K = \mathbb{C}(s_1, \ldots, s_n)$, where the $s_j$ are the elementary symmetric polynomials in the $t_j$.

To delete 'Ruffini' we need:

**Definition 8.12.** An extension $L/K$ in $\mathbb{C}$ is *radical* if $L = K(\alpha_1, \ldots, \alpha_m)$ where for each $j = 1, \ldots, m$ there exists an integer $n_j$ such that

$$\alpha_j^{n_j} \in K(\alpha_1, \ldots, \alpha_{j-1})$$

The elements $\alpha_j$ form a *radical sequence* for $L/K$. The *radical degree* of the radical $\alpha_j$ is $n_j$.

The essential point is:

**Theorem 8.13.** *If the general polynomial equation $F(t) = 0$ can be solved by radicals, then it can be solved by Ruffini radicals.*

**Corollary 8.14.** *The general polynomial equation $F(t) = 0$ is insoluble by radicals if $n \geq 5$.*

To prove the above, all we need is the so-called 'Theorem on Natural Irrationalities', which states that extraneous radicals like $\sqrt[5]{s_1}$ cannot help in the solution of $F(t) = 0$. More precisely:

**Theorem 8.15 (Natural Irrationalities).** *If $L$ contains an element $x$ that lies in some radical extension $R$ of $K$, then there exists a radical extension $R'$ of $K$ with $x \in R'$ and $R' \subseteq L$.*

Once we have proved Theorem 8.15, any solution of $F(t) = 0$ by radicals can be converted into one by Ruffini radicals. Theorem 8.13 and Corollary 8.14 are then immediate.

It remains to prove Theorem 8.15. A proof using Galois theory is straightforward, see Exercise 15.11. With what we know at the moment, we have to work a little harder—but, following Abel's strategic insights, not much harder. We need several lemmas, and a technical definition.

**Definition 8.16.** Let $R/K$ be a radical extension. The *height* of $R/K$ is the smallest integer $h$ such that there exist elements $\alpha_1, \ldots, \alpha_h \in R$ and primes $p_1, \ldots, p_h$ such that $R = K(\alpha_1, \ldots, \alpha_h)$ and

$$\alpha_j^{p_j} \in K(\alpha_1, \ldots, \alpha_{j-1}) \qquad 1 \leq j \leq h$$

where when $j = 1$ we interpret $K(\alpha_1, \ldots, \alpha_{j-1})$ as $K$.

Proposition 8.9 shows that the height of every radical extension is defined.

We prove Theorem 8.15 by induction on the height of a radical extension $R$ that contains $x$. The key step is extensions of height 1, and this is where all the work is put in.

**Lemma 8.17.** *Let $M$ be a subfield of $L$ such that $K \subseteq M$, and let $a \in M$, where $a$ is not a $p$th power in $M$. Then*

(1) *$a^k$ is not a $p$th power in $M$ for $k = 1, 2, \ldots, p-1$.*

(2) *The polynomial $m(t) = t^p - a$ is irreducible over $M$.*

*Proof.* (1) Since $k$ is prime to $p$ there exist integers $q, l$ such that $qp + lk = 1$. If $a^k = b^p$ with $b \in M$, then

$$(a^q b^l)^p = a^{qp} b^{lp} = a^{qp} a^{kl} = a$$

contrary to $a$ not being a $p$th power in $M$.
(2) Assume for a contradiction that $t^p - a$ is reducible over $M$. Suppose that $P(t)$ is a monic irreducible factor of $m(t) = t^p - a$ over $M$. For $0 \leq j \leq p-1$ let $P_j(t) = P(\zeta^j t)$, where $\zeta \in \mathbb{C} \subseteq K \subseteq M$ is a primitive $p$th root of unity. Then $P_0 = P$, and $P_j$ is irreducible for all $j$, for if $P(\zeta^j t) = g(t)h(t)$ then $P(t) = g(\zeta^{-j}t)h(\zeta^{-j}t)$. Moreover, $m(\zeta^j t) = (\zeta^j t)^p - a = t^p - a = m(t)$, so $P_j$ divides $m$ for all $j = 0, \ldots, p-1$ by Lemma 5.5.

We claim that $P_k$ and $P_j$ are coprime whenever $0 \leq j < k \leq p-1$. If not, by irreducibility

$$P_j(t) = cP_k(t) \qquad c \in M$$

Let

$$P(t) = p_0 + p_1 t + \cdots + p_{r-1}t^{r-1} + t^r$$

where $r \leq p$. By irreducibility, $p_0 \neq 0$. Then

$$\begin{aligned} P_j(t) &= p_0 + p_1\zeta^j t + \cdots + p_{r-1}\zeta^{j(r-1)}t^{r-1} + \zeta^{jr}t^r \\ P_k(t) &= p_0 + p_1\zeta^k t + \cdots + p_{r-1}\zeta^{k(r-1)}t^{r-1} + \zeta^{kr}t^r \end{aligned}$$

so $c = \zeta^{(j-k)r}$ from the coefficient of $t^r$. But then $p_0 = \zeta^{(j-k)r}p_0$. Since $p_0 \neq 0$, we must have $\zeta^{(j-k)r} = 1$, so $r = p$. But this implies that $\partial P = \partial m$, so $m$ is irreducible over $M$.

Thus we may assume that the $P_j$ are pairwise coprime. We know that $P_j | m$ for all $j$, so

$$P_0 P_1 \dots P_{p-1} \,|\, m$$

Since $\partial p = r$, it follows that $pr \leq p$, so $r = 1$. Thus $P$ is linear, so there exists $b \in M$ such that $(t - b)|m(t)$. But this implies that $b^p = a$, contradicting the assumption that $a$ is not a $p$th power. Thus $t^p - a$ is irreducible. □

Now suppose that $R$ is a radical extension of height 1 over $M$. Then $R = M(\alpha)$ where $\alpha^p \in M$, $\alpha \notin M$. Therefore every $x \in R \setminus M$ is uniquely expressible as

$$x = x_0 + x_1\alpha + x_2\alpha^2 + \cdots x_{p-1}\alpha^{p-1} \tag{8.8}$$

where the $x_j \in M$. This follows since $[M(\alpha) : M] = p$ by irreducibility of $m$. We want to put $x$ into a more convenient form, and for this we need the following result:

**Lemma 8.18.** *Let $L \subseteq M$ be fields, and let $p$ be a prime such that $L$ contains a primitive $p$th root of unity $\zeta$. Suppose that $\alpha, x_0, \dots, x_{p-1} \in M$ with $\alpha \neq 0$, and $L$ contains all of the elements*

$$X_r = x_0 + (\zeta^r\alpha)x_1 + (\zeta^r\alpha)^2 x_2 + \cdots + (\zeta^r\alpha)^{p-1}x_{p-1} \tag{8.9}$$

*for $0 \leq r \leq p-1$. Then each of the elements $x_0, \alpha x_1, \alpha^2 x_2, \dots, \alpha^{p-1}x_{p-1}$ also lies in $L$. Hence, if $x_1 = 1$, then $\alpha$ and each $x_j$ $(0 \leq j \leq p-1)$ lies in $L$.*

*Proof.* For any $m$ with $0 \leq m \leq p-1$, consider the sum

$$X_0 + \zeta^{-m}X_1 + \zeta^{-2m}X_2 + \cdots + \zeta^{-(p-1)m}X_{p-1}$$

Since $1 + \zeta + \zeta^2 + \cdots + \zeta^{p-1} = 0$, all terms vanish except for those in which the power of $\zeta$ is zero. These terms sum to $p\alpha^m x_m$. Therefore $p\alpha^m x_m \in L$, so $\alpha^m x_m \in L$.

If $x_1 = 1$ then the case $m = 1$ shows that $\alpha \in L$, so now $x_m \in L$ for all $m$ with $0 \leq m \leq p-1$. □

We can also prove:

**Lemma 8.19.** *With the above notation, for a given $x \in R$, there exist $\beta \in M(\alpha)$ and $b \in M$ with $b = \beta^p$, such that $b$ is not the $p$th power of an element of $M$, and*

$$x = y_0 + \beta + y_2\beta^2 + \cdots y_{p-1}\beta^{p-1}$$

*where the $y_j \in M$.*

*Proof.* We know that $x \notin M$, so in (8.8) some $x_s \neq 0$ for $1 \leq s \leq p-1$. Let $\beta = x_s\alpha^s$, and let $b = \beta^p$. Then $b = x_s^p\alpha^{sp} = x_s^p a^s$, and if $b$ is a $p$th power of an element of $M$ then $a^s$ is a $p$th power of an element of $M$, contrary to Lemma 8.17(2). Therefore $b$ is not the $p$th power of an element of $M$.

Now $s$ is prime to $p$, and the additive group $\mathbb{Z}_p$ is cyclic of prime order $p$, so $s$ generates $\mathbb{Z}_p$. Therefore, up to multiplication by nonzero elements of $M$, the powers $\beta^j$ of $\beta$ run through the powers of $\alpha$ precisely once as $j$ runs from 0 to $p-1$. Since $\beta^0 = 1, \beta^1 = x_s\alpha^s$, we have

$$x = y_0 + \beta + y_2\beta^2 + \cdots + y_{p-1}\beta^{p-1}$$

for suitable $y_j \in M$, where in fact $y_0 = x_0$. □

**Lemma 8.20.** *Let $q \in L$. Then the minimal polynomial of $q$ over $K$ splits into linear factors over $L$.*

*Proof.* The element $q$ is a rational expression $q(t_1, \ldots, t_n) \in \mathbb{C}(t_1, \ldots, t_n)$. The polynomial

$$f_q(t) = \prod_{\sigma \in \mathbb{S}_n} (t - q(t_{\sigma(1)}, \ldots, t_{\sigma(n)}))$$

has $q$ as a zero. Symmetry under $\mathbb{S}_n$ implies that $f_q(t) \in K[t]$. The minimal polynomial $m_q$ of $q$ over $K$ divides $f_q$, and $f_q$ is a product of linear factors; therefore $m_q$ is the product of some subset of those linear factors. □

We are now ready for the climax of Diet Galois:

*Proof of Theorem* 8.15. We prove the theorem by induction on the height $h$ of $R$.

If $h = 0$ then the theorem is obvious.

Suppose that $h \geq 1$. Then $R = R_1(\alpha)$ where $R_1$ is a radical extension of $K$ of height $h-1$, and $\alpha^p \in R_1$, $\alpha \notin R_1$, with $p$ prime. Let $\alpha^p = a \in R_1$.

By Lemma 8.19 we may assume without loss of generality that

$$x = x_0 + \alpha + x_2\alpha^2 + \cdots + x_{p-1}\alpha^{p-1}$$

where the $x_j \in R_1$. (Replace $\alpha$ by $\beta$ as in the lemma, and then change notation back to $\alpha$.) The mimimum polynomial $m(t)$ of $x$ over $K$ splits into linear factors in $L$ by Lemma 8.20. In particular, $x$ is a zero of $m(t)$, while all zeros of $m(t)$ lie in $L$.

Take the equation $m(\alpha) = 0$, write $x$ as above in terms of powers of $\alpha$ with coefficients in $R_1$, and consider the result as an equation satisfied by $\alpha$. The equation has the form $f(\alpha) = 0$ where $f(t) \in R_1[t]$. Therefore $f(t)$ is divisible by the minimal polynomial of $\alpha$, which is $t^p - a$. Hence all the roots of that equation, namely $\zeta^r\alpha$ for $0 \leq r \leq p-1$, are also roots of $f(t)$. Therefore all the elements $X_r$ in (8.9) are roots of $m(t)$, so they lie in $L$. Lemma 18.4 now shows that $\alpha, x_0, x_2, \ldots x_{p-1} \in L$.

Also, $\alpha^p, x_0, x_2, \ldots x_{p-1} \in R_1$. The height of $R_1$ is $h-1$, so by induction, each of these elements lies in some radical extension of $K$ that is contained in $L$. The subfield $J$ generated by all of these radical extensions is clearly radical (Exercise 8.12), and contains $\alpha^p, x_0, x_2, \ldots x_{p-1}$. Then $x \in J(\alpha) \subseteq L$, and $J(\alpha)$ is radical. This completes the induction step, and with it, the proof. □

So much for the general quintic. We have used virtually everything that led up to Galois theory, but instead of thinking of a group of automorphisms of a field extension, we have used a group of permutations of the roots of a polynomial. Indeed, we have used only the group $\mathbb{S}_n$, which permutes the roots $t_j$ of the general polynomial $F(t)$. It would be possible to stop here, with a splendid application of group theory to the insolubility of the 'general' quintic. But for Galois, and for us, there is much more to do. The general quintic is not general *enough*, and it would be nice to find out why the various tricks used above actually *work*. At the moment, they seem to be fortunate accidents. In fact, they conceal an elegant theory (which, in particular, makes the Theorem on Natural Irrationalities entirely obvious; so much so that we can ignore it altogether). That theory is, of course, Galois theory. Now motivated up to the hilt, we can start to develop it in earnest.

---

## EXERCISES

8.1 Prove that in Section 8.2, the permutations $R$ and $S$ of equations (8.1, 8.2) preserve every valid polynomial equation over $Q$ relating $\alpha$, $\beta$, $\gamma$, and $\delta$. (*Hint*: The permutation $R$ has the same effect as complex conjugation. For the permutation $S$, observe that any polynomial equation in $\alpha, \beta, \gamma, \delta$ can be expressed as

$$p\gamma + q\delta = 0$$

where $p, q \in \mathbb{Q}(\mathrm{i})$. Substitute $\gamma = \sqrt{5}, \delta = -\sqrt{5}$ to derive a condition on $p$ and $q$. Show that this condition also implies that the equation holds if we change the values so that $\gamma = -\sqrt{5}, \delta = \sqrt{5}$.)

8.2 Show that the only subfields of $\mathbb{Q}(\mathrm{i}, \sqrt{5})$ are $\mathbb{Q}$, $\mathbb{Q}(\mathrm{i})$, $\mathbb{Q}(\sqrt{5})$, $\mathbb{Q}(\mathrm{i}\sqrt{5})$, and $\mathbb{Q}(\mathrm{i}, \sqrt{5})$.

8.3 Express the following in terms of elementary symmetric polynomials of $\alpha, \beta, \gamma$.

(a) $\alpha^2 + \beta^2 + \gamma^2$

(b) $\alpha^3 + \beta^3 + \gamma^3$

(c) $\alpha^2\beta + \alpha^2\gamma + \beta^2\alpha + \beta^2\gamma + \gamma^2\alpha + \gamma^2\beta$

(d) $(\alpha - \beta)^2 + (\beta - \gamma)^2 + (\gamma - \alpha)^2$

8.4 Prove that every symmetric polynomial $p(x, y) \in \mathbb{Q}[x, y]$ can be written as a polynomial in $xy$ and $x + y$, as follows. If $p$ contains a term $ax^i y^j$, with $i \neq j \in \mathbb{N}$ and $a \in \mathbb{Q}$, show that it must also contain the term $ax^j y^i$. Use this to write $p$ as a sum of terms of the form $a(x^i y^j + x^j y^i)$ or $ax^i y^i$. Observe that

$$\begin{aligned} x^i y^j + x^j y^i &= x^i y^i (x^{j-i} + y^{j-i}) \quad \text{if } i < j \\ x^i y^i &= (xy)^i \\ (x^i + y^i) &= (x + y)(x^{i-1} + y^{i-1}) - xy(x^{i-2} + y^{i-2}). \end{aligned}$$

Hence show that $p$ is a sum of terms that are polynomials in $x + y, xy$.

8.5* This exercise generalises Exercise 8.3 to $n$ variables. Suppose that $p(t_1, \ldots, t_n) \in K[t_1, \ldots, t_n]$ is symmetric and let the $s_i$ be the elementary symmetric polynomials in the $t_j$. Define the *rank* of a monomial $t_1^{a_1} t_2^{a_2} \ldots t_n^{a_n}$ to be $a_1 + 2a_2 + \cdots na_n$. Define the *rank* of $p$ to be the: maximum of the ranks of all monomials that occur in $p$, and let its part of highest rank be the sum of the terms whose ranks attain this maximum value. Find a polynomial $q$ composed of terms of the form $ks_1^{b_1} s_2^{b_2} \ldots s_n^{b_n}$, where $k \in K$, such that the part of $q$ of highest rank equals that of $p$. Observe that $p - q$ has smaller rank than $p$, and use induction on the rank to prove that $p$ is a polynomial in the $s_i$.

8.6 Suppose that $f(t) = a_n t^n + \cdots + a_0 \in K[t]$, and suppose that in some subfield $L$ of $\mathbb{C}$ such that $K \subset L$ we can factorise $f$ as

$$f(t) = a_n(t - \alpha_1) \ldots (t - \alpha_n)$$

Define

$$\lambda_j = \alpha_1^j + \cdots + \alpha_n^j$$

Prove *Newton's identities*

$$\begin{aligned} a_{n-1} + a_n \lambda_1 &= 0 \\ 2a_{n-2} + a_{n-1}\lambda_1 + a_n \lambda_2 &= 0 \\ &\vdots \\ na_0 + a_1\lambda_1 + \cdots + a_{n-1}\lambda_{n-1} + a_n\lambda_n &= 0 \\ &\vdots \\ a_0\lambda_k + a_1\lambda_{k+1} + \cdots + a_{n-1}\lambda_{k+n-1} + a_n\lambda_{k+n} &= 0 \quad (k \geq 1) \end{aligned}$$

Show how to use these identities inductively to obtain formulas for the $\lambda_j$.

8.7 Prove that the alternating group $\mathbb{A}_n$ is generated by 3-cycles.

8.8 Prove that every element of $\mathbb{A}_5$ is the product of two 5-cycles. Deduce that $\mathbb{A}_5$ is simple.

8.9 Solve the general quadratic by Ruffini radicals. (*Hint:* If the roots are $\alpha_1, \alpha_2$, show that $\alpha_1 - \alpha_2$ is a Ruffini radical.)

8.10 Solve the general cubic by Ruffini radicals. (*Hint:* If the roots are $\alpha_1, \alpha_2, \alpha_3$ show that $\alpha_1 + \omega\alpha_2 + \omega^2\alpha_3$ and $\alpha_1 + \omega^2\alpha_2 + \omega\alpha_3$ are Ruffini radicals.)

8.11 Suppose that $I \subseteq J$ are subfields of $\mathbb{C}(t_1, \ldots, t_n)$ (that is, subsets closed under the operations $+, -, \times, \div$), and $J$ is generated by $J_1, \ldots, J_r$ where $I \subseteq J_j \subseteq J$ for each $j$ and $J_j/I$ is radical. By induction on $r$, prove that $J/I$ is radical.

8.12 Using computer algebra, find an explicit formula for the discriminant (8.7) of any fourth degree polynomial of the form $t^4 - 2t^2 + a$ where $a \in \mathbb{C}$.

8.13 Mark the following true or false.

(a) The $K$-automorphisms of a field extension $L/K$ form a subfield of $\mathbb{C}$.

(b) The $K$-automorphisms of a field extension $L/K$ form a group.

(c) The fixed field of the Galois group of any finite extension $L/K$ contains $K$.

(d) The fixed field of the Galois group of any finite extension $L/K$ equals $K$.

(e) The alternating group $\mathbb{A}_5$ has a normal subgroup $H$ with quotient isomorphic to $\mathbb{Z}_5$.

(f) The alternating group $\mathbb{A}_5$ has a normal subgroup $H$ with quotient isomorphic to $\mathbb{Z}_3$.

(g) The alternating group $\mathbb{A}_5$ has a normal subgroup $H$ with quotient isomorphic to $\mathbb{Z}_2$.

(h) The general quintic equation can be solved using radicals, but it cannot be solved using Ruffini radicals.

# Chapter 9

# Normality and Separability

In this chapter we define the important concepts of *normality* and *separability* for field extensions, and develop some of their key properties.

Suppose that $K$ is a subfield of $\mathbb{C}$. Often a polynomial $p(t) \in K[t]$ has no zeros in $K$. But it must have zeros in $\mathbb{C}$, by the Fundamental Theorem of Algebra, Theorem 2.4. Therefore it may have at least some zeros in a given extension field $L$ of $K$. For example $t^2 + 1 \in \mathbb{R}[t]$ has no zeros in $\mathbb{R}$, but it has zeros $\pm\mathrm{i} \in \mathbb{C}$, in $\mathbb{Q}(\mathrm{i})$, and for that matter in any subfield containing $\mathbb{Q}(\mathrm{i})$. We shall study this phenomenon in detail, showing that every polynomial can be resolved into a product of linear factors (and hence has its full complement of zeros) if the ground field $K$ is extended to a suitable 'splitting field' $N$, which has finite degree over $K$. An extension $N/K$ is normal if any irreducible polynomial over $K$ with at least one zero in $N$ splits into linear factors in $N$. We show that a finite extension is normal if and only if it is a splitting field.

Separability is a complementary property to normality. An irreducible polynomial is separable if its zeros in its splitting field are simple. It turns out that over $\mathbb{C}$, this property is automatic. We make it explicit because it is *not* automatic for more general fields, see Chapter 16.

## 9.1 Splitting Fields

The most tractable polynomials are products of linear ones, so we are led to single this property out:

**Definition 9.1.** If $K$ is a subfield of $\mathbb{C}$, the subfield $L$ is an extension of $K$, and $f$ is a nonzero polynomial over $K$, then $f$ *splits* over $L$ if it can be expressed as a product of linear factors

$$f(t) = k(t - \alpha_1) \ldots (t - \alpha_n)$$

where $k \in K$ and $\alpha_1, \ldots, \alpha_n \in L$.

(Necessarily $k \in K$ here since $f(t) \in K[t]$ with leading coefficient $k$.) If $f$ splits in this manner, the zeros of $f$ in $K$ are precisely $\alpha_1, \ldots, \alpha_n$. The Fundamental Theorem of Algebra, Theorem 2.4, implies that $f$ splits over $K$

DOI: 10.1201/9781003213949-9

if and only if all of its zeros in $\mathbb{C}$ actually lie in $K$. Equivalently, $K$ contains the subfield generated by all the zeros of $f$.

**Examples 9.2.** (1) The polynomial $f(t) = t^3 - 1 \in \mathbb{Q}[t]$ splits over $\mathbb{C}$, because it can be written as

$$f(t) = (t-1)(t-\omega)(t-\omega^2)$$

where $\omega = e^{2\pi i/3} \in \mathbb{C}$. Similarly, $f$ splits over the subfield $\mathbb{Q}(i, \sqrt{3})$ since $\omega \in \mathbb{Q}(i, \sqrt{3})$, and indeed $f$ splits over $\mathbb{Q}(\omega)$, the smallest subfield of $\mathbb{C}$ with that property.

(2) The polynomial $f(t) = t^4 - 4t^2 - 5$ splits over $\mathbb{Q}(i, \sqrt{5})$, because

$$f(t) = (t-i)(t+i)(t-\sqrt{5})(t+\sqrt{5})$$

However, over $\mathbb{Q}(i)$ the best we can do is factorise it as

$$(t-i)(t+i)(t^2-5)$$

with an irreducible factor $t^2 - 5$ of degree greater than 1. (It is easy to show that 5 is not a square in $\mathbb{Q}(i)$. If $(p + iq)^2 = 5$ with $p, q \in \mathbb{Q}$, then $p^2 - q^2 = 5$ and $pq = 0$. If $p = 0$ then $-q^2 = 5$, which is impossible. If $q = 0$ then $p^2 = 5$, but $\sqrt{5}$ is irrational.) This proves that the polynomial $f$ does not split over $\mathbb{Q}(i)$. We see that even if a polynomial $f(t)$ has some linear factors in an extension field $L$, it need not split over $L$.

We show that given $K$ and $f$, we can always construct an extension $\Sigma$ of $K$ such that $f$ splits over $\Sigma$. It is convenient to require in addition that $f$ does not split over any smaller field, so that $\Sigma$ is as economical as possible.

**Definition 9.3.** A subfield $\Sigma$ of $\mathbb{C}$ is a *splitting field* for the nonzero polynomial $f$ over the subfield $K$ of $\mathbb{C}$ if $K \subseteq \Sigma$ and

(1) $f$ splits over $\Sigma$.

(2) If $K \subseteq \Sigma' \subseteq \Sigma$ and $f$ splits over $\Sigma'$ then $\Sigma' = \Sigma$.

The second condition is clearly equivalent to:

(2′) $\Sigma = K(\sigma_1, \ldots, \sigma_n)$ where $\sigma_1, \ldots, \sigma_n$ are the zeros of $f$ in $\Sigma$.

Clearly every polynomial over a subfield $K$ of $\mathbb{C}$ has a splitting field:

**Theorem 9.4.** *If $K$ is any subfield of $\mathbb{C}$ and $f$ is any nonzero polynomial over $K$, then there exists a unique splitting field $\Sigma$ for $f$ over $K$. Moreover, $[\Sigma : K]$ is finite.*

*Proof.* We can take $\Sigma = K(\sigma_1, \ldots, \sigma_n)$, where the $\sigma_j$ are the zeros of $f$ in $\mathbb{C}$. In fact, this is the only possibility, so $\Sigma$ is unique. The degree $[\Sigma : K]$ is finite since $K(\sigma_1, \ldots, \sigma_n)$ is finitely generated and algebraic, so Lemma 6.11 applies. □

Polynomials over isomorphic subfields of $\mathbb{C}$ have isomorphic splitting fields, in the following strong sense:

**Lemma 9.5.** *Suppose that $\iota : K \to K'$ is an isomorphism of subfields of $\mathbb{C}$. Let $f$ be a nonzero polynomial over $K$ and let $\Sigma \supseteq K$ be the splitting field for $f$. Let $L$ be any extension field of $K'$ such that $\iota(f)$ splits over $L$. Then there exists a monomorphism $j : \Sigma \to L$ such that $j|_K = \iota$.*

*Proof.* We have the following situation:

$$\begin{array}{ccc} K & \longrightarrow & \Sigma \\ \downarrow{\scriptstyle \iota} & & \downarrow{\scriptstyle j} \\ K' & \longrightarrow & L \end{array}$$

where $j$ has yet to be found. We construct $j$ using induction on $\partial f$. We can view $f$ as a polynomial over $\Sigma$, in which case

$$f(t) = k(t - \sigma_1)\dots(t - \sigma_n)$$

for $\sigma_1 \dots \sigma_n \in \Sigma$. The minimal polynomial $m$ of $\sigma_1$ over $K$ is an irreducible factor of $f$. Now $\iota(m)$ divides $\iota(f)$ which splits over $L$, so that over $L$

$$\iota(m) = (t - \alpha_1)\dots(t - \alpha_r)$$

where $\alpha_1, \dots, \alpha_r \in L$. Since $\iota(m)$ is irreducible over $K'$ it must be the minimal polynomial of $\alpha_1$ over $K'$. So by Theorem 5.16 there is an isomorphism

$$j_1 : K(\sigma_1) \to K'(\alpha_1)$$

such that $j_1|_K = \iota$ and $j_1(\sigma_1) = \alpha_1$. Now $\Sigma$ is a splitting field over $K(\sigma_1)$ of the polynomial $g = f/(t - \sigma_1)$. By induction there exists a monomorphism $j : \Sigma \to L$ such that $j|_{K(\sigma_1)} = j_1$. But then $j|_K = \iota$ and we are finished. □

This lets us prove the uniqueness theorem:

**Theorem 9.6.** *Let $\iota : K \to K'$ be an isomorphism. Let $\Sigma$ be the splitting field for $f$ over $K$, and let $\Sigma'$ be the splitting field for $\iota(f)$ over $K'$. Then there is an isomorphism $j : \Sigma \to \Sigma'$ such that $j|_K = \iota$. In other words, the extensions $\Sigma/K$ and $\Sigma'/K'$ are isomorphic.*

*Proof.* Consider the following diagram:

$$\begin{array}{ccc} K & \longrightarrow & \Sigma \\ \downarrow{\scriptstyle \iota} & & \downarrow{\scriptstyle j} \\ K' & \longrightarrow & \Sigma' \end{array}$$

We must find $j$ to make the diagram commute, given the rest of the diagram.

By Lemma 9.5 there is a monomorphism $j : \Sigma \to \Sigma'$ such that $j|_K = \iota$. But $j(\Sigma)$ is clearly the splitting field for $\iota(f)$ over $K'$ and is contained in $\Sigma'$. Since $\Sigma'$ is also the splitting field for $\iota(f)$ over $K'$, we have $j(\Sigma) = \Sigma'$, so that $j$ is onto. Hence $j$ is an isomorphism, and the theorem follows. □

**Examples 9.7.** (1) Let $f(t) = t^5 - 3t^3 + t^2 - 3 = (t^2 - 3)(t^3 + 1)$ over $\mathbb{Q}$. We can construct a splitting field for $f$ as follows: over $\mathbb{C}$ the polynomial $f$ splits into linear factors

$$f(t) = (t + \sqrt{3})(t - \sqrt{3})(t + 1)\left(t - \frac{1 + \mathrm{i}\sqrt{3}}{2}\right)\left(t - \frac{1 - \mathrm{i}\sqrt{3}}{2}\right)$$

so there exists a splitting field in $\mathbb{C}$, namely

$$\mathbb{Q}\left(\sqrt{3}, \frac{1 + \mathrm{i}\sqrt{3}}{2}\right)$$

This is clearly the same as $\mathbb{Q}(\sqrt{3}, \mathrm{i})$.
(2) Let $f(t) = t^4 - 2t^3 - t^2 - 2t - 2 = (t^2 - 2t - 2)(t^2 + 1)$ over $\mathbb{Q}$. The zeros of $f$ in $\mathbb{C}$ are $1 \pm \sqrt{3}, \pm\mathrm{i}$, so a splitting field is afforded by $\mathbb{Q}(1 + \sqrt{3}, \mathrm{i})$ which equals $\mathbb{Q}(\sqrt{3}, \mathrm{i})$. This is the same field as in the previous example, although the two polynomials involved are different.
(3) It is even possible to have two distinct irreducible polynomials with the same splitting field. For example $t^2 - 3$ and $t^2 - 2t - 2$ are both irreducible over $\mathbb{Q}$, and both have $\mathbb{Q}(\sqrt{3})$ as their splitting field over $\mathbb{Q}$.

---

## 9.2 Normality

The idea of a normal extension was explicitly recognised by Galois (but, as always, in terms of polynomials over $\mathbb{C}$). In the modern treatment it takes the following form:

**Definition 9.8.** An algebraic field extension $L/K$ is *normal* if every irreducible polynomial $f$ over $K$ that has at least one zero in $L$ splits in $L$.

For example, $\mathbb{C}/\mathbb{R}$ is normal since every polynomial (irreducible or not) splits in $\mathbb{C}$. On the other hand, we can find extensions that are not normal. Let $\alpha$ be the real cube root of 2 and consider $\mathbb{Q}(\alpha)/\mathbb{Q}$. The irreducible polynomial $t^3 - 2$ has a zero, namely $\alpha$, in $\mathbb{Q}(\alpha)$, but it does not split in $\mathbb{Q}(\alpha)$. If it did, then there would be three real cube roots of 2, not all equal. This is absurd.

Compare with the examples of Galois groups given in Chapter 8. The normal extension $\mathbb{C}/\mathbb{R}$ has a well-behaved Galois group, in the sense that the Galois correspondence is a bijection. The same goes for $\mathbb{Q}(\sqrt{2}, \sqrt{3}, \sqrt{5})/\mathbb{Q}$. In contrast, the non-normal extension $\mathbb{Q}(\alpha)/\mathbb{Q}$ has a badly behaved Galois group. Although this is not the whole story, it illustrates the importance of normality.

There is a close connection between normal extensions and splitting fields which provides a wide range of normal extensions:

**Theorem 9.9.** *A field extension $L/K$ is normal and finite if and only if $L$ is a splitting field for some polynomial over $K$.*

*Proof.* Suppose $L/K$ is normal and finite. By Lemma 6.11, $L = K(\alpha_1, \ldots, \alpha_s)$ for certain $\alpha_j$ algebraic over $K$. Let $m_j$ be the minimal polynomial of $\alpha_j$ over $K$ and let $f = m_1 \ldots m_s$. Each $m_j$ is irreducible over $K$ and has a zero $\alpha_j \in L$, so by normality each $m_j$ splits over $L$. Hence $f$ splits over $L$. Since $L$ is generated by $K$ and the zeros of $f$, it is the splitting field for $f$ over $K$.

To prove the converse, suppose that $L$ is the splitting field for some polynomial $g$ over $K$. The extension $L/K$ is then obviously finite; we must show it is normal. To do this we must take an irreducible polynomial $f$ over $K$ with a zero in $L$ and show that it splits in $L$. Let $M \supseteq L$ be a splitting field for $fg$ over $K$. Suppose that $\theta_1$ and $\theta_2$ are zeros of $f$ in $M$. By irreducibility, $f$ is the minimal polynomial of both $\theta_1$ and $\theta_2$ over $K$.

We claim that

$$[L(\theta_1) : L] = [L(\theta_2) : L]$$

This is proved by an interesting trick. We look at several subfields of $M$, namely $K, L, K(\theta_1), L(\theta_1), K(\theta_2), L(\theta_2)$. There are two towers

$$K \subseteq K(\theta_1) \subseteq L(\theta_1) \subseteq M$$
$$K \subseteq K(\theta_2) \subseteq L(\theta_2) \subseteq M$$

The claim follows from a simple computation of degrees. For $j = 1$ or $2$

$$[L(\theta_j) : L][L : K] = [L(\theta_j) : K] = [L(\theta_j) : K(\theta_j)][K(\theta_j) : K] \qquad (9.1)$$

By Proposition 6.7, $[K(\theta_1){:}K] = [K(\theta_2){:}K]$. Clearly $L(\theta_j)$ is the splitting field for $g$ over $K(\theta_j)$, and by Corollary 5.13 $K(\theta_1)$ is isomorphic to $K(\theta_2)$. Therefore by Theorem 9.6 the extensions $L(\theta_j)/K(\theta_j)$ are isomorphic for $j = 1, 2$, so they have the same degree. Substituting in (9.1) and cancelling,

$$[L(\theta_1) : L] = [L(\theta_2) : L]$$

as claimed. From this point on, the rest is easy. If $\theta_1 \in L$ then $[L(\theta_1) : L] = 1$, so $[L(\theta_2) : L] = 1$ and $\theta_2 \in L$ also. Hence $L/K$ is normal. □

---

## 9.3 Separability

Galois did not explicitly recognise the concept of separability, since he worked only with the complex field, where, as we shall see, separability is automatic. However, the concept is implicit in his work and must be invoked when studying more general fields.

**Definition 9.10.** An irreducible polynomial $f$ over a subfield $K$ of $\mathbb{C}$ is *separable* over $K$ if it has simple zeros in $\mathbb{C}$, or equivalently, simple zeros in its splitting field.

This means that over its splitting field, or over $\mathbb{C}$, $f$ takes the form

$$f(t) = k(t - \sigma_1) \dots (t - \sigma_n)$$

where the $\sigma_j$ are all different.

**Example 9.11.** The polynomial $t^4 + t^3 + t^2 + t + 1$ is separable over $\mathbb{Q}$, since its zeros in $\mathbb{C}$ are $e^{2\pi i/5}$, $e^{4\pi i/5}$, $e^{6\pi i/5}$, $e^{8\pi i/5}$, which are all different.

For polynomials over $\mathbb{R}$ there is a standard method for detecting multiple zeros by differentiation, used in Lemma 6.12. To obtain maximum generality later, we redefine the derivative in a purely formal manner.

**Definition 9.12.** Suppose that $K$ is a subfield of $\mathbb{C}$, and let

$$f(t) = a_0 + a_1 t + \cdots + a_n t^n \in K[t]$$

Then the *formal derivative* of $f$ is the polynomial

$$Df = a_1 + 2a_2 t + \cdots + n a_n t^{n-1} \in K[t]$$

For $K = \mathbb{R}$ (and indeed for $K = \mathbb{C}$) this is the usual derivative. Several useful properties of the derivative carry over to $D$. In particular, simple computations (Exercise 9.3) show that for all polynomials $f$ and $g$ over $K$,

$$\begin{aligned} D(f+g) &= Df + Dg \\ D(fg) &= (Df)g + f(Dg) \end{aligned}$$

Also, if $\lambda \in K$ then $D(\lambda) = 0$, so

$$D(\lambda f) = \lambda(Df)$$

These properties of $D$ let us state a criterion for the existence of multiple zeros without knowing what the zeros are. It extends Lemma 6.12 in different notation.

**Lemma 9.13.** *Let $f \neq 0$ be a polynomial over a subfield $K$ of $\mathbb{C}$, and let $\Sigma$ be its splitting field. Then $f$ has a multiple zero (in $\mathbb{C}$ or $\Sigma$) if and only if $f$ and $Df$ have a common factor of degree $\geq 1$ in $K[t]$.*

*Proof.* Suppose $f$ has a repeated zero in $\Sigma$, so that over $\Sigma$

$$f(t) = (t - \alpha)^2 g(t)$$

where $\alpha \in \Sigma$. Then

$$Df = (t - \alpha)[(t - \alpha)Dg + 2g]$$

so $f$ and $Df$ have a common factor $(t-\alpha)$ in $\Sigma[t]$. Hence $f$ and $Df$ have a common factor in $K[t]$, namely the minimal polynomial of $\alpha$ over $K$.

Now suppose that $f$ has no repeated zeros. Suppose that $f$ and $Df$ have a common factor, and let $\alpha$ be a zero of that factor. Then $f=(t-\alpha)g$ and $Df=(t-\alpha)h$. Differentiate the former to get $(t-\alpha)h = Df = g+(t-\alpha)Dg$, so $(t-\alpha)$ divides $g$, hence $(t-\alpha)^2$ divides $f$, a contradiction. □

We now prove that separability of an irreducible polynomial is automatic over subfields of $\mathbb{C}$.

**Proposition 9.14.** *If $K$ is a subfield of $\mathbb{C}$ then every irreducible polynomial over $K$ is separable.*

*Proof.* An irreducible polynomial $f$ over $K$ is inseparable if and only if $f$ and $Df$ have a common factor of degree $\geq 1$. If so, then since $f$ is irreducible the common factor must be $f$, but $Df$ has smaller degree than $f$, and the only multiple of $f$ having smaller degree is 0, so $Df = 0$. Thus if

$$f(t) = a_0 + \cdots + a_m t^m$$

then this is equivalent to $na_n = 0$ for all integers $n > 0$. For subfields of $\mathbb{C}$, this is equivalent to $a_n = 0$ for all $n > 0$. □

---

## EXERCISES

9.1 Determine splitting fields over $\mathbb{Q}$ for the polynomials $t^3-1, t^4+5t^2+6, t^6-8$, in the form $\mathbb{Q}(\alpha_1,\ldots,\alpha_k)$ for explicit $\alpha_j$.

9.2 Find the degrees of these fields as extensions of $\mathbb{Q}$.

9.3 Prove that the formal derivative $D$ has the following properties:

(a) $D(f+g) = Df + Dg$

(b) $D(fg) = (Df)g + f(Dg)$

(c) If $f(t) = t^n$, then $Df(t) = nt^{n-1}$

9.4 Show that we can extend the definition of the formal derivative to $K(t)$ by defining

$$D(f/g) = (Df \cdot g - f \cdot Dg)/g^2$$

when $g \neq \mathbf{0}$. Verify the relevant properties of $D$.

9.5 Which of the following extensions are normal?

(a) $\mathbb{Q}(t)/\mathbb{Q}$

(b) $\mathbb{Q}(\sqrt{-5})/\mathbb{Q}$

(c) $\mathbb{Q}(\alpha)/\mathbb{Q}$ where $\alpha$ is the real seventh root of 5

(d) $\mathbb{Q}(\sqrt{5}, \alpha)/\mathbb{Q}(\alpha)$, where $\alpha$ is as in (c)

(e) $\mathbb{R}(\sqrt{-7})/\mathbb{R}$

9.6 Show that every extension in $\mathbb{C}$, of degree 2, is normal. Is this true if the degree is greater than 2?

9.7 If $\Sigma$ is the splitting field for $f$ over $K$ and $K \subseteq L \subseteq \Sigma$, show that $\Sigma$ is the splitting field for $f$ over $L$.

9.8* Let $f$ be a polynomial of degree $n$ over $K$, and let $\Sigma$ be the splitting field for $f$ over $K$. Show that $[\Sigma : K]$ divides $n!$ (*Hint:* Use induction on $n$. Consider separately the cases when $f$ is reducible or irreducible. Note that $a!b!$ divides $(a+b)!$ (why?).)

9.9 Mark the following true or false.

(a) Every polynomial over $\mathbb{Q}$ splits over some subfield of $\mathbb{C}$.

(b) Splitting fields in $\mathbb{C}$ are unique.

(c) Every finite extension is normal.

(d) $\mathbb{Q}(\sqrt{19})/\mathbb{Q}$ is a normal extension.

(e) $\mathbb{Q}(\sqrt[4]{19})/\mathbb{Q}$ is a normal extension.

(f) $\mathbb{Q}(\sqrt[4]{19})/\mathbb{Q}(\sqrt{19})$ is a normal extension.

(g) A normal extension of a normal extension is a normal extension.

# Chapter 10

# Counting Principles

When proving the Fundamental Theorem of Galois theory in Chapter 12, we will need to show that if $H$ is a subgroup of the Galois group of a finite normal extension $L/K$, then $H^{\dagger *} = H$. Here the maps $*$ and $\dagger$ are as defined in Section 8.6. Our method will be to show that $H$ and $H^{\dagger *}$ are finite groups and have the same order. Since we already know that $H \subseteq H^{\dagger *}$, the two groups must be equal. This is an archetypal application of a *counting principle*: showing that two finite sets, one contained in the other, are identical, by counting how many elements they have, and showing that the two numbers are the same.

It is largely for this reason that we need to restrict attention to finite extensions and finite groups. If an infinite set is contained in another of the same cardinality, they need not be equal—for example, $\mathbb{Z} \subseteq \mathbb{Q}$ and both sets are countable, but $\mathbb{Z} \neq \mathbb{Q}$. So counting principles may fail for infinite sets.

The object of this chapter is to perform part of the calculation of the order of $H^{\dagger *}$. Namely, we find the degree $[H^\dagger : K]$ in terms of the order of $H$. In Chapter 11 we find the order of $H^{\dagger *}$ in terms of this degree; putting the pieces together will give the desired result.

## 10.1 Linear Independence of Monomorphisms

We begin with a theorem of Dedekind, who was the first to make a systematic study of field monomorphisms.

To motivate the theorem and its proof, we consider a special case. Suppose that $K$ and $L$ are subfields of $\mathbb{C}$, and let $\lambda$ and $\mu$ be monomorphisms $K \to L$. We claim that $\lambda$ cannot be a constant multiple of $\mu$ unless $\lambda = \mu$. By 'constant' here we mean an element of $L$. Suppose that there exists $a \in L$ such that

$$\mu(x) = a\lambda(x) \tag{10.1}$$

for all $x \in K$. Replace $x$ by $yx$, where $y \in K$, to get

$$\mu(yx) = a\lambda(yx)$$

DOI: 10.1201/9781003213949-10

Since $\lambda$ and $\mu$ are monomorphisms,

$$\mu(y)\mu(x) = a\lambda(y)\lambda(x)$$

Multiplying (10.1) by $\lambda(y)$, we also have

$$\lambda(y)\mu(x) = a\lambda(y)\lambda(x)$$

Comparing the two, $\lambda(y) = \mu(y)$ for all $y$, so $\lambda = \mu$.

In other words, if $\lambda$ and $\mu$ are distinct monomorphisms $K \to L$, they must be *linearly independent* over $L$.

Next, suppose that $\lambda_1, \lambda_2, \lambda_3$ are three distinct monomorphisms $K \to L$, and assume that they are linearly dependent over $L$. That is,

$$a_1\lambda_1 + a_2\lambda_2 + a_3\lambda_3 = 0$$

for $a_j \in L$. In more detail,

$$a_1\lambda_1(x) + a_2\lambda_2(x) + a_3\lambda_3(x) = 0 \tag{10.2}$$

for all $x \in K$. If some $a_j = 0$ then we reduce to the previous case, so we may assume all $a_j \neq 0$.

Substitute $yx$ for $x$ in (10.2) to get

$$a_1\lambda_1(yx) + a_2\lambda_2(yx) + a_3\lambda_3(yx) = 0 \tag{10.3}$$

That is,

$$[a_1\lambda_1(y)]\lambda_1(x) + [a_2\lambda_2(y)]\lambda_2(x) + [a_3\lambda_3(y)]\lambda_3(x) = 0 \tag{10.4}$$

Relations (10.2) and (10.4) are independent—that is, they are not scalar multiples of each other—unless $\lambda_1(y) = \lambda_2(y) = \lambda_3(y)$, and we can choose $y$ to prevent this. Therefore we may eliminate one of the $\lambda_j$ to deduce a linear relation between at most two of them, contrary to the previous case. Specifically, there exists $y \in K$ such that $\lambda_1(y) \neq \lambda_3(y)$. Multiply (10.2) by $\lambda_3(y)$ and subtract from (10.4) to get

$$[a_1\lambda_1(y) - a_1\lambda_3(y)]\lambda_1(x) + [a_2\lambda_2(y) - a_2\lambda_3(y)]\lambda_2(x) = 0$$

Then the coefficient of $\lambda_1(x)$ is $a_1(\lambda_1(y) - \lambda_3(y)) \neq 0$, a contradiction.

Dedekind realised that this approach can be used inductively to prove:

**Lemma 10.1 (Dedekind).** *If $K$ and $L$ are subfields of $\mathbb{C}$, then every set of distinct monomorphisms $K \to L$ is linearly independent over $L$.*

*Proof.* Let $\lambda_1, \ldots, \lambda_n$ be distinct monomorphisms $K \to L$. To say these are linearly independent over $L$ is to say that there do not exist elements $a_1, \ldots, a_n \in L$ such that

$$a_1\lambda_1(x) + \cdots + a_n\lambda_n(x) = 0 \tag{10.5}$$

for all $x \in K$, unless all the $a_j$ are 0.

Assume the contrary, so that (10.5) holds. At least one of the $a_i$ is non-zero. Among all the valid equations of the form (10.5) with all $a_i \neq 0$, there must be at least one for which the number $n$ of non-zero terms is least. (This 'minimal criminal' approach is essentially a version of induction.) Since all $\lambda_j$ are non-zero, $n \neq 1$. We choose notation so that equation (10.5) is such as expression. Hence we may assume that there does not exist an equation like (10.5) with fewer than $n$ terms. From this we deduce a contradiction.

Since $\lambda_1 \neq \lambda_n$, there exists $y \in K$ such that $\lambda_1(y) \neq \lambda_n(y)$. Therefore $y \neq 0$. Now (10.5) holds with $yx$ in place of $x$, so

$$a_1\lambda_1(yx) + \cdots + a_n\lambda_n(yx) = 0$$

for all $x \in K$, whence

$$a_1\lambda_1(y)\lambda_1(x) + \cdots + a_n\lambda_n(y)\lambda_n(x) = 0 \tag{10.6}$$

for all $x \in K$. Multiply (10.5) by $\lambda_1(y)$ and subtract (10.6), so that the first terms cancel: we obtain

$$a_2[\lambda_2(x)\lambda_1(y) - \lambda_2(x)\lambda_2(y)] + \cdots + a_n[\lambda_n(x)\lambda_1(y) - \lambda_n(x)\lambda_n(y)] = 0$$

The coefficient of $\lambda_n(x)$ is $a_n[\lambda_1(y) - \lambda_n(y)] \neq 0$, so we have an equation of the form (10.5) with fewer terms. Deleting any zero terms does not alter this statement. This contradicts the italicised assumption above.

Consequently no equation of the form (10.5) exists, so the monomorphisms are linearly independent. □

**Example 10.2.** Let $K = \mathbb{Q}(\alpha)$ where $\alpha = \sqrt[3]{2} \in \mathbb{R}$. There are three monomorphisms $K \to \mathbb{C}$, namely

$$\begin{aligned}
\lambda_1(p + q\alpha + r\alpha^2) &= p + q\alpha + r\alpha^2 \\
\lambda_2(p + q\alpha + r\alpha^2) &= p + q\omega\alpha + r\omega^2\alpha^2 \\
\lambda_3(p + q\alpha + r\alpha^2) &= p + q\omega^2\alpha + r\omega\alpha^2
\end{aligned}$$

where $p, q, r \in \mathbb{Q}$ and $\omega$ is a primitive cube root of unity. We prove by 'bare hands' methods that the $\lambda_j$ are linearly independent. Suppose that $a_1\lambda_1(x) + a_2\lambda_2(x) + a_3\lambda_3(x) = 0$ for all $x \in K$. Set $x = 1, \alpha, \alpha^2$ respectively to get

$$\begin{aligned}
a_1 + a_2 + a_3 &= 0 \\
a_1 + \omega a_2 + \omega^2 a_3 &= 0 \\
a_1 + \omega^2 a_2 + \omega a_3 &= 0
\end{aligned}$$

The determinant is non-zero (Exercise 10.5), so the only solution of this system of linear equations is $a_1 = a_2 = a_3 = 0$.

For our next result we need two lemmas. The first is a standard theorem of linear algebra, which we quote without proof.

**Lemma 10.3.** *If $n > m$ then a system of $m$ homogeneous linear equations*

$$a_{i1}x_1 + \cdots + a_{in}x_n = 0 \quad 1 \le i \le m$$

*in $n$ unknowns $x_1, \ldots, x_n$, with coefficients $a_{ij}$ in a field $K$, has a solution in which the $x_i$ are all in $K$ and are not all zero.* □

This lemma is proved in most first-year undergraduate linear algebra courses and can be found in any text of linear algebra, for example Anton (1987). The second lemma states a useful general principle.

**Lemma 10.4.** *If $G$ is a group whose distinct elements are $g_1, \ldots, g_n$, and if $g \in G$, then as $j$ varies from 1 to $n$ the elements $gg_j$ run through the whole of $G$, each element of $G$ occurring precisely once.*

*Proof.* If $h \in G$ then $g^{-1}h = g_j$ for some $j$ and $h = gg_j$. If $gg_i = gg_j$ then $g_i = g^{-1}gg_i = g^{-1}gg_j = g_j$. Thus the map $g_i \mapsto gg_i$ is a bijection $G \to G$, and the result follows. □

We also recall some standard notation. We denote the cardinality of a set $S$ by $|S|$. Thus if $G$ is a group, then $|G|$ is the *order* of $G$. For example, $|\mathbb{S}_n| = n!$ and $|\mathbb{A}_n| = n!/2$.

We now come to the main theorem of this chapter, whose proof is similar to that of Lemma 10.1, and which can be motivated in a similar manner.

**Theorem 10.5.** *Let $G$ be a finite subgroup of the group of automorphisms of a field $K$, and let $K_0$ be the fixed field of $G$. Then $[K : K_0] = |G|$.*

*Proof.* Let $n = |G|$, and suppose that the elements of $G$ are $g_1, \ldots, g_n$, where $g_1 = 1$. We prove separately that $[K : K_0] < n$ and $[K : K_0] > n$ are impossible.

(1) Suppose that $[K : K_0] = m < n$. Let $\{x_1, \ldots, x_m\}$ be a basis for $K$ over $K_0$. By Lemma 10.3 there exist $y_1, \ldots, y_n \in K$, not all zero, such that

$$y_1 g_1(x_i) + \cdots + y_n g_n(x_i) = 0 \tag{10.7}$$

for $i = 1, \ldots, m$. Let $x$ be any element of $K$. Then

$$x = \alpha_1 x_1 + \cdots + \alpha_m x_m$$

where $\alpha_1, \ldots, \alpha_m \in K_0$. Hence

$$\begin{aligned} y_1 g_1(x) + \cdots + y_n g_n(x) &= y_1 g_1\left(\sum_l \alpha_l x_l\right) + \cdots + y_n g_n\left(\sum_l \alpha_l x_l\right) \\ &= \sum_l \alpha_l [y_1 g_1(x_l) + \cdots + y_n g_n(x_l)] \\ &= 0 \end{aligned}$$

using (10.7). Hence the distinct monomorphisms $g_1, \ldots, g_n$ are linearly dependent, contrary to Lemma 10.1. Therefore $m \geq n$.
(2) Next, suppose for a contradiction that $[K : K_0] > n$. Then there exists a set of $n+1$ elements of $K$ that are linearly independent over $K_0$; let such a set be $\{x_1, \ldots, x_{n+1}\}$. By Lemma 10.3 there exist $y_1, \ldots, y_{n+1} \in K$, not all zero, such that for $j = 1, \ldots, n$

$$y_1 g_j(x_1) + \cdots + y_{n+1} g_j(x_{n+1}) = 0 \tag{10.8}$$

We subject this equation to a combinatorial attack, similar to that used in proving Lemma 10.1. Choose $y_1, \ldots, y_{n+1}$ so that as few as possible are non-zero, and renumber so that

$$y_1, \ldots, y_r \neq 0, \quad y_{r+1}, \ldots, y_{n+1} = 0$$

Equation (10.8) now becomes

$$y_1 g_j(x_1) + \cdots + y_r g_j(x_r) = 0 \tag{10.9}$$

Let $g \in G$, and operate on (10.9) with $g$. This gives a system of equations

$$g(y_1) g g_j(x_1) + \cdots + g(y_r) g g_j(x_r) = 0$$

By Lemma 10.4, as $j$ varies, this system of equations is equivalent to the system

$$g(y_1) g_j(x_1) + \cdots + g(y_r) g_j(x_r) = 0 \tag{10.10}$$

Multiply (10.9) by $g(y_1)$ and (10.10) by $y_1$ and subtract, to get

$$[y_2 g(y_1) - g(y_2) y_1] g_j(x_2) + \cdots + [y_r g(y_1) - g(y_r) y_1] g_j(x_r) = 0$$

This is a system of equations like (10.9) but with fewer terms, which gives a contradiction unless all the coefficients

$$y_i g(y_1) - y_1 g(y_i)$$

are zero. If this happens then

$$y_i y_1^{-1} = g(y_i y_1^{-1})$$

for all $g \in G$, so that $y_i y_1^{-1} \in K_0$. Thus there exist $z_1, \ldots, z_r \in K_0$ and an element $k \in K$ such that $y_i = k z_i$ for all $i$. Then (10.9), with $j = 1$, becomes

$$x_1 k z_1 + \cdots + x_r k z_r = 0$$

and since $k \neq 0$ we may divide by $k$, which shows that the $x_i$ are linearly dependent over $K_0$. This is a contradiction.

Therefore $[K : K_0]$ is not less than $n$ and not greater than $n$, so $[K : K_0] = n = |G|$ as required. □

**Corollary 10.6.** *If $G$ is the Galois group of the finite extension $L/K$, and $H$ is a finite subgroup of $G$, then*

$$[H^\dagger : K] = [L : K]/|H|$$

*Proof.* By the Tower Law, $[L : K] = [L : H^\dagger][H^\dagger : K]$, so $[H^\dagger : K] = [L : K]/[L : H^\dagger]$. But this equals $[L : K]/|H|$ by Theorem 10.5. □

**Examples 10.7.** We illustrate Theorem 10.5 by two examples, one simple, the other more intricate.

(1) Let $G$ be the group of automorphisms of $\mathbb{C}$ consisting of the identity and complex conjugation. The fixed field of $G$ is $\mathbb{R}$, for if $x - \mathrm{i}y = x + \mathrm{i}y$ $(x, y \in \mathbb{R})$ then $y = 0$, and conversely. Hence $[\mathbb{C} : \mathbb{R}] = |G| = 2$, a conclusion that is manifestly correct.

(2) Let $K = \mathbb{Q}(\zeta)$ where $\zeta = \exp(2\pi\mathrm{i}/5) \in \mathbb{C}$. Now $\zeta^5 = 1$ and $\mathbb{Q}(\zeta)$ consists of all elements

$$p + q\zeta + r\zeta^2 + s\zeta^3 + t\zeta^4 \tag{10.11}$$

where $p, q, r, s, t \in \mathbb{Q}$. The Galois group of $\mathbb{Q}(\zeta)/\mathbb{Q}$ is easy to find, for if $\alpha$ is a $\mathbb{Q}$-automorphism of $\mathbb{Q}(\zeta)$ then

$$(\alpha(\zeta))^5 = \alpha(\zeta^5) = \alpha(1) = 1,$$

so that $\alpha(\zeta) = \zeta, \zeta^2, \zeta^3$, or $\zeta^4$. This gives four candidates for $\mathbb{Q}$-automorphisms:

$$\begin{array}{ll} \alpha_1 : p + q\zeta + r\zeta^2 + s\zeta^3 + t\zeta^4 & \mapsto p + q\zeta + r\zeta^2 + s\zeta^3 + t\zeta^4 \\ \alpha_2 : & \mapsto p + s\zeta + q\zeta^2 + t\zeta^3 + r\zeta^4 \\ \alpha_3 : & \mapsto p + r\zeta + t\zeta^2 + q\zeta^3 + s\zeta^4 \\ \alpha_4 : & \mapsto p + t\zeta + s\zeta^2 + r\zeta^3 + q\zeta^4 \end{array}$$

It is easy to check that all of these are $\mathbb{Q}$-automorphisms. The only point to bear in mind is that $1, \zeta, \zeta^2, \zeta^3, \zeta^4$ are *not* linearly independent over $\mathbb{Q}$. However, their linear relations are generated by just one: $\zeta + \zeta^2 + \zeta^3 + \zeta^4 = -1$, and this relation is preserved by all of the candidate $\mathbb{Q}$-automorphisms.

Alternatively, observe that $\zeta, \zeta^2, \zeta^3, \zeta^4$ all have the same minimal polynomial $t^4 + t^3 + t^2 + t + 1$ and use Corollary 5.13.

We deduce that the Galois group of $\mathbb{Q}(\zeta)/\mathbb{Q}$ has order 4. It is easy to find the fixed field of this group: it turns out to be $\mathbb{Q}$. Therefore, by Theorem 10.5, $[\mathbb{Q}(\zeta) : \mathbb{Q}] = 4$. At first sight this might seem wrong, for (10.11) expresses each element in terms of five basic elements; the degree should be 5. In support of this contention, $\zeta$ is a zero of $t^5 - 1$. The astute reader will already have seen the source of this dilemma: $t^5 - 1$ is not the minimal polynomial of $\zeta$ over $\mathbb{Q}$, since it is reducible. The minimal polynomial is, as we have seen, $t^4 + t^3 + t^2 + t + 1$, which has degree 4. Equation (10.11) holds, but the elements of the supposed 'basis' are linearly dependent. Every element of $\mathbb{Q}(\zeta)$ can be expressed *uniquely* in the form $p + q\zeta + r\zeta^2 + s\zeta^3$ where $p, q, r, s \in \mathbb{Q}$. We did not use this expression because it lacks symmetry, making the computations formless and therefore harder.

## EXERCISES

10.1 Check Theorem 10.5 for the extension $\mathbb{C}(t_1,\ldots,t_n)/\mathbb{C}(s_1,\ldots,s_n)$ of Chapter 8 Section 8.7.

10.2 Find the fixed field of the subgroup $\{\alpha_1,\alpha_4\}$ for Example 10.7(2). Check that Theorem 10.5 holds.

10.3 Parallel the argument of Example 10.7(2) when $\zeta = e^{2\pi i/7}$.

10.4 Find all monomorphisms $\mathbb{Q}\to\mathbb{C}$.

10.5 To fill in details in Example 10.2, calculate the determinant

$$\begin{vmatrix} 1 & 1 & 1 \\ 1 & \omega & \omega^2 \\ 1 & \omega^2 & \omega \end{vmatrix}$$

where $\omega = e^{2\pi i/3}$, and show that it is non-zero.

10.6 Verify directly that the $\mathbb{Q}$-automorphisms $\alpha_j$ in Example 10.7(2) are linearly independent over $\mathbb{Q}$.

10.7 Let $\zeta$ be a primitive 6th root of unity. Find all $\mathbb{Q}$-automorphisms of $\mathbb{Q}(\zeta)$ and prove directly that they are linearly independent over $\mathbb{Q}$.

10.8* Generalise the results of Exercises 10.6 and 10.7 to primitive $n$th roots of unity.

10.9 Mark the following true or false.

(a) If $S\subseteq T$ is a finite set and $|S|=|T|$, then $S=T$.

(b) The same is true of infinite sets.

(c) There is only one monomorphism $\mathbb{Q}\to\mathbb{Q}$.

(d) If $K$ and $L$ are subfields of $\mathbb{C}$, then there exists at least one monomorphism $K\to L$.

(e) Distinct automorphisms of a field $K$ are linearly independent over $K$.

(f) Linearly independent monomorphisms are distinct.

# Chapter 11

# Field Automorphisms

The theme of this chapter is the construction of automorphisms to given specifications. We begin with a generalisation of a $K$-automorphism, known as a $K$-monomorphism. For normal extensions we shall use $K$-monomorphisms to build up $K$-automorphisms. Using this technique, we can calculate the order of the Galois group of any finite normal extension, which combines with the result of Chapter 10 to give a crucial part of the fundamental theorem of Chapter 12.

We also introduce the concept of a normal closure of a finite extension. This useful device enables us to steer around some of the technical obstructions caused by non-normal extensions.

## 11.1 $K$-Monomorphisms

We begin by generalising the concept of a $K$-automorphism of a subfield $L$ of $\mathbb{C}$, by relaxing the condition that the map should be onto. We continue to require it to be one-to-one.

**Definition 11.1.** Suppose that $K$ is a subfield of each of the subfields $M$ and $L$ of $\mathbb{C}$. Then a *$K$-monomorphism* of $M$ into $L$ is a field monomorphism $\phi : M \to L$ such that $\phi(k) = k$ for every $k \in K$. We say that $\phi$ *fixes* $K$.

**Example 11.2.** Suppose that $K = \mathbb{Q}, M = \mathbb{Q}(\alpha)$ where $\alpha$ is a real cube root of 2, and $L = \mathbb{C}$. We can define a $K$-monomorphism $\phi : M \to L$ by insisting that $\phi(\alpha) = \omega\alpha$, where as usual $\omega = \mathrm{e}^{2\pi \mathrm{i}/3}$. In more detail, every element of $M$ is of the form $p + q\alpha + r\alpha^2$ where $p, q, r \in \mathbb{Q}$, and

$$\phi(p + q\alpha + r\alpha^2) = p + q\omega\alpha + r\omega^2\alpha^2$$

Since $\alpha$ and $\omega\alpha$ have the same minimal polynomial, namely $t^3 - 2$, Corollary 5.13 implies that $\phi$ is a $K$-monomorphism.

There are two other $K$-monomorphisms $M \to L$ in this case. One is the identity, and the other takes $\alpha$ to $\omega^2\alpha$ (see Figure 18).

In general if $K \subseteq M \subseteq L$ then any $K$-automorphism of $L$ restricts to a $K$-monomorphism $M \to L$. We are particularly interested in when this process can be reversed.

DOI: 10.1201/9781003213949-11

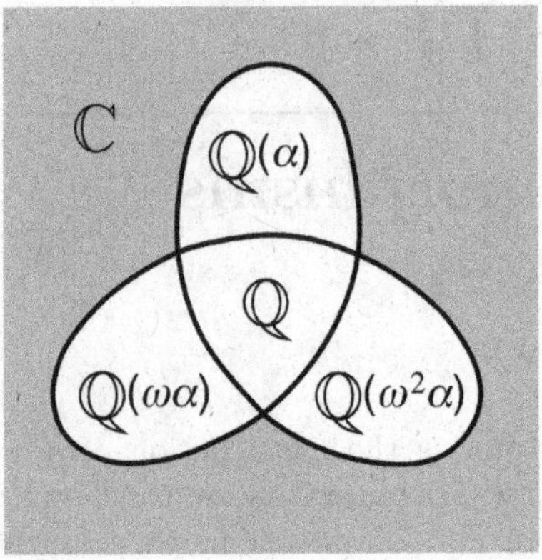


FIGURE 18: Images of $\mathbb{Q}$-monomorphisms of $\mathbb{Q}(\alpha)$ when $\alpha = \sqrt[3]{2}$.

**Theorem 11.3.** *Suppose that $L/K$ is a finite normal extension and $K \subseteq M \subseteq L$. Let $\tau$ be any $K$-monomorphism $M \to L$. Then there exists a $K$-automorphism $\sigma$ of $L$ such that $\sigma|_M = \tau$.*

*Proof.* By Theorem 9.9, $L$ is the splitting field over $K$ of some polynomial $f$ over $K$. Hence it is simultaneously the splitting field over $M$ for $f$ and over $\tau(M)$ for $\tau(f)$. But $\tau|_K$ is the identity, so $\tau(f) = f$. Consider the diagram

$$\begin{array}{ccc} M & \longrightarrow & L \\ \downarrow{\scriptstyle \tau} & & \downarrow{\scriptstyle \sigma} \\ \tau(M) & \longrightarrow & L \end{array}$$

where $\sigma$ is yet to be found. By Theorem 9.6 there is an isomorphism $\sigma : L \to L$ such that $\sigma|_M = \tau$. Therefore $\sigma$ is an automorphism of $L$, and since $\sigma|_K = \tau|_K$ is the identity, $\sigma$ is a $K$-automorphism of $L$. □

This result can be used to construct $K$-automorphisms:

**Proposition 11.4.** *Suppose that $L/K$ is a finite normal extension, and $\alpha, \beta$ are zeros in $L$ of the irreducible polynomial $p$ over $K$. Then there exists a $K$-automorphism $\sigma$ of $L$ such that $\sigma(\alpha) = \beta$.*

*Proof.* By Corollary 5.13 there is an isomorphism $\tau : K(\alpha) \to K(\beta)$ such that $\tau|_K$ is the identity and $\tau(\alpha) = \beta$. By Theorem 11.3 $\tau$ extends to a $K$-automorphism $\sigma$ of $L$. □

---

## 11.2 Normal Closures

When extensions are not normal, we can try to recover normality by making the extensions larger.

**Definition 11.5.** Let $L$ be a finite extension of $K$. A *normal closure* of $L/K$ is an extension $N$ of $L$ such that

(1) $N/K$ is normal;

(2) If $L \subseteq M \subseteq N$ and $M/K$ is normal, then $M = N$.

Thus $N$ is the smallest extension of $L$ that is normal over $K$.

The next theorem assures us of a sufficient supply of normal closures, and shows that (working inside $\mathbb{C}$) they are unique.

**Theorem 11.6.** *If $L/K$ is a finite extension in $\mathbb{C}$, then there exists a unique normal closure $N \subseteq \mathbb{C}$ of $L/K$, which is a finite extension of $K$.*

*Proof.* Let $x_1, \ldots, x_r$ be a basis for $L$ over $K$, and let $m_j$ be the minimal polynomial of $x_j$ over $K$. Let $N$ be the splitting field for $f = m_1 m_2 \ldots m_r$ over $L$. Then $N$ is also the splitting field for $f$ over $K$, so $N/K$ is normal and finite by Theorem 9.9. Suppose that $L \subseteq P \subseteq N$ where $P/K$ is normal. Each polynomial $m_j$ has a zero $x_j \in P$, so by normality $f$ splits in $P$. Since $N$ is the splitting field for $f$, we have $P = N$. Therefore $N$ is a normal closure.

Now suppose that $M$ and $N$ are both normal closures. The above polynomial $f$ splits in $M$ and in $N$, so each of $M$ and $N$ contain the splitting field for $f$ over $K$. This splitting field contains $L$ and is normal over $K$, so it must be equal to both $M$ and $N$. □

**Example 11.7.** Consider $\mathbb{Q}(\alpha)/\mathbb{Q}$ where $\alpha$ is the real cube root of 2. This extension is not normal, as we have seen. If we let $K$ be the splitting field for $t^3 - 2$ over $\mathbb{Q}$, contained in $\mathbb{C}$, then $K = \mathbb{Q}(\alpha, \alpha\omega, \alpha\omega^2)$ where $\omega = (-1 + \mathrm{i}\sqrt{3})/2$. This is the same as $\mathbb{Q}(\alpha, \omega)$. Now $K$ is the normal closure for $\mathbb{Q}(\alpha)/\mathbb{Q}$. So here we obtain the normal closure by adjoining all the 'missing' zeros.

Normal closures let us place restrictions on the image of a monomorphism.

**Lemma 11.8.** *Suppose that $K \subseteq L \subseteq N \subseteq M$ where $L/K$ is finite and $N$ is the normal closure of $L/K$. Let $\tau$ be any $K$-monomorphism $L \to M$. Then $\tau(L) \subseteq N$.*

*Proof.* Let $\alpha \in L$. Let $m$ be the minimal polynomial of $\alpha$ over $K$. Then $m(\alpha) = 0$ so $\tau(m(\alpha)) = 0$. But $\tau(m(\alpha)) = m(\tau(\alpha))$ since $\tau$ is a $K$-monomorphism, so $m(\tau(\alpha)) = 0$ and $\tau(\alpha)$ is a zero of $m$. Therefore $\tau(\alpha)$ lies in $N$ since $N/K$ is normal. Therefore $\tau(L) \subseteq N$. □

This result often lets us to restrict attention to the normal closure of a given extension when discussing monomorphisms. The next theorem provides a sort of converse.

**Theorem 11.9.** *For a finite extension $L/K$ the following are equivalent:*

(1) *$L/K$ is normal.*

(2) *There exists a finite normal extension $N$ of $K$ containing $L$ such that every $K$-monomorphism $\tau : L \to N$ is a $K$-automorphism of $L$.*

(3) *For every finite extension $M$ of $K$ containing $L$, every $K$-monomorphism $\tau : L \to M$ is a $K$-automorphism of $L$.*

*Proof.* We show that (1) $\Rightarrow$ (3) $\Rightarrow$ (2) $\Rightarrow$ (1).

(1) $\Rightarrow$ (3). If $L/K$ is normal then $L$ is the normal closure of $L/K$, so by Lemma 11.8, $\tau(L) \subseteq L$. But $\tau$ is a $K$-linear map defined on the finite-dimensional vector space $L$ over $K$ and is a monomorphism. Therefore $\tau(L)$ has the same dimension as $L$, whence $\tau(L) = L$ and $\tau$ is a $K$-automorphism of $L$.

(3) $\Rightarrow$ (2). Let $N$ be the normal closure for $L/K$. Then $N$ exists by Theorem 11.6, and has the requisite properties by (3).

(2) $\Rightarrow$ (1). Suppose that $f$ is any irreducible polynomial over $K$ with a zero $\alpha \in L$. Then $f$ splits over $N$ by normality, and if $\beta$ is any zero of $f$ in $N$, then by Proposition 11.4 there exists an automorphism $\sigma$ of $N$ such that $\sigma(\alpha) = \beta$. By hypothesis, $\sigma$ is a $K$-automorphism of $L$, so $\beta = \sigma(\alpha) \in \sigma(L) = L$. Therefore $f$ splits over $L$ and $L/K$ is normal. □

Our next result is of a more computational nature.

**Theorem 11.10.** *Suppose that $L/K$ is a finite extension of degree $n$. Then there are precisely $n$ distinct $K$-monomorphisms of $L$ into the normal closure $N$ of $L/K$, and hence into any given normal extension $M$ of $K$ containing $L$.*

*Proof.* Use induction on $[L : K]$. If $[L : K] = 1$ the result is clear. Suppose that $[L : K] = k > 1$. Let $\alpha \in L \backslash K$ with minimal polynomial $m$ over $K$. Then

$$\partial m = [K(\alpha) : K] = r > 1$$

Now $m$ is an irreducible polynomial over a subfield of $\mathbb{C}$ with one zero in the normal extension $N$, so $m$ splits in $N$ and its zeros $\alpha_1, \ldots, \alpha_r$ are distinct. By induction there are precisely $s$ distinct $K(\alpha)$-monomorphisms $\rho_1, \ldots, \rho_s : L \to N$, where $s = [L : K(\alpha)] = k/r$. By Proposition 11.4, there are $r$ distinct $K$-automorphisms $\tau_1, \ldots, \tau_r$ of $N$ such that $\tau_i(\alpha) = \alpha_i$. The maps

$$\phi_{ij} = \tau_i \rho_j \quad (1 \leq i \leq r, 1 \leq j \leq s)$$

are $K$-monomorphisms $L \to N$.

We claim they are distinct. Suppose $\phi_{ij} = \phi_{kl}$. Then $\tau_k^{-1}\tau_i = \rho_l \rho_j^{-1}$. The $\rho_j$ fix $K(\alpha)$, so they map $\alpha$ to itself. But $\rho_j$ is defined by its action on $\alpha$, so $\rho_l \rho_j^{-1}$ is the identity. That is, $\rho_l = \rho_j$. So $\tau_k^{-1}\tau_i$ is the identity, and $\tau_k = \tau_i$. Therefore $i = k, j = l$, so the $\phi_{ij}$ are distinct. They therefore provide $rs = k$ distinct $K$-monomorphisms $L \to N$.

Finally, we show that these are all of the $K$-monomorphisms $L \to N$. Let $\tau : L \to N$ be a $K$-monomorphism. Then $\tau(\alpha)$ is a zero of $m$ in $N$, so $\tau(\alpha) = \alpha_i$ for some $i$. The map $\phi = \tau_i^{-1}\tau$ is a $K(\alpha)$-monomorphism $L \to N$, so by induction $\phi = \rho_j$ for some $j$. Hence $\tau = \tau_i \rho_j = \phi_{ij}$ and the theorem is proved. □

We can now calculate the order of the Galois group of a finite normal extension, a result of fundamental importance.

**Corollary 11.11.** *If $L/K$ is a finite normal extension inside $\mathbb{C}$, then there are precisely $[L : K]$ distinct $K$-automorphisms of $L$. That is,*

$$|\Gamma(L/K)| = [L : K]$$

*Proof.* Use Theorem 11.10. □

From this we easily deduce the important:

**Theorem 11.12.** *Let $L/K$ be a finite extension with Galois group $G$. If $L/K$ is normal, then $K$ is the fixed field of $G$.*

*Proof.* Let $K_0$ be the fixed field of $G$, and let $[L : K] = n$. Corollary 11.11 implies that $|G| = n$. By Theorem 10.5, $[L : K_0] = n$. Since $K \subseteq K_0$ we must have $K = K_0$. □

An alternative and in some ways simpler approach to Corollary 11.11 and Theorem 11.12 can be found in Geck (2014). There is a converse to Theorem 11.12, which shows why we must consider normal extensions in order to make the Galois correspondence a bijection. Before we can prove the converse, we need a theorem whose statement and proof closely resemble those of Theorem 11.10.

**Theorem 11.13.** *Suppose that $K \subseteq L \subseteq M$ and $M/K$ is finite. Then the number of distinct $K$-monomorphisms $L \to M$ is at most $[L : K]$.*

*Proof.* Let $N$ be a normal closure of $M/K$. Then the set of $K$-monomorphisms $L \to M$ is contained in the set of $K$-monomorphisms $L \to N$, and by Theorem 11.10 there are precisely $[L : K]$ of those. □

**Theorem 11.14.** *If $L$ is any field, $G$ any finite group of automorphisms of $L$, and $K$ is its fixed field, then $L/K$ is finite and normal, with Galois group $G$.*

*Proof.* By Theorem 10.5, $[L : K] = |G| = n$, say. There are exactly $n$ distinct $K$-monomorphisms $L \to L$, namely, the elements of the Galois group.

We prove normality using Theorem 11.9. Thus let $N$ be an extension of $K$ containing $L$, and let $\tau$ be a $K$-monomorphism $L \to N$. Since every element of the Galois group of $L/K$ defines a $K$-monomorphism $L \to N$, the Galois group provides $n$ distinct $K$-monomorphisms $L \to N$, and these are automorphisms of $L$. But by Theorem 11.13 there are at most $n$ distinct $K$-monomorphisms $L \to N$, so $\tau$ must be one of these monomorphisms. Hence $\tau$ is an automorphism of $L$. Finally, $L/K$ is normal by Theorem 11.9. □

If the Galois correspondence is a bijection, then $K$ must be the fixed field of the Galois group of $L/K$, so by the above $L/K$ must be normal. That these hypotheses are also sufficient to make the Galois correspondence bijective (for subfields of $\mathbb{C}$) will be proved in Chapter 12. For general fields we need the additional concept of 'separability', see Chapter 17.

## EXERCISES

11.1 Suppose that $L/K$ is finite. Show that every $K$-monomorphism $L \to L$ is an automorphism. Does this result hold if the extension is not finite?

11.2 Construct the normal closure $N$ for the following extensions:

(a) $\mathbb{Q}(\alpha)/\mathbb{Q}$ where $\alpha$ is the real fifth root of 3

(b) $\mathbb{Q}(\beta)/\mathbb{Q}$ where $\beta$ is the real seventh root of 2

(c) $\mathbb{Q}(\sqrt{2}, \sqrt{3})/\mathbb{Q}$

(d) $\mathbb{Q}(\alpha, \sqrt{2})/\mathbb{Q}$ where $\alpha$ is the real cube root of 2

(e) $\mathbb{Q}(\gamma)/\mathbb{Q}$ where $\gamma$ is a zero of $t^3 - 3t^2 + 3$

11.3 Find the Galois groups of the extensions (a), (b), (c), (d) in Exercise 11.2.

11.4 Find the Galois groups of the extensions $N/\mathbb{Q}$ for their normal closures $N$.

11.5 Show that Lemma 11.8 fails if we do not assume that $N/K$ is normal but is true for any extension $N$ of $L$ such that $N/K$ is normal, rather than just for a normal closure.

11.6 Use Corollary 11.11 to find the order of the Galois group of the extension

$$\mathbb{Q}(\sqrt{3}, \sqrt{5}, \sqrt{7})/\mathbb{Q}$$

(*Hint:* Argue as in Example 6.8.)

11.7 Mark the following true or false.

(a) Every $K$-monomorphism is a $K$-automorphism.

(b) Every finite extension has a normal closure.

(c) If $K \subseteq L \subseteq M$ and $\sigma$ is a $K$-automorphism of $M$, then the restriction $\sigma|_L$ is a $K$-automorphism of $L$.

(d) An extension having Galois group of order 1 is normal.

(e) A finite normal extension has finite Galois group.

(f) Every Galois group is abelian (commutative).

(g) The Galois correspondence fails to be bijective for non-normal extensions.

(h) A finite normal extension inside $\mathbb{C}$, of degree $n$, has Galois group of order $n$.

(i) The Galois group of a normal extension is cyclic.

# Chapter 12

# The Galois Correspondence

We are at last in a position to establish the fundamental properties of the Galois correspondence between a field extension and its Galois group. Most of the work has already been done, and all that remains is to put the pieces together.

## 12.1 The Fundamental Theorem of Galois Theory

Recall a few points of notation from Chapter 8. Let $L/K$ be a field extension in $\mathbb{C}$ with Galois group $G$, which consists of all $K$-automorphisms of $L$. Let $\mathcal{F}$ be the set of intermediate fields, that is, subfields $M$ such that $K \subseteq M \subseteq L$, and let $\mathcal{G}$ be the set of all subgroups $H$ of $G$. We defined two maps

$$^* : \mathcal{F} \to \mathcal{G} \qquad ^\dagger : \mathcal{G} \to \mathcal{F}$$

as follows: if $M \in \mathcal{F}$, then $M^*$ is the group of all $M$-automorphisms of $L$. If $H \in \mathcal{G}$, then $H^\dagger$ is the fixed field of $H$. We observed in (8.4) that the maps $^*$ and $^\dagger$ reverse inclusions.

Before proceeding to the main theorem, we need a lemma:

**Lemma 12.1.** *Suppose that $L/K$ is a field extension, $M$ is an intermediate field, and $\tau$ is a $K$-automorphism of $L$. Then $\tau(M)^* = \tau M^* \tau^{-1}$.*

*Proof.* Let $M' = \tau(M)$, and take $\gamma \in M^*, x_1 \in M'$. Then $x_1 = \tau(x)$ for some $x \in M$. Compute:

$$(\tau\gamma\tau^{-1})(x_1) = \tau\gamma(x) = \tau(x) = x_1$$

so $\tau M^* \tau^{-1} \subseteq M'^*$. Similarly $\tau^{-1} M'^* \tau \subseteq M^*$, so $\tau M^* \tau^{-1} \supseteq M'^*$, and the lemma is proved. □

We are now ready to prove the main result:

**Theorem 12.2 (Fundamental Theorem of Galois Theory).** *If $L/K$ is a finite normal field extension inside $\mathbb{C}$, with Galois group $G$, and if $\mathcal{F}, \mathcal{G}, ^*, ^\dagger$ are defined as above, then:*

DOI: 10.1201/9781003213949-12

(1) *The Galois group $G$ has order $[L : K]$.*

(2) *The maps $^*$ and $^\dagger$ are mutual inverses, and set up an order-reversing one-to-one correspondence between $\mathcal{F}$ and $\mathcal{G}$.*

(3) *If $M$ is an intermediate field, then*

$$[L : M] = |M^*| \qquad [M : K] = |G|/|M^*|$$

(4) *An intermediate field $M$ is a normal extension of $K$ if and only if $M^*$ is a normal subgroup of $G$.*

(5) *If an intermediate field $M$ is a normal extension of $K$, then the Galois group of $M/K$ is isomorphic to the quotient group $G/M^*$.*

*Proof.* Part (1) is a restatement of Corollary 11.11.

For part (2), suppose that $M$ is an intermediate field, and let $[L : M] = d$. Then $|M^*| = d$ by Theorem 10.5. On the other hand, if $H$ is a subgroup of $G$ of order $d$, then $[L : H^\dagger] = d$ by Corollary 11.11. Hence the composite operators $*\dagger$ and $\dagger*$ preserve $[L : M]$ and $|H|$ respectively. (Because these operators are written on the right, the natural order for composing them is from left to right, so $M^{*\dagger} = (M^*)^\dagger$ and $H^{\dagger *} = (H^\dagger)^*$.)

From their definitions, $M^{*\dagger} \supseteq M$ and $H^{\dagger *} \supseteq H$. Therefore these inclusions are equalities.

For part (3), again note that $L/M$ is normal. Corollary 11.11 states that $[L : M] = |M^*|$, and the other equality follows immediately.

We now prove part (4). If $M/K$ is normal, let $\tau \in G$. Then $\tau|_M$ is a $K$-monomorphism $M \to L$, so is a $K$-automorphism of $M$ by Theorem 11.9. Hence $\tau(M) = M$. By Lemma 12.1, $\tau M^* \tau^{-1} = M^*$, so $M^*$ is a normal subgroup of $G$.

Conversely, suppose that $M^*$ is a normal subgroup of $G$. Let $\sigma$ be any $K$-monomorphism $M \to L$. By Theorem 11.3, there is a $K$-automorphism $\tau$ of $L$ such that $\tau|_M = \sigma$. Now $\tau M^* \tau^{-1} = M^*$ since $M^*$ is a normal subgroup of $G$, so by Lemma 12.1, $\tau(M)^* = M^*$. By part 2 of Theorem 12.2, $\tau(M) = M$. Hence $\sigma(M) = M$ and $\sigma$ is a $K$-automorphism of $M$. By Theorem 11.9, $M/K$ is normal.

Finally we prove part (5). Let $G'$ be the Galois group of $M/K$. We can define a map $\phi : G \to G'$ by

$$\phi(\tau) = \tau|_M \qquad \tau \in G$$

This is clearly a group homomorphism $G \to G'$, for by Theorem 11.9 $\tau|_M$ is a $K$-automorphism of $M$. By Theorem 11.3, $\phi$ is onto. The kernel of $\phi$ is obviously $M^*$, so by standard group theory

$$G' = \mathrm{im}(\phi) \cong G/\ker(\phi) = G/M^*$$

where im is the image and ker the kernel. □

The use of Theorem 10.5 in the proof of part (2) of Theorem 12.2 is crucial. Many of the most beautiful results in mathematics hang by equally slender threads.

Parts (4) and (5) of Theorem 12.2 can be generalised: see Exercise 12.2. Note that the proof of part (5) provides an explicit isomorphism between $\Gamma(M/K)$ and $G/M^*$, namely, restriction to $M$.

The importance of the Fundamental Theorem of Galois Theory derives from its potential as a tool rather than its intrinsic merit. It enables us to apply group theory to otherwise intractable problems about polynomials over $\mathbb{C}$ and associated subfields of $\mathbb{C}$, and we shall spend most of the remaining chapters exploiting such applications.

---

## EXERCISES

12.1 Work out the details of the Galois correspondence for the extension

$$\mathbb{Q}(\mathrm{i}, \sqrt{5})/\mathbb{Q}$$

whose Galois group is $G = \{I, R, S, T\}$ as in Chapter 8.

12.2 Let $L/K$ be a finite normal extension in $\mathbb{C}$ with Galois group $G$. Suppose that $M, N$ are intermediate fields with $M \subseteq N$. Prove that $N/M$ is normal if and only if $N^*$ is a normal subgroup of $M^*$. In this case prove that the Galois group of $N/M$ is isomorphic to $M^*/N^*$.

12.3* Let $\gamma = \sqrt{2+\sqrt{2}}$. Show that $\mathbb{Q}(\gamma)/\ \mathbb{Q}$ is normal, with cyclic Galois group. Show that $\mathbb{Q}(\gamma, \mathrm{i})=\mathbb{Q}(\mu)$ where $\mu^4 = \mathrm{i}$.

12.4* Find the Galois group of $t^6 - 7$ over $\mathbb{Q}$.

12.5* Find the Galois group of $t^6 - 2t^3 - 1$ over $\mathbb{Q}$.

12.6 Let $\zeta = \mathrm{e}^{\pi \mathrm{i}/6}$ be a primitive 12th root of unity. Find the Galois group $\Gamma(\mathbb{Q}(\zeta)/\mathbb{Q})$ as follows.

(a) Prove that $\zeta$ is a zero of the polynomial $t^4 - t^2 + 1$, and that the other zeros are $\zeta^5, \zeta^7, \zeta^{11}$.

(b) Prove that $t^4 - t^2 + 1$ is irreducible over $\mathbb{Q}$ and is the minimal polynomial of $\zeta$ over $\mathbb{Q}$.

(c) Prove that $\Gamma(\mathbb{Q}(\zeta)/\mathbb{Q})$ consists of four $\mathbb{Q}$-automorphisms $\phi_j$, defined by

$$\phi_j(\zeta) = \zeta^j \qquad j = 1, 5, 7, 11$$

(d) Prove that $\Gamma(\mathbb{Q}(\zeta)/\mathbb{Q}) \cong \mathbb{Z}_2 \times \mathbb{Z}_2$.

12.7 Using the subgroup structure of $\mathbb{Z}_2 \times \mathbb{Z}_2$ as in Exercise 12.6, find all intermediate fields between $\mathbb{Q}$ and $\mathbb{Q}(\zeta)$. (*Hint*: Calculate the fixed fields of the subgroups.)

12.8 Mark the following true or false.

(a) If $L/K$ is a finite normal extension inside $\mathbb{C}$, then the order of the Galois group of $L/K$ is equal to the dimension of $L$ considered as a vector space over $K$.

(b) If $M$ is any intermediate field of a finite normal extension inside $\mathbb{C}$, then $M^{\dagger *} = M$.

(c) If $M$ is any intermediate field of a finite normal extension inside $\mathbb{C}$, then $M^{*\dagger} = M$.

(d) If $M$ is any intermediate field of a finite normal extension $L/K$ inside $\mathbb{C}$, then the Galois group of $M/K$ is a subgroup of the Galois group of $L/K$.

(e) If $M$ is any intermediate field of a finite normal extension $L/K$ inside $\mathbb{C}$, then the Galois group of $L/M$ is a quotient of the Galois group of $L/K$.

# Chapter 13

# Worked Examples

The Fundamental Theorem of Galois theory is quite a lot to take in at one go, so it is worth spending some time thinking it through. We therefore analyse how the Galois correspondence works out on some extended examples. With the machinery currently at our disposal, these examples have to be relatively simple. In particular, most of these examples use polynomials whose zeros can be calculated explicitly, either in terms of radicals or the complex exponential function. In all cases we calculate the Galois group over the rationals $\mathbb{Q}$.

## 13.1 Examples of Galois Groups

**Example 13.1.** The first example is a simple warm-up problem. Let $f(t) = t^2 - 5$ over $\mathbb{Q}$. The zeros are $\pm\sqrt{5}$, so the splitting field over $\mathbb{Q}$ is $K = \mathbb{Q}(\sqrt{5})$, consisting of all $p + q\sqrt{5}$ where $p, q \in \mathbb{Q}$. That is, a basis for $K$ considered as a vector space over $\mathbb{Q}$ is $\{1, \sqrt{5}\}$. Since $K$ is the splitting field, $K/\mathbb{Q}$ is finite and normal. We are working in $\mathbb{C}$, so separability is automatic. The degree $[K : \mathbb{Q}] = 2$, so the Galois group $G = \Gamma(K/\mathbb{Q})$ has order 2. It must therefore be $\mathbb{S}_2 \cong \mathbb{Z}_2$, with elements

$$\begin{aligned} 1 : p + q\sqrt{5} &\mapsto p + q\sqrt{5} \\ \alpha : p + q\sqrt{5} &\mapsto p - q\sqrt{5} \end{aligned}$$

The subgroups of $G$ are $\mathbf{1}$ and $G$, with fixed fields $K$ and $\mathbb{Q}$ respectively. The order is reversed: $\mathbf{1} \subseteq G$ but $K \supseteq \mathbb{Q}$.

**Example 13.2.** Let $f(t) = t^2 - q$ over $\mathbb{Q}$, where $q$ is any element of $q$. The zeros are $\pm\sqrt{q}$, the splitting field is $K = \mathbb{Q}(\sqrt{q})$, and the Galois group $G \subseteq \mathbb{S}_2$. Therefore $G = \mathbf{1}$ or $G = \mathbb{S}_2$. Which subgroup occurs depends on a number-theoretic property of $q$. If $q$ is not the square of a rational number, then $\sqrt{q}$ is irrational, and $G = \mathbb{S}_2$. But if $q$ is the square of a rational number, then $\sqrt{q}$ is rational, and $G = \mathbf{1}$.

**Example 13.3.** Let $f(t) = t^4 - 5t^2 + 6$ over $\mathbb{Q}$. This factorises as $(t^2 - 2)(t^2 - 3)$ so its zeros are $\pm\sqrt{2}, \pm\sqrt{3}$. Therefore its splitting field over $\mathbb{Q}$ is $K = \mathbb{Q}(\sqrt{2}, \sqrt{3})$. Since $K$ is the splitting field, $K/\mathbb{Q}$ is finite and normal.

DOI: 10.1201/9781003213949-13

We are working in $\mathbb{C}$, so separability is automatic. The degree $[K:\mathbb{Q}] = 4$, so the Galois group $G = \Gamma(K/\mathbb{Q})$ has order 4. A basis for $K$ over $\mathbb{Q}$ is $\{1, \sqrt{2}, \sqrt{3}, \sqrt{6}\}$.

The minimal polynomial of $\sqrt{2}$ is $t^2 - 2$, so any $\gamma \in \Gamma$ maps $\sqrt{2}$ either to itself or to $-\sqrt{2}$. Similarly $\gamma$ maps $\sqrt{3}$ either to itself or to $-\sqrt{3}$. Combining these with all possible signs gives four distinct maps, which must all be $\mathbb{Q}$-automorphisms of $K$ because $|G| = 4$:

| Automorphism | Effect on $\sqrt{2}$ | Effect on $\sqrt{3}$ |
|---|---|---|
| 1 | $\sqrt{2}$ | $\sqrt{3}$ |
| $\alpha$ | $-\sqrt{2}$ | $\sqrt{3}$ |
| $\beta$ | $\sqrt{2}$ | $-\sqrt{3}$ |
| $\alpha\beta$ | $-\sqrt{2}$ | $-\sqrt{3}$ |

The effects on 1 and $\sqrt{6}$ can be deduced because these maps are $\mathbb{Q}$-automorphisms. For example,

$$\alpha(\sqrt{6}) = \alpha(\sqrt{2}\sqrt{3}) = \alpha(\sqrt{2})\alpha(\sqrt{3}) = -\sqrt{2}\sqrt{3} = -\sqrt{6}$$

Clearly $\alpha^2 = \beta^2 = 1$ and $\alpha, \beta$ commute, so $G \cong \mathbb{Z}_2 \times \mathbb{Z}_2$.

We now investigate the Galois correspondence. The group $G$ has five subgroups:

| | | |
|---|---|---|
| Order 4: | $G$ | $G \cong \mathbb{D}_4$ |
| Order 2: | $\{1, \alpha\}$ | $A \cong \mathbb{Z}_2$ |
| | $\{1, \beta\}$ | $B \cong \mathbb{Z}_2$ |
| | $\{1, \alpha\beta\}$ | $C \cong \mathbb{Z}_2$ |
| Order 1: | $\{1\}$ | $I \cong 1$ |

The inclusion relations between the subgroups of $G$ can be summed up by the *lattice diagram* of Figure 19 (left). In such diagrams, $X \subseteq Y$ if there is a sequence of upward-sloping lines from $X$ to $Y$.

Under the Galois correspondence we obtain the intermediate fields. Since the correspondence reverses inclusions, we obtain the lattice diagram in Figure 19 (right).

It is easy to spot the fixed fields:

$$I^\dagger = K \qquad A^\dagger = \mathbb{Q}(\sqrt{3}) \qquad B^\dagger = \mathbb{Q}(\sqrt{2}) \qquad C^\dagger = \mathbb{Q}(\sqrt{6}) \qquad G^\dagger = \mathbb{Q}$$

To find them systematically, rather than by inspired guesswork, we can apply the relevant elements $\gamma$ of the Galois group to a general field element

$$x = a + b\sqrt{2} + c\sqrt{3} + d\sqrt{6}$$

and solve the equation $\gamma(x) = x$ for $a, b, c, d$. We omit these calculations.

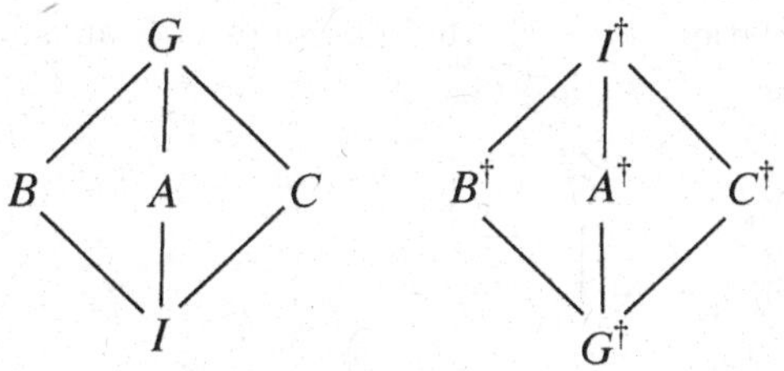


FIGURE 19: *Left*: Lattice of subgroups of $G = \mathbb{Z}_2 \times \mathbb{Z}_2$. *Right*: Lattice of fixed fields.

**Example 13.4.** Let $f(t) = t^3 - 1$ over $\mathbb{Q}$, and let $K \subseteq \mathbb{C}$ be the splitting field for $f$. The Galois group $G$ is a subgroup of $\mathbb{S}_3$. Since $f$ is cubic we might expect $G$ to be either $\mathbb{Z}_3$ or $\mathbb{S}_3$. However, $f$ is reducible: $f(t) = (t-1)(t^2+t+1)$. The zero $t = 1$ is rational, hence fixed by any element of the Galois group. The other two zeros are $\frac{1}{2}(-1 \pm \mathrm{i}\sqrt{3})$, so the splitting field is actually $K = \mathbb{Q}(\mathrm{i}\sqrt{3})$ and $[K : Q] = 2$. Thus $G = \mathbb{S}_2$, generated by the permutation $(2\,3)$ in cycle notation (assuming we number the zeros so that the first is 1).

**Example 13.5.** Let $f(t) = t^4 + t^3 + t^2 + t + 1$ over $\mathbb{Q}$. This is irreducible by Lemma 3.24. Its zeros are $\zeta, \zeta^2, \zeta^3, \zeta^4$ where $\zeta = \mathrm{e}^{2\pi\mathrm{i}/5}$ is a primitive 5th root of 1. In particular $\zeta^5 = 1$, so the splitting field of $f$ over $\mathbb{Q}$ is $K = \mathbb{Q}(\zeta)$. The degree $[K : \mathbb{Q}] = 4$, so the Galois group $G = \Gamma(K/\mathbb{Q})$ has order 4. A basis for $K$ over $\mathbb{Q}$ is $\{1, \zeta, \zeta^2, \zeta^3\}$.

The Galois group permutes these zeros, and it must also preserve products, so each $\mathbb{Q}$-automorphism is uniquely determined by its effect on $\zeta$. Thus there are four $\mathbb{Q}$-automorphisms:

| Automorphism | Effect on $\zeta$ |
|---|---|
| 1 | $\zeta$ |
| $\alpha$ | $\zeta^2$ |
| $\beta$ | $\zeta^3$ |
| $\gamma$ | $\zeta^4$ |

Apply $\alpha$ repeatedly to $\zeta$. What happens is:

$$\zeta \mapsto \zeta^2 \mapsto \zeta^4 \mapsto \zeta^8 = \zeta^3 \mapsto \zeta^6 = \zeta$$

so $\alpha^4 = 1$ and $G \cong \mathbb{Z}_4$ is cyclic. Moreover, $\alpha^2 = \gamma$ and this has order 2. Therefore $G$ has three subgroups:

| | | |
|---|---|---|
| Order 4: | $G$ | $G \cong \mathbb{Z}_4$ |
| Order 2: | $\{1, \gamma\}$ | $A \cong \mathbb{Z}_2$ |
| Order 1: | $\{1\}$ | $I \cong \mathbf{1}$ |

The inclusion relations between the subgroups of $G$ are shown in Figure 20 (left).

FIGURE 20: *Left*: Lattice of subgroups of $G = \mathbb{Z}_4$. *Right*: Lattice of fixed fields.

The Galois correspondence gives the intermediate fields, see Figure 20 (right). The fixed fields are obvious except perhaps for $A$. Consider a general element of $K$:

$$x = a + b\zeta + c\zeta^2 + d\zeta^3$$

Then

$$\gamma(x) = a + b\zeta^4 + c\zeta^3 + d\zeta$$

so $x = \gamma(x)$ if and only if $a = d, b = c$. Therefore $A^\dagger$ is spanned by $\zeta + \zeta^4$ and $\zeta^2 + \zeta^3$. Since $1 + \zeta + \zeta^2 + \zeta^3 + \zeta^4 = 0$, we have

$$\zeta^2 + \zeta^3 = 1 - (\zeta + \zeta^4)$$

so $A^\dagger = \mathbb{Q}(\zeta + \zeta^4)$. In summary:

$$\begin{aligned} I^\dagger &= K \\ A^\dagger &= \mathbb{Q}(\zeta + \zeta^4) \\ G^\dagger &= \mathbb{Q} \end{aligned}$$

**Example 13.6.** This example is very similar to the previous one. Let $f(t) = t^6 + t^5 + t^3 + t^2 + t + 1$ over $\mathbb{Q}$, and let $K \subseteq \mathbb{C}$ be the splitting field for $f$. The Galois group $G$ is a subgroup of $\mathbb{S}_6$. Let $\zeta = e^{2\pi i/7}$ be a primitive 7th root of unity. The zeros of $f$ are $\zeta, \zeta^2, \zeta^3, \zeta^4, \zeta^5, \zeta^6$. Since $\zeta^a\zeta^b = \zeta^{a+b}$, the splitting field is $K = \mathbb{Q}(\zeta)$, and $[K : \mathbb{Q}] = 6$.

If $\sigma$ is a $\mathbb{Q}$-automorphism of $K$ then $(\sigma(\zeta))^7 = \sigma(\zeta^7) = \sigma(1) = 1$. Clearly $\sigma(\zeta) \neq 1$, so $\sigma(\zeta) = \zeta^a$ where $1 \leq a \leq 6$. The automorphism property implies that $\sigma(\zeta^b) = \zeta^{ab}$. Define permutations $\sigma_a$ of the zeros by $\sigma_a(\zeta) = \zeta^a$, so $\sigma_a(\zeta^b) = \zeta^{ab}$.

Since $f$ is irreducible by Lemma 3.24, the conjugates of $\zeta$ are the six powers $\zeta^a$ where $1 \leq a \leq 6$, and the six maps $\sigma_a$ are $\mathbb{Q}$-automorphisms. Therefore $G = \{\sigma_a : 1 \leq a \leq 6\}$. This group is generated by $\sigma_3$, which has order 6, so $G \cong \mathbb{Z}_6$.

We leave finding the intermediate fields as an exercise: see Exercise 13.12.

**Example 13.7.** Let $f(t) = t^3 + 12t^2 + 3t + 6$ over $\mathbb{Q}$, and let $K$ be the splitting field for $f$ such that $K \subseteq \mathbb{C}$. We find the Galois group of $f$ without using explicit formulas for the zeros.

The coefficients are all positive so $f$ has exactly one real zero, which is simple. Let this zero be $a$. Eisenstein's criterion shows that $f$ is irreducible over $\mathbb{Q}$, so it is the minimal polynomial of $a$. Therefore $\{1, a, a^2\}$ is linearly independent over $\mathbb{Q}$.

The Galois group $G$ is a subgroup of $\mathbb{S}_3$, so it is (conjugate to) one of the subgroups $\mathbf{1}, \mathbb{Z}_2, \mathbb{A}_3 \cong \mathbb{Z}_3$, or $\mathbb{S}_3$. If $G = \mathbf{1}$ all zeros would be rational, which is false; if $G = \mathbb{Z}_2$ one zero would be rational, which is also false. If $G = \mathbb{Z}_3$ then $[K : \mathbb{Q}] = 3$ so $K$ has basis $\{1, a, a^2\}$. Every zero lies in $K$, so is of the form $p + qa + ra^2$ for $p, q, r \in \mathbb{Q}$. Therefore all three zeros would be real, which is false. The only possibility remaining is $G = \mathbb{S}_3$.

**Example 13.8.** Let $f(t) = t^5 + 15t^4 + 85t^3 + 225t^2 + 274t + 120$. This is a quintic, so the Galois group $G \subseteq \mathbb{S}_5$. However, $f$ factorises completely as

$$f(t) = (t+1)(t+2)(t+3)(t+4)(t+5)$$

and the zeros $-1, -2, -3, -4, -5$ are all rational. Therefore $G$ fixes all five zeros, and $G = \mathbf{1}$.

**Example 13.9.** This example is a favourite with writers on Galois theory, because of its archetypal quality: the Galois group of the splitting field of $t^4 - 2$ over $\mathbb{Q}$. We describe the calculations in stages.

(1) Let $f(t) = t^4 - 2$ over $\mathbb{Q}$, and let $K$ be the splitting field for $f$ such that $K \subseteq \mathbb{C}$. We can factorise $f$ as follows:

$$f(t) = (t - \xi)(t + \xi)(t - \mathrm{i}\xi)(t + \mathrm{i}\xi)$$

where $\xi = \sqrt[4]{2}$ is real and positive. Therefore $K = \mathbb{Q}(\xi, \mathrm{i})$. Since $K$ is the splitting field, $K/\mathbb{Q}$ is finite and normal; it is separable because we are working in $\mathbb{C}$.

(2) We find the degree of $K/\mathbb{Q}$. By the Tower Law,

$$[K : \mathbb{Q}] = [\mathbb{Q}(\xi, \mathrm{i}) : \mathbb{Q}(\xi)][\mathbb{Q}(\xi) : \mathbb{Q}]$$

The minimal polynomial of i over $\mathbb{Q}(\xi)$ is $t^2 + 1$, since $\mathrm{i}^2 + 1 = 0$ but $\mathrm{i} \notin \mathbb{R} \supseteq \mathbb{Q}(\xi)$. So $[\mathbb{Q}(\xi, \mathrm{i}) : \mathbb{Q}(\xi)] = 2$.

Now $\xi$ is a zero of $f$ over $\mathbb{Q}$, and $f$ is irreducible by Eisenstein's Criterion, Theorem 3.21. Hence $f$ is the minimal polynomial of $\xi$ over $\mathbb{Q}$, and $[\mathbb{Q}(\xi) : \mathbb{Q}] = 4$. Therefore

$$[K : \mathbb{Q}] = 2.4 = 8$$

(3) We find the elements of the Galois group of $K/\mathbb{Q}$. By a direct check, or by Corollary 5.13, there are $\mathbb{Q}$-automorphisms $\sigma, \tau$ of $K$ such that

$$\begin{aligned} \sigma(i) &= \mathrm{i} & \sigma(\xi) &= \mathrm{i}\xi \\ \tau(i) &= -\mathrm{i} & \tau(\xi) &= \xi \end{aligned}$$

Products of these yield eight distinct $\mathbb{Q}$-automorphisms of $K$:

| Automorphism | Effect on $\xi$ | Effect on i |
|---|---|---|
| $1$ | $\xi$ | i |
| $\sigma$ | $i\xi$ | i |
| $\sigma^2$ | $-\xi$ | i |
| $\sigma^3$ | $-i\xi$ | i |
| $\tau$ | $\xi$ | $-$i |
| $\sigma\tau$ | $i\xi$ | $-$i |
| $\sigma^2\tau$ | $-\xi$ | $-$i |
| $\sigma^3\tau$ | $-i\xi$ | $-$i |

Other products do not give new automorphisms, since $\sigma^4 = 1, \tau^2 = 1$, $\tau\sigma = \sigma^3\tau$, $\tau\sigma^2 = \sigma^2\tau$, $\tau\sigma^3 = \sigma\tau$. (The last two relations follows from the first three.) Any $\mathbb{Q}$-automorphism of $K$ sends i to some zero of $t^2+1$, so $i \mapsto \pm i$; similarly $\xi$ is mapped to $\xi, i\xi, -\xi$, or $-i\xi$. All possible combinations of these (eight in number) appear in the above list, so these are precisely the $\mathbb{Q}$-automorphisms of $K$.

(4) The abstract structure of the Galois group $G$ can be found. The generator-relation presentation

$$G = \langle \sigma, \tau : \sigma^4 = \tau^2 = 1, \ \tau\sigma = \sigma^3\tau \rangle$$

shows that $G$ is the dihedral group of order 8, which we write as $\mathbb{D}_4$. (In some books the notation $\mathbb{D}_8$ is used instead. It depends on what you think is important: the order is 8 or there is a normal subgroup $\mathbb{Z}_4$.)

The group $\mathbb{D}_4$ has a geometric interpretation as the symmetry group of a square. In fact we can label the four vertices of a square with the zeros of $t^4-2$, in such a way that the geometric symmetries are precisely the permutations of the zeros that occur in the Galois group (Figure 21).

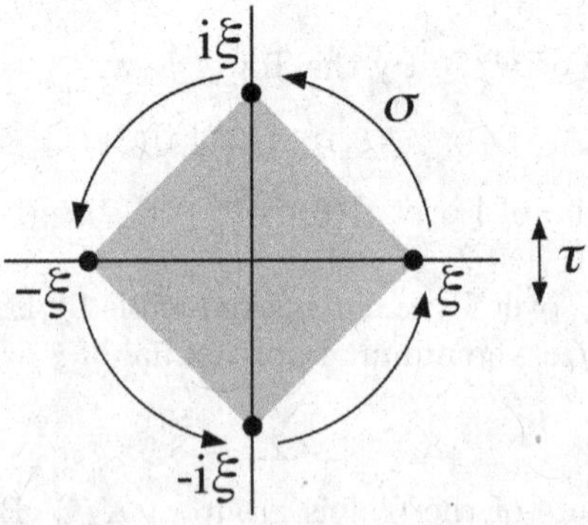


FIGURE 21: The Galois group $\mathbb{D}_4$ interpreted as the symmetry group of a square.

(5) It is a routine but lengthy exercise to find the subgroups of $G$:

| | | |
|---|---|---|
| Order 8: | $G$ | $G \cong \mathbb{D}_4$ |
| Order 4: | $\{1, \sigma, \sigma^2, \sigma^3\}$ | $S \cong \mathbb{Z}_4$ |
| | $\{1, \sigma^2, \tau, \sigma^2\tau\}$ | $T \cong \mathbb{Z}_2 \times \mathbb{Z}_2$ |
| | $\{1, \sigma^2, \sigma\tau, \sigma^3\tau\}$ | $U \cong \mathbb{Z}_2 \times \mathbb{Z}_2$ |
| Order 2: | $\{1, \sigma^2\}$ | $A \cong \mathbb{Z}_2$ |
| | $\{1, \tau\}$ | $B \cong \mathbb{Z}_2$ |
| | $\{1, \sigma\tau\}$ | $C \cong \mathbb{Z}_2$ |
| | $\{1, \sigma^2\tau\}$ | $D \cong \mathbb{Z}_2$ |
| | $\{1, \sigma^3\tau\}$ | $E \cong \mathbb{Z}_2$ |
| Order 1: | $\{1\}$ | $I \cong 1$ |

(6) The inclusion relations between the subgroups of $G$ can be summed up by the lattice diagram of Figure 22 (left).

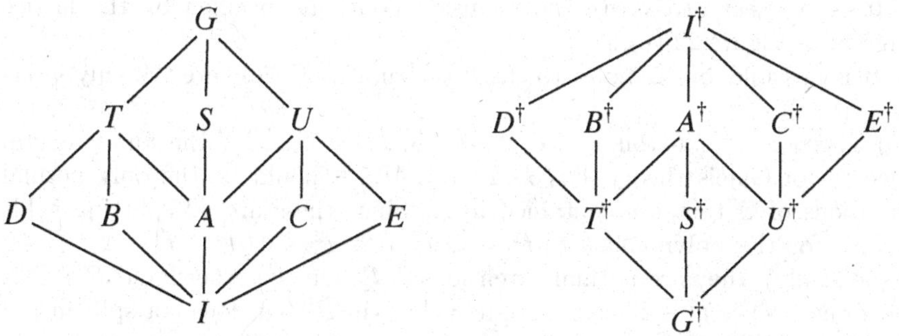


FIGURE 22: *Left*: Lattice of subgroups. *Right*: Lattice of subfields.

(7) Under the Galois correspondence we obtain the intermediate fields. Since the correspondence reverses inclusions, we obtain the lattice diagram in Figure 22 (right).

(8) We now describe the elements of these intermediate fields. There are three obvious subfields of $K$ of degree 2 over $\mathbb{Q}$, namely $\mathbb{Q}(\mathrm{i})$, $\mathbb{Q}(\sqrt{2})$, $\mathbb{Q}(\mathrm{i}\sqrt{2})$. These are clearly the fixed fields $S^\dagger, T^\dagger$, and $U^\dagger$, respectively. The other fixed fields are less obvious. To illustrate a possible approach we shall find $C^\dagger$. Any element of $K$ can be expressed uniquely in the form

$$x = a_0 + a_1\xi + a_2\xi^2 + a_3\xi^3 + a_4\mathrm{i} + a_5\mathrm{i}\xi + a_6\mathrm{i}\xi^2 + a_7\mathrm{i}\xi^3$$

where $a_0, \ldots, a_7 \in \mathbb{Q}$. Then

$$\begin{aligned} \sigma\tau(x) &= a_0 + a_1\mathrm{i}\xi - a_2\xi^2 - a_3\mathrm{i}\xi^3 - a_4\mathrm{i} + a_5(-\mathrm{i})\mathrm{i}\xi - a_6\mathrm{i}(\mathrm{i}\xi)^2 - a_7\mathrm{i}(\mathrm{i}\xi)^3 \\ &= a_0 + a_5\xi - a_2\xi^2 - a_7\xi^3 - a_4\mathrm{i} + a_1\mathrm{i}\xi + a_6\mathrm{i}\xi^2 - a_3\mathrm{i}\xi^3 \end{aligned}$$

The element $x$ is fixed by $\sigma\tau$ (and hence by $C$) if and only if

$$\begin{array}{llll} a_0 = a_0 & a_1 = a_5 & a_2 = -a_2 & a_3 = -a_7 \\ a_4 = -a_4 & a_5 = a_1 & a_6 = a_6 & a_7 = -a_3 \end{array}$$

Therefore $a_0$ and $a_6$ are arbitrary, while

$$a_2 = 0 = a_4 \qquad a_1 = a_5 \qquad a_3 = -a_7$$

It follows that

$$\begin{aligned} x &= a_0 + a_1(1+\mathrm{i})\xi + a_6\mathrm{i}\xi^2 + a_3(1-\mathrm{i})\xi^3 \\ &= a_0 + a_1[(1+\mathrm{i})\xi] + \frac{a_6}{2}[(1+\mathrm{i})\xi]^2 - \frac{a_3}{2}[(1+\mathrm{i})\xi]^3 \end{aligned}$$

which shows that

$$C^\dagger = \mathbb{Q}((1+\mathrm{i})\xi)$$

Similarly,

$$A^\dagger = \mathbb{Q}(\mathrm{i}, \sqrt{2}) \qquad B^\dagger = \mathbb{Q}(\xi) \qquad D^\dagger = \mathbb{Q}(\mathrm{i}\xi) \qquad E^\dagger = \mathbb{Q}((1-\mathrm{i})\xi)$$

It is now easy to verify the inclusion relations specified by the lattice diagram in Figure 22 (right).

(9) It is possible, but tedious, to check by hand that these are the only intermediate fields.

(10) The normal subgroups of $G$ are $G$, $S$, $T$, $U$, $A$, $I$. By the Fundamental Theorem of Galois theory, $G^\dagger, S^\dagger, T^\dagger, U^\dagger, A^\dagger, I^\dagger$ should be the only normal extensions of $\mathbb{Q}$ that are contained in $K$. Since these are all splitting fields over $\mathbb{Q}$, for the polynomials $t$, $t^2+1$, $t^2-2$, $t^2+2$, $t^4-t^2-2$, $t^4-2$ (respectively), they are normal extensions of $\mathbb{Q}$. On the other hand $B^\dagger/\mathbb{Q}$ is not normal, since $t^4-2$ has a zero, namely $\xi$, in $B^\dagger$ but does not split in $B^\dagger$. Similarly $C^\dagger, D^\dagger, E^\dagger$ are not normal extensions of $\mathbb{Q}$.

(11) According to the Fundamental Theorem of Galois theory, the Galois group of $A^\dagger/\mathbb{Q}$ is isomorphic to $G/A$. Now $G/A$ is isomorphic to $\mathbb{Z}_2 \times \mathbb{Z}_2$. We calculate directly the Galois group of $A^\dagger/\mathbb{Q}$. Since $A^\dagger = \mathbb{Q}(\mathrm{i}, \sqrt{2})$ there are four $\mathbb{Q}$-automorphisms:

| Automorphism | Effect on i | Effect on $\sqrt{2}$ |
|---|---|---|
| 1 | i | $\sqrt{2}$ |
| $\alpha$ | i | $-\sqrt{2}$ |
| $\beta$ | $-$i | $\sqrt{2}$ |
| $\alpha\beta$ | $-$i | $-\sqrt{2}$ |

and since $\alpha^2 = \beta^2 = 1$ and $\alpha\beta = \beta\alpha$, this group is $\mathbb{Z}_2 \times \mathbb{Z}_2$ as expected.

(12) The lattice diagrams for $\mathcal{F}$ and $\mathcal{G}$ do *not* look the same unless one of them is turned upside-down. Hence there does not exist a correspondence like the Galois correspondence but preserving inclusion relations. It may seem a little odd at first that the Galois correspondence reverses inclusions, but in fact it is entirely natural, and quite as useful a property as preservation of inclusions.

**Example 13.10.** The final example illustrates that calculating Galois groups can be a bit of a scramble, even when we have simple explicit formulas for the zeros. We omit some details and do not deal with all cases.

Let $c \in \mathbb{Q}$ and let $f(t) = t^4 - 2t^2 + c \in \mathbb{Q}[t]$. Since $f$ is a quadratic in $t^2$, The zeros of $f$ can be calculated using the quadratic formula and then taking a further square root. This gives the zeros in the form

$$\sqrt{1+\sqrt{1-c}} \quad \sqrt{1-\sqrt{1-c}} \quad -\sqrt{1+\sqrt{1-c}} \quad -\sqrt{1-\sqrt{1-c}} \tag{13.1}$$

We choose one of the two possible values for each $\sqrt{\phantom{x}}$; the other one is automatically minus this.

Suppose that $c$ is chosen to make $f(t)$ irreducible over $\mathbb{Q}$. (At this stage we do not describe which $c$ have this property, but Eisenstein's criterion shows that such $c$ exist: let $c$ be twice any odd integer.) By Proposition 9.14 the zeros of $f$ are distinct.

The nature of the zeros (13.1) depends on $c$. Routine calculations show that:

If $c < 0$ there are two real zeros and a complex conjugate pair.
If $0 \leq c \leq 1$ all zeros are real.
If $c > 1$ there are two complex conjugate pairs of zeros.

Since $f$ is assumed irreducible, Proposition 11.4 implies that $G$ is transitive (that is, it maps any zero to any other zero). Therefore the listed zeros are conjugates of each other. We can now make an educated guess at some of the $\mathbb{Q}$-automorphisms: change the signs of either square root in (13.1). That is, $\sqrt{1+\sqrt{1-c}}$ maps to any of the four possible zeros, and the others change sign in the same manner. To prove that these permutations specify $\mathbb{Q}$-automorphisms, observe that $G$ acts transitively on the zeros and consider possible algebraic relations among the zeros. It follows that $G$ has a subgroup isomorphic to $\mathbb{Z}_2 \times \mathbb{Z}_2$, where the first $\mathbb{Z}_2$ changes the sign of the innermost square root and the second $\mathbb{Z}_2$ changes the sign of the outermost square root.

There remains the possibility that other $\mathbb{Q}$-automorphisms exist. Indeed, since the coefficients of $f$ are real, there is at least one other 'obvious' $\mathbb{Q}$-automorphism: complex conjugation. Denote this permutation by $\kappa$. How $\kappa$ acts depends on $c$. If all roots are real, it acts trivially; if any are complex, it acts nontrivially. For simplicity, assume that $c < 0$ so two zeros are real and the other two form a complex conjugate pair. In this case, $\kappa$ fixes the real zeros and swaps the complex ones, so it is a 2-cycle.

It is an easy exercise to prove that the only subgroup of $\mathbb{S}_4$ that is isomorphic to $\mathbb{Z}_2 \times \mathbb{Z}_2$ is the Klein four-group $\mathbb{V}$. Therefore $G$ contains $\mathbb{V}$, but it might be larger. The only subgroups of $\mathbb{S}_4$, up to conjugacy, that contain $\mathbb{V}$ are $\mathbb{V}$ itself, the conjugates of $\mathbb{D}_4$ (which differ only in the numbering of the zeros), $\mathbb{A}_4$, and $\mathbb{S}_4$. We know that $G$ contains $\mathbb{V}$ and $\kappa$, and $\kappa$ is a 2-cycle. Since $\mathbb{V}$ contains no 2-cycles, $G$ must be larger than $\mathbb{V}$. The order of $G$ is divisible by 8 because $\mathbb{V}$ is a normal subgroup and $\kappa^2 = 1$, so $G$ is either (a conjugate of) $\mathbb{D}_4$ or it is the whole of $\mathbb{S}_4$.

The case $\mathbb{S}_4$ can be ruled out using results discussed in Chapter 22, exercise 22.8. (Specifically, the discriminant of $f$ is not a square in $\mathbb{Q}$, and the resolvent cubic of $f$ is reducible.) This implies that $G \cong \mathbb{D}_4$. The lattices of subgroups and fixed fields are again those shown in Figure 19. We leave the calculation of the three nontrivial fixed fields as an exercise, and note that the case $c \geq 0$ has not been dealt with.

---

## 13.2 Discussion

The examples above were chosen to illustrate various features of the Galois group.

Example 13.1 is simple and routine, but even then, we have to do some work to check that the obvious answer is correct.

Example 13.2 demonstrates that the Galois group can depend in an intricate manner on number-theoretic properties of the coefficients. It also shows that—even though we are working with concrete polynomials over $\mathbb{C}$—the geometry of $\mathbb{C}$ often plays a very minor role. Both the squares and the non-squares are dense in the non-negative rationals $\mathbb{Q}^+$, so the Galois group $G$ varies in a complicated and discontinuous manner as $q$ runs through $\mathbb{Q}^+$. When $q > 0$ the zeros $z_1 = \pm z_2$ lie on the real line $\mathbb{R}$, and $G$ is either trivial, or acts as reflection of $\mathbb{R}$ through the origin: $z_j \mapsto -z_j$. In contrast, when $q < 0$ the Galois group is always $\mathbb{S}_2$, which acts via complex conjugation $z_j \mapsto -\bar{z}_j$.

In Example 13.3 it is easy to guess the $\mathbb{Q}$-automorphisms, but even in this simple example we have to do some work to prove that these maps actually are $\mathbb{Q}$-automorphisms, and that there are no others. In this example the polynomial is reducible, and the Galois group splits up as a direct product of subgroups acting on the corresponding subsets of zeros.

Example 13.4 warns us not to jump to conclusions. This cubic equation has a Galois group of order 2, not 3 or 6.

Examples 13.4-13.6 illustrate the very special behaviour of 'cyclotomic polynomials', which we investigate in Chapter 21. These polynomials have a very small Galois group compared to $\mathbb{S}_m$ when the degree $m$ is large enough.

Example 13.7 shows that we can sometimes find the Galois group without having explicit formulas for the zeros. (Since this example is a cubic, such formulas exist, but we make no use of them and they would not be much help.) Trickery of this kind is useful for constructing interesting examples. In fact, we use a slightly subtler argument in Theorem 15.10 below to exhibit an explicit quintic equation that cannot be solved by radicals. However, such tricks are few and far between.

Example 13.8 is another warning not to jump to conclusions. It is a complicated-looking quintic, but its zeros are all rational and its Galois group is trivial.

Example 13.9 illustrates the need to check carefully that we have not missed out any $\mathbb{Q}$-automorphisms. Roots of unity, solutions of $t^n - 1 = 0$, lead to abelian Galois groups (for a general proof, see Theorem 21.9). It would be easy to assume that the same goes for roots of other complex numbers, solutions of $t^n - a = 0$ for $a \neq 1$. This example shows that even when $a$ is an integer, the Galois group can be non-abelian. In fact, if $f$ is a polynomial with real coefficients, complex conjugation is always an element of the Galois group. However, it leaves all real zeros fixed, so it acts trivially unless some zeros are non-real.

We might then wonder whether we missed complex conjugation out in Examples 13.5 and 13.6. Where has it gone? The answer is that when $\zeta$ is a $k$th root of unity, its complex conjugate $\bar{\zeta} = \zeta^{-1} = \zeta^{k-1}$ is a power of $\zeta$. So complex conjugation is already included.

Example 13.10 shows how complicated the calculation of the Galois group can be when the formula for the zeros involves nested radicals. Again, it is unwise to assume that our initial guess is correct. We have to consider the issue of reducibility, remember that $k$th roots have many values, ensure that what we think ought to be $\mathbb{Q}$-automorphisms really are, and make sure that we have not omitted any.

As a final remark: all of the above examples are somewhat contrived, because it is in general a difficult problem to compute the Galois group of a given polynomial—especially when there is no explicit representation of its zeros. Indeed, this is one reason why the mathematicians of the Paris Academy were unimpressed by Galois's memoir when he submitted it for a prize. However, a general theoretical method does exist, although it is not very practical for hand calculations. More recently, several computer algorithms have been developed that can deal with fairly complicated polynomials of high degree. See Chapter 22.

---

## EXERCISES

13.1 Find the Galois groups of the following extensions:

(a) $\mathbb{Q}(\sqrt{2}, \sqrt{5})/\mathbb{Q}$.

(b) $\mathbb{Q}(\omega)/\mathbb{Q}$ where $\omega = e^{2\pi i/3}$.

(c) $K/\mathbb{Q}$ where $K$ is the splitting field over $\mathbb{Q}$ for $t^4 - 3t^2 + 4$.

13.2 Find all subgroups of these Galois groups.

13.3 Find the corresponding fixed fields.

13.4 Find all normal subgroups of the above Galois groups.

13.5 Check that the corresponding extensions are normal.

13.6 Verify that the Galois groups of these normal extensions are the relevant quotient groups.

13.7* Consider the Galois group of $t^6 - 7$ over $\mathbb{Q}$, found in exercise 12.4. Use the Galois correspondence to find all intermediate fields.

13.8* Consider the Galois group of $t^6 - 2t^3 - 1$ over $\mathbb{Q}$, found in exercise 12.5. Use the Galois correspondence to find all intermediate fields.

13.9 Find the Galois group of $t^8 - \mathrm{i}$ over $\mathbb{Q}(\mathrm{i})$.

13.10 Find the Galois group of $t^8 + t^4 + 1$ over $\mathbb{Q}(\mathrm{i})$.

13.11 Use the Galois group $\mathbb{Z}_2 \times \mathbb{Z}_2 \times \mathbb{Z}_2$ of $\mathbb{Q}(\sqrt{2}, \sqrt{3}, \sqrt{5})/\mathbb{Q}$ to find all intermediate fields. Which of these are normal over $\mathbb{Q}$?

13.12 Find the intermediate fields in Example 13.6.

13.13* Let $f(t) = t^6 + 3t^4 - 4t^3 + 3t^2 + 12t + 5$. Using a computer algebra package if you wish, find the zeros of $f$ in $\mathbb{C}$ in terms of radicals. Show that the Galois group of $f$ over $\mathbb{Q}$ is isomorphic to $\mathbb{D}_3$. Use the Galois correspondence to calculate the fixed fields. (*Hint*: One of the zeros is $\sqrt[3]{2} + \mathrm{i}$.)

13.14 Let $f(t) \in \mathbb{Q}[t]$. If the Galois group of $f$ has odd order, prove that all zeros of $f$ are real.

13.15 Mark the following true or false.

(a) A $3 \times 3$ square has exactly 9 distinct symmetries.

(b) The symmetry group of a square is isomorphic to $\mathbb{Z}_8$.

(c) The symmetry group of a square is isomorphic to $\mathbb{S}_8$.

(d) The symmetry group of a square is isomorphic to a subgroup of $\mathbb{S}_8$.

(e) The group $\mathbb{D}_4$ has 10 distinct subgroups.

(f) The Galois correspondence preserves inclusion relations.

(g) The Galois correspondence reverses inclusion relations.

# Chapter 14

# Solubility and Simplicity

In order to apply the Galois correspondence, in particular to solving equations by radicals, we need to have at our fingertips a number of group-theoretic concepts and theorems. We have already assumed familiarity with elementary group theory: subgroups, normal subgroups, quotient groups, conjugates, permutations (up to cycle decomposition): to these we now add the standard isomorphism theorems. The relevant theory, along with most of the material in this chapter, can be found in any basic textbook on group theory, for example Fraleigh (1989), Humphreys (1996), or Neumann, Stoy, and Thompson (1994).

We start by defining soluble groups and proving some basic properties. These groups are of cardinal importance for the theory of the solution of equations by radicals. Next, we discuss simple groups, the main target being a proof of the simplicity of the alternating group of degree 5 or more. We end by proving Cauchy's Theorem: if a prime $p$ divides the order of a finite group, then the group has an element of order $p$.

## 14.1 Soluble Groups

Soluble groups were first defined and studied (though not in the current abstract way) by Galois in his work on the solution of equations by radicals. They have since proved extremely important in many branches of mathematics.

In the following definition, and thereafter, the notation $H \lhd G$ means that $H$ is a normal subgroup of the group $G$. Recall also that an *abelian* (or *commutative*) group is one in which $gh = hg$ for all elements $g, h$.

**Definition 14.1.** A group $G$ is *soluble* (in the US: *solvable*) if it has a finite series of subgroups

$$1 = G_0 \subseteq G_1 \subseteq \ldots \subseteq G_n = G \tag{14.1}$$

such that

(1) $G_i \lhd G_{i+1}$ for $i = 0, \ldots, n-1$.

(2) $G_{i+1}/G_i$ is abelian for $i = 0, \ldots, n-1$.

DOI: 10.1201/9781003213949-14

Condition (14.1) does not imply that $G_i \lhd G$, since $G_i \lhd G_{i+1} \lhd G_{i+2}$ does not imply $G_i \lhd G_{i+2}$. See Exercise 14.10.

**Examples 14.2.** (1) Every abelian group $G$ is soluble, with series $1 \lhd G$.
(2) The symmetric group $\mathbb{S}_3$ of degree 3 is soluble, since it has a cyclic normal subgroup of order 3 generated by the cycle $(1\,2\,3)$ whose quotient is cyclic of order 2. All cyclic groups are abelian.
(3) The dihedral group $\mathbb{D}_4$ of order 8 is soluble. In the notation of Chapter 13, it has a normal subgroup $S \cong \mathbb{Z}_4$ of order 4 whose quotient has order 2, and $S$ is abelian.
(4) The symmetric group $\mathbb{S}_4$ of degree 4 is soluble, having a series

$$1 \lhd \mathbb{V} \lhd \mathbb{A}_4 \lhd \mathbb{S}_4$$

where $\mathbb{A}_4$ is the alternating group of order 12, and $\mathbb{V}$ is the Klein four-group, which we recall consists of the permutations 1, (12)(34),(13)(24), (14)(23) and hence is a direct product of two cyclic groups of order 2. The quotient groups are

$$\begin{array}{ll} \mathbb{V}/1 \cong \mathbb{V} & \text{abelian of order 4} \\ \mathbb{A}_4/\mathbb{V} \cong \mathbb{Z}_3 & \text{abelian of order 3} \\ \mathbb{S}_4/\mathbb{A}_4 \cong \mathbb{Z}_2 & \text{abelian of order 2.} \end{array}$$

(5) The symmetric group $\mathbb{S}_5$ of degree 5 is not soluble. This follows from Lemma 8.11 with a bit of extra work. See Corollary 14.8.

We recall the following isomorphism theorems:

**Lemma 14.3.** *Let $G, H$, and $A$ be groups.*

(1) *If $H \lhd G$ and $A \subseteq G$ then $H \cap A \lhd A$ and*

$$\frac{A}{H \cap A} \cong \frac{HA}{H}$$

(2) *If $H \lhd G$, and $H \subseteq A \lhd G$ then $H \lhd A$, $A/H \lhd G/H$ and*

$$\frac{G/H}{A/H} \cong \frac{G}{A}$$

(3) *If $H \lhd G$ and $A/H \lhd G/H$ then $A \lhd G$.*

Parts (1) and (2) are respectively the *First* and *Second Isomorphism Theorems.* They are the translation into normal subgroup language of two straightforward facts: restricting a homomorphism to a subgroup yields a homomorphism, and composing two homomorphisms yields a homomorphism. See Exercise 14.11. Part (3) is a converse to part (2) and is easy to prove.

Judicious use of these isomorphism theorems lets us prove that soluble groups persist in being soluble even when subjected to quite drastic treatment.

**Theorem 14.4.** *Let $G$ be a group, $H$ a subgroup of $G$, and $N$ a normal subgroup of $G$.*

(1) *If $G$ is soluble, then $H$ is soluble.*

(2) *If $G$ is soluble, then $G/N$ is soluble.*

(3) *If $N$ and $G/N$ are soluble, then $G$ is soluble.*

*Proof.* (1) Let

$$1 = G_0 \lhd G_1 \lhd \ldots \lhd G_r = G$$

be a series for $G$ with abelian quotients $G_{i+1}/G_i$. Let $H_i = G_i \cap H$. Then $H$ has a series

$$1 = H_0 \lhd \ldots \lhd H_r = H$$

We show the quotients are abelian. Now

$$\frac{H_{i+1}}{H_i} = \frac{G_{i+1} \cap H}{G_i \cap H} = \frac{G_{i+1} \cap H}{G_i \cap (G_{i+1} \cap H)} \cong \frac{G_i(G_{i+1} \cap H)}{G_i}$$

by the first isomorphism theorem. But this latter group is a subgroup of $G_{i+1}/G_i$, which is abelian. Hence $H_{i+1}/H_i$ is abelian for all $i$, and $H$ is soluble.
(2) Take $G_i$ as before. Then $G/N$ has a series

$$N/N = G_0N/N \lhd G_1N/N \lhd \ldots \lhd G_rN/N = G/N$$

A typical quotient is

$$\frac{G_{i+1}N/N}{G_iN/N}$$

which by the second isomorphism theorem is isomorphic to

$$\frac{G_{i+1}N}{G_iN} = \frac{G_{i+1}(G_iN)}{G_iN} \cong \frac{G_{i+1}}{G_{i+1} \cap (G_iN)} \cong \frac{G_{i+1}/G_i}{(G_{i+1} \cap (G_iN))/G_i}$$

which is a quotient of the abelian group $G_{i+1}/G_i$, so is abelian. Therefore $G/N$ is soluble.
(3) There exist two series

$$1 = N_0 \lhd N_1 \lhd \ldots \lhd N_r = N$$
$$N/N = G_0/N \lhd G_1/N \lhd \ldots \lhd G_s/N = G/N$$

with abelian quotients. Consider the series of $G$ given by combining them:

$$1 = N_0 \lhd N_1 \lhd \ldots \lhd N_r = N = G_0 \lhd G_1 \lhd \ldots \lhd G_s = G$$

The quotients are either $N_{i+1}/N_i$ (which is abelian) or $G_{i+1}/G_i$, which is isomorphic to

$$\frac{G_{i+1}/N}{G_i/N}$$

and again is abelian. Therefore $G$ is soluble. □

A group $G$ is an *extension* of a group $A$ by a group $B$ if $G$ has a normal subgroup $N$ isomorphic to $A$ such that $G/N$ is isomorphic to $B$. We may sum up the three properties of the above theorem as: the class of soluble groups is closed under taking subgroups, quotients, and extensions. In contrast, the class of abelian groups is closed under taking subgroups and quotients, but not extensions. It is largely for this reason that Galois was led to define soluble groups.

---

## 14.2 Simple Groups

We turn to groups that are, in a sense, the opposite of soluble.

**Definition 14.5.** A group $G$ is *simple* if it is nontrivial and its only normal subgroups are 1 and $G$.

Every cyclic group $\mathbb{Z}_p$ of prime order is simple, since it has no subgroups other than 1 and $\mathbb{Z}_p$, hence in particular no other normal subgroups. These groups are also abelian, hence soluble. They are in fact the only soluble simple groups:

**Theorem 14.6.** *A soluble group is simple if and only if it is cyclic of prime order.*

*Proof.* Since $G$ is soluble group, it has a series

$$1 = G_0 \lhd G_1 \lhd \ldots \lhd G_n = G$$

where by deleting repeats we may assume $G_{i+1} \neq G_i$. Then $G_{n-1}$ is a proper normal subgroup of $G$. However, $G$ is simple, so $G_{n-1} = 1$ and $G = G_n/G_{n-1}$, which is abelian. Since every subgroup of an abelian group is normal, and every element of $G$ generates a cyclic subgroup, $G$ must be cyclic with no non-trivial proper subgroups. Hence $G$ has prime order.

The converse is trivial. □

Simple groups play an important role in finite group theory. They are in a sense the fundamental units from which all finite groups are made. Indeed the Jordan–Hölder theorem, which we do not prove, states that every finite group has a series of subgroups like (14.1) whose quotients are simple, and these simple groups depend only on the group and not on the series chosen.

We do not need to know much about simple groups, intriguing as they are. We require just one result:

**Theorem 14.7.** *If $n \geq 5$, then the alternating group $\mathbb{A}_n$ of degree $n$ is simple.*

*Proof.* We use much the same strategy as in Lemma 8.11, but we are proving a rather stronger property, so we have to work a bit harder.

Suppose that $1 \neq N \lhd \mathbb{A}_n$. Our strategy will be as follows: first, observe that if $N$ contains a 3-cycle then it contains all 3-cycles, and since the 3-cycles generate $\mathbb{A}_n$, we must have $N = \mathbb{A}_n$. Second, prove that $N$ must contain a 3-cycle. It is here that we need $n \geq 5$.

Suppose then, that $N$ contains a 3-cycle; without loss of generality $N$ contains $(1\,2\,3)$. Now for any $k > 3$ the cycle $(3\,2\,k)$ is an even permutation, so lies in $\mathbb{A}_n$, and therefore

$$(3\,2\,k)(1\,2\,3)(3\,2\,k)^{-1} = (1\,k\,2)$$

lies in $N$. Hence $N$ contains $(1\,k\,2)^2 = (1\,2\,k)$ for all $k \geq 3$. We claim that $\mathbb{A}_n$ is generated by all 3-cycles of the form $(1\,2\,k)$. If $n = 3$ then we are done. If $n > 3$ then for all $a, b > 2$ the permutation $(1\,a)(2\,b)$ is even, so lies in $\mathbb{A}_n$, and then $\mathbb{A}_n$ contains

$$(1\,a)(2\,b)(1\,2\,k)((1\,a)(2\,b))^{-1} = (abk)$$

if $k \neq a, b$. Since $\mathbb{A}_n$ is generated by all 3-cycles (Exercise 8.7), it follows that $N = \mathbb{A}_n$.

It remains to show that $N$ must contain at least one 3-cycle. We do this by an analysis into cases.

(1) Suppose that $N$ contains an element $x = abc\ldots$, where $a, b, c, \ldots$ are disjoint cycles and

$$a = (a_1 \ldots a_m) \quad (m \geq 4)$$

Let $t = (a_1\,a_2\,a_3)$. Then $N$ contains $t^{-1}xt$. Since $t$ commutes with $b, c, \ldots$ (disjointness of cycles) it follows that

$$t^{-1}xt = (t^{-1}at)bc\ldots = z \quad \text{(say)}$$

so that $N$ contains

$$zx^{-1} = (a_1\,a_3\,a_m)$$

which is a 3-cycle.

(2) Now suppose $N$ contains an element involving at least two 3-cycles. Without loss of generality $N$ contains

$$x = (1\,2\,3)(4\,5\,6)y$$

where $y$ is a permutation fixing 1, 2, 3, 4, 5, 6. Let $t = (2\,3\,4)$. Then $N$ contains

$$(t^{-1}xt)x^{-1} = (1\,2\,4\,3\,6)$$

Then by case (1) $N$ contains a 3-cycle.

(3) Now suppose that $N$ contains an element $x$ of the form $(i\,j\,k)p$, where $p$ is a product of 2-cycles disjoint from each other and from $(i\,j\,k)$. Then $N$ contains $x^2 = (i\,k\,j)$, which is a 3-cycle.

(4) There remains the case when every element of $N$ is a product of disjoint 2-cycles. (This actually occurs when $n = 4$, giving the four-group $\mathbb{V}$.) But as $n \geq 5$, we can assume that $N$ contains

$$x = (1\,2)(3\,4)p$$

where $p$ fixes 1, 2, 3, 4. If we let $t = (2\,3\,4)$ then $N$ contains

$$(t^{-1}xt)x^{-1} = (1\,4)(2\,3)$$

and if $u = (1\,4\,5)$ then $N$ contains

$$u^{-1}(t^{-1}xtx^{-1})u = (4\,5)(2\,3)$$

so that $N$ contains

$$(4\,5)(2\,3)(1\,4)(2\,3) = (1\,4\,5)$$

contradicting the assumption that every element of $N$ is a product of disjoint 2-cycles.

Hence $\mathbb{A}_n$ is simple if $n \geq 5$. □

In fact $\mathbb{A}_5$ is the smallest non-abelian simple group. As a historical aside, this result is often attributed to Galois. However, Neumann (2011), in his translation of Galois's mathematical writings, points out on pages 384–385 that Galois does not mention alternating groups in any significant work. In particular, his results that imply that the quintic cannot be solved by radicals use other special features of the Galois group of an equation of prime degree: see Neumann (2011) chapter IV. Neumann also suggests that the methods available to Galois are inadequate to eliminate various orders for a potential simple group, such as 56.

However, Galois did mention alternating groups in unpublished documents, and his letter to Chevalier states that the smallest simple group (excluding groups of prime order, that is, the simple abelian ones) has order 5.4.3, which is 60. So he definitely knew this result. The question is: could he have proved it? We discuss this point further in Chapter 25, and give a proof that no non-cyclic simple group has order less than 60, using methods that Galois definitely knew or could easily have proved. Order 56 is the trickiest case, but it can be done with a little ingenuity. Shedding some light on a historical puzzle, we also prove that any simple group of order 60 is isomorphic to $\mathbb{A}_5$, again using only methods that Galois knew or could easily have proved.

From Theorem 14.7 we deduce:

**Corollary 14.8.** *The symmetric group $\mathbb{S}_n$ of degree $n$ is not soluble if $n \geq 5$.*

*Proof.* If $\mathbb{S}_n$ were soluble then $\mathbb{A}_n$ would be soluble by Theorem 14.4, and simple by Theorem 14.7, hence of prime order by Theorem 14.6. But $|\mathbb{A}_n| = \frac{1}{2}(n!)$ is not prime if $n \geq 5$. □

## 14.3 Cauchy's Theorem

We next prove Cauchy's Theorem: if a prime $p$ divides the order of a finite group, then the group has an element of order $p$. We begin by recalling several ideas from group theory.

**Definition 14.9.** Elements $a$ and $b$ of a group $G$ are *conjugate* in $G$ if there exists $g \in G$ such that $a = g^{-1}bg$.

Conjugacy is an equivalence relation; the equivalence classes are the *conjugacy classes* of $G$.

If the conjugacy classes of $G$ are $C_1, \ldots, C_r$, then one of them, say $C_1$, contains only the identity element of $G$. Therefore $|C_1| = 1$. Since the conjugacy classes form a partition of $G$, we have

$$|G| = 1 + |C_2| + \cdots + |C_r| \tag{14.2}$$

which is the *class equation* for $G$.

**Definition 14.10.** If $G$ is a group and $x \in G$, then the *centraliser* $C_G(x)$ of $x$ in $G$ is the set of all $g \in G$ for which $xg = gx$. It is always a subgroup of $G$.

There is a useful connection between centralisers and conjugacy classes.

**Lemma 14.11.** *If $G$ is a group and $x \in G$, then the number of elements in the conjugacy class of $x$ is the index of $C_G(x)$ in $G$.*

*Proof.* The equation $g^{-1}xg = h^{-1}xh$ holds if and only if $hg^{-1}x = xhg^{-1}$, which means that $hg^{-1} \in C_G(x)$, that *is*, $h$ and $g$ lie in the same coset of $C_G(x)$ in $G$. The number of these cosets is the index of $C_G(x)$ in $G$, so the lemma is proved. □

**Corollary 14.12.** *The number of elements in any conjugacy class of a finite group $G$ divides the order of $G$.*

**Definition 14.13.** The *centre* $Z(G)$ of a group $G$ is the set of all elements $x \in G$ such that $xg = gx$ for all $g \in G$.

The centre of $G$ is a normal subgroup of $G$. Many groups have trivial centre, for example $Z(\mathbb{S}_3) = 1$. Abelian groups go to the other extreme and have $Z(G) = G$.

**Lemma 14.14.** *If $A$ is a finite abelian group whose order is divisible by a prime $p$, then $A$ has an element of order $p$.*

*Proof.* Use induction on $|A|$. If $|A|$ is prime the result follows. Otherwise take a proper subgroup $M$ of $A$ whose order $m$ is maximal. If $p$ divides $m$ we are home by induction, so we may assume that $p$ does not divide $m$. Let $b$ be in $A$

but not in $M$, and let $B$ be the cyclic subgroup generated by $b$. Then $MB$ is a subgroup of $A$, larger than $M$, so by maximality $A = MB$. From the First Isomorphism Theorem, Lemma 14.3(1),

$$|MB| = |M||B|/|M \cap B|$$

so $p$ divides the order $r$ of $B$. Since $B$ is cyclic, the element $b^{r/p}$ has order $p$. □

From this result we can derive a more general theorem of Cauchy in which the group need not be abelian:

**Theorem 14.15 (Cauchy's Theorem).** *If a prime $p$ divides the order of a finite group $G$, then $G$ has an element of order $p$.*

*Proof.* We prove the theorem by induction on the order $|G|$. The first few cases $|G| = 1, 2, 3$ are obvious. For the induction step, start with the class equation

$$|G| = 1 + |C_2| + \cdots + |C_r|$$

Since $p \,|\, |G|$, we must have $p \nmid |C_j|$ for some $j \geq 2$. If $x \in C_j$ it follows that $p \,|\, |C_G(x)|$, since $|C_j| = |G|/|C_G(x)|$.

If $C_G(x) \neq G$ then by induction $C_G(x)$ contains an element of order $p$, and this element also belongs to $G$.

Otherwise $C_G(x) = G$, which implies that $x \in Z(G)$, and by choice $x \neq 1$, so $Z(G) \neq 1$.

Either $p \,|\, |Z(G)|$ or $p \nmid |Z(G)|$. In the first case the proof reduces to the abelian case, Lemma 14.14. In the second case, by induction there exists $x \in G$ such that the image $\bar{x} \in G/Z(G)$ has order $p$. That is, $x^p \in Z(G)$ but $x \notin Z(G)$. Let $X$ be the cyclic group generated by $x$. Now $XZ(G)$ is abelian and has order divisible by $p$, so by Lemma 14.14 it has an element of order $p$, and again this element also belongs to $G$.

This completes the induction step, and with it the proof. □

Cauchy's Theorem does not work for composite divisors of $|G|$. See Exercise 14.6.

---

## EXERCISES

14.1 Show that the general dihedral group

$$\mathbb{D}_n = \langle a, b : a^n = b^2 = 1, b^{-1}ab = a^{-1} \rangle$$

is a soluble group. Here $a, b$ are generators and the equalities are relations between them.

14.2 Prove that $\mathbb{S}_n$ is not soluble for $n \geq 5$, using only the simplicity of $\mathbb{A}_5$.

14.3 Prove that a normal subgroup of a group is a union of conjugacy classes. Find the conjugacy classes of $\mathbb{A}_5$, using the cycle type of the permutations, and hence show that $\mathbb{A}_5$ is simple.

14.4 Prove that $\mathbb{S}_n$ is generated by the 2-cycles $(1\,2), \ldots, (1\,n)$.

14.5 If the point $\alpha \in \mathbb{C}$ is constructible by ruler and compasses, show that the Galois group of $\mathbb{Q}(\alpha) : \mathbb{Q}$ is soluble.

14.6 Show that $\mathbb{A}_5$ has no subgroup of order 15, even though 15 divides its order.

14.7 Show that $\mathbb{S}_n$ has trivial centre if $n \geq 3$.

14.8 Find the conjugacy classes of the dihedral group $\mathbb{D}_n$ defined in Exercise 14.1. Work out the centralisers of selected elements, one from each conjugacy class, and check Lemma 14.11.

14.9 If $G$ is a group and $x, g \in G$, show that $C_G(g^{-1}xg) = g^{-1}C_G(x)g$.

14.10 Show that the relation 'normal subgroup of' is not transitive. (*Hint:* Consider the subgroup $G \subseteq \mathbb{V} \subseteq \mathbb{S}_4$ generated by the element $(1\,2)(3\,4)$.)

14.11 There are (at least) two distinct ways to think about a group homomorphism. One is the definition as a structure-preserving mapping, the other is in terms of a quotient group by a normal subgroup. The relation between these is as follows. If $\phi : G \to H$ is a homomorphism then

$$\ker(\phi) \lhd G \quad \text{and} \quad G/\ker(\phi) \cong \operatorname{im}(\phi)$$

If $N \lhd G$ then there is a natural surjective homomorphism

$$\phi : G \to G/N \quad \text{with } \ker(\phi) = N$$

Show that the first and second isomorphism theorems are the translations into 'quotient group' language of two facts that are trivial in 'structure-preserving mapping' language:

(1) The restriction of a homomorphism to a subgroup is a homomorphism.

(2) The composition of two homomorphisms is a homomorphism.

14.12* By counting the sizes of conjugacy classes, prove that the group of rotational symmetries of a regular icosahedron is simple. Show that it is isomorphic to $\mathbb{A}_5$.

14.13 Mark the following true or false.

(a) The direct product of two soluble groups is soluble.

(b) Every simple soluble group is cyclic.

(c) Every cyclic group is simple.

(d) The symmetric group $\mathbb{S}_n$ is simple if $n \geq 5$.

(e) Every conjugacy class of a group $G$ is a subgroup of $G$.

# Chapter 15

## Solution by Radicals

The historical aspects of the problem of solving polynomial equations by radicals have been discussed in the introduction. The object of this chapter is to use the Galois correspondence to derive a condition that must be satisfied by any polynomial equation that is soluble by radicals, namely: the associated Galois group must be a soluble group. We then construct a quintic polynomial equation whose Galois group is not soluble, namely the disarmingly straightforward-looking $t^5 - 6t + 3 = 0$, which shows that the quintic equation cannot be solved by radicals.

Solubility of the Galois group is also a sufficient condition for an equation to be soluble by radicals, but we defer this result to Chapter 18.

### 15.1 Radical Extensions

Some care is needed in formalising the idea of 'solubility by radicals'. We begin from the point of view of field extensions.

Informally, a radical extension is obtained by a sequence of adjunctions of $n$th roots, for various $n$. For example, the following expression is radical:

$$\sqrt[3]{11}\sqrt[5]{\frac{7+\sqrt{3}}{2}} + \sqrt[4]{1+\sqrt[3]{4}} \tag{15.1}$$

To find an extension of $\mathbb{Q}$ that contains this element we may adjoin in turn elements

$$\alpha = \sqrt[3]{11} \qquad \beta = \sqrt{3} \qquad \gamma = \sqrt[5]{(7+\beta)/2} \qquad \delta = \sqrt[3]{4} \qquad \varepsilon = \sqrt[4]{1+\delta}$$

Recall Definition 8.12, which formalises the idea of a radical extension: $L/K$ is radical if $L = K(\alpha_1, \ldots, \alpha_m)$ where for each $j = 1, \ldots, m$ there exists $n_j$ such that

$$\alpha_j^{n_j} \in K(\alpha_1, \ldots, \alpha_{j-1}) \qquad (j \geq 1)$$

The elements $\alpha_j$ form a radical sequence for $L/K$, and the *radical degree* of $\alpha_j$ is $n_j$.

DOI: 10.1201/9781003213949-15

FIGURE 23: Galois thought he had solved the quintic...but changed his mind.

For example, the expression (15.1) is contained in a radical extension of the form $\mathbb{Q}(\alpha, \beta, \gamma, \delta, \varepsilon)$ of $\mathbb{Q}$, where $\alpha^3 = 11$, $\beta^2 = 3$, $\gamma^5 = (7 + \beta)/2$, $\delta^3 = 4$, $\varepsilon^4 = 1 + \delta$.

It is clear that any radical expression, in the sense of the introduction, is contained in some radical extension.

A polynomial should be considered soluble by radicals provided *all* of its zeros are radical expressions over the ground field.

**Definition 15.1.** Let $f$ be a polynomial over a subfield $K$ of $\mathbb{C}$, and let $\Sigma$ be the splitting field for $f$ over $K$. Then $f$ is *soluble by radicals* if there exists a field $M$ containing $\Sigma$ such that $M/K$ is a radical extension.

We emphasise that in the definition, we do not require the splitting field extension $\Sigma/K$ to be radical. There is a good reason for this. We want everything in the splitting field $\Sigma$ to be expressible by radicals, but it is pointless to expect everything expressible by the same radicals to be inside the splitting field. Moreover, if $M/K$ is radical and $L$ is an intermediate field, then $L/K$ need not be radical: see Exercise 15.6.

An essential feature is that we require *all* zeros of $f$ to be expressible by radicals. It is possible for some zeros to be expressible by radicals, while others are not—simply take a product of two polynomials, one soluble by radicals and one not. However, if an irreducible polynomial $f$ has one zero expressible by radicals, then all the zeros must be so expressible, by a simple argument based on Corollary 5.13.

The main theorem of this chapter is:

**Theorem 15.2.** *If $K$ is a subfield of $\mathbb{C}$ and $K \subseteq L \subseteq M \subseteq \mathbb{C}$ where $M/K$ is a radical extension, then the Galois group of $L/K$ is soluble.*

The otherwise curious word 'soluble' for groups arises in this context: a soluble (by radicals) polynomial has a soluble Galois group (of its splitting field over the base field).

The proof of this result is not entirely straightforward, and we must spend some time on preliminaries.

**Lemma 15.3.** *If $L/K$ is a radical extension in $\mathbb{C}$ and $M$ is the normal closure of $L/K$, then $M/K$ is radical.*

*Proof.* Let $L = K(\alpha_1, \ldots, \alpha_r)$ with $\alpha_i^{n_i} \in K(\alpha_1, \ldots, \alpha_{i-1})$. Let $f_i$ be the minimal polynomial of $\alpha_i$ over $K$. Then $M \supseteq L$ is clearly the splitting field of $\prod_{i=1}^{r} f_i$. For every zero $\beta_{ij}$ of $f_i$ in $M$ there exists an isomorphism $\sigma / K(\alpha_i) \to K(\beta_{ij})$ by Corollary 5.13. By Proposition 11.4, $\sigma$ extends to a $K$-automorphism $\tau : M \to M$. Since $\alpha_i$ is a member of a radical sequence for a subfield of $M$, so is $\beta_{ij}$. Combining the sequences yields a radical sequence for $M$. □

The next two lemmas show that certain Galois groups are abelian.

**Lemma 15.4.** *Let $K$ be a subfield of $\mathbb{C}$, and let $L$ be the splitting field for $t^p - 1$ over $K$, where $p$ is prime. Then the Galois group of $L/K$ is abelian.*

*Proof.* The derivative of $t^p - 1$ is $pt^{p-1}$, which is prime to $t^p - 1$, so by Lemma 9.13 the polynomial $t^p - 1$ has no multiple zeros in $L$. Clearly its zeros form a group under multiplication; this group has prime order $p$ since the zeros are distinct, so it is cyclic. Let $\varepsilon$ be a generator of this group. Then $L = K(\varepsilon)$,which implies that any $K$-automorphism of $L$ is determined by its effect on $\varepsilon$. Further, $K$-automorphisms permute the zeros of $t^p - 1$. Hence any $K$-automorphism of $L$ is of the form

$$\alpha_j : \varepsilon \mapsto \varepsilon^j$$

and is uniquely determined by this condition.

But then $\alpha_i \alpha_j$ and $\alpha_j \alpha_i$ both map $\varepsilon$ to $\varepsilon^{ij}$, so the Galois group is abelian. □

It is possible to determine the precise structure of the above Galois group, and to remove the condition that $p$ be prime. However, this involves extra work and is not needed at this stage. We return to this topic later in Theorem 21.9.

**Lemma 15.5.** *Let $K$ be a subfield of $\mathbb{C}$ in which $t^n - 1$ splits. Let $a \in K$, and let $L$ be a splitting field for $t^n - a$ over $K$. Then the Galois group of $L/K$ is abelian.*

*Proof.* Let $\alpha$ be any zero of $t^n - a$. Since $t^n - 1$ splits in $K$, the general zero of $t^n - a$ is $\varepsilon\alpha$ where $\varepsilon$ is a zero of $t^n - 1$ in $K$. Since $L = K(\alpha)$, any $K$-automorphism of $L$ is determined by its effect on $\alpha$. Given two $K$-automorphisms

$$\phi : \alpha \mapsto \varepsilon\alpha \qquad \psi : \alpha \mapsto \eta\alpha$$

where $\varepsilon$ and $\eta \in K$ are zeros of $t^n - 1$, then

$$\phi\psi(\alpha) = \varepsilon\eta\alpha = \eta\varepsilon\alpha = \psi\phi(\alpha)$$

As before, the Galois group is abelian. □

The main work in proving Theorem 15.2 is done in the next lemma.

**Lemma 15.6.** *If $K$ is a subfield of $\mathbb{C}$ and $L/K$ is normal and radical, then $\Gamma(L/K)$ is soluble.*

*Proof.* Suppose that $L = K(\alpha_1, \ldots, \alpha_n)$ with $\alpha_j^{n_j} \in K(\alpha_1, \ldots, \alpha_{j-1})$. By Proposition 8.9 we may assume that $n_j$ *is prime for all* $j$. In particular there is a prime $p$ such that $\alpha_1^p \in K$.

We prove the result by induction on $n$, using the additional hypothesis that all $n_j$ are prime. The case $n = 0$ is trivial, which gets the induction started.

If $\alpha_1 \in K$, then $L = K(\alpha_2, \ldots, \alpha_n)$ and $\Gamma(L/K)$ is soluble by induction.

We may therefore assume that $\alpha_1 \notin K$. Let $f$ be the minimal polynomial of $\alpha_1$ over $K$. Since $L/K$ is normal, $f$ splits in $L$; since $K \subseteq \mathbb{C}$, $f$ has no repeated zeros. Since $\alpha_1 \notin K$, the degree of $f$ is at least 2. Let $\beta$ be a zero of $f$ different from $\alpha_1$, and put $\varepsilon = \alpha_1/\beta$. Then $\varepsilon^p = 1$ and $\varepsilon \neq 1$. Thus $\varepsilon$ has order $p$ in the multiplicative group of $L$, so the elements $1, \varepsilon, \varepsilon^2, \ldots, \varepsilon^{p-1}$ are distinct $p$th roots of unity in $L$. Therefore $t^p - 1$ splits in $L$.

Let $M \subseteq L$ be the splitting field for $t^p - 1$ over $K$, that is, let $M = K(\varepsilon)$. Consider the chain of subfields $K \subseteq M \subseteq M(\alpha_1) \subseteq L$. The strategy of the remainder of the proof is illustrated in the following diagram:

$L$

$\longleftarrow \Gamma(L/M(\alpha_1))$ soluble by induction

$M(\alpha_1)$

$\longleftarrow \Gamma(M(\alpha_1)/M)$ abelian by Lemma 15.5

$M$

$\longleftarrow \Gamma(M/K)$ abelian by Lemma 15.4

$K$

Observe that $L/K$ is finite and normal, hence so is $L/M$, therefore Theorem 12.2 applies to $L/K$ and to $L/M$.

Since $t^p - 1$ splits in $M$ and $\alpha_1^p \in M$, the proof of Lemma 15.5 implies that $M(\alpha_1)$ is a splitting field for $t^p - \alpha_1^p$ over $M$. Thus $M(\alpha_1)/M$ is normal, and by Lemma 15.5 $\Gamma(M(\alpha_1)/M)$ is abelian. Apply Theorem 12.2 to $L/M$ to deduce that

$$\Gamma(M(\alpha_1)/M) \cong \Gamma(L/M)/\Gamma(L/M(\alpha_1))$$

Now

$$L = M(\alpha_1)(\alpha_2, \ldots, \alpha_n)$$

so that $L/M(\alpha_1)$ is a normal radical extension. By induction $\Gamma(L/M(\alpha_1))$ is soluble. Hence by Theorem 14.4(3), $\Gamma(L/M)$ is soluble.

Since $M$ is the splitting field for $t^p - 1$ over $K$, the extension $M/K$ is normal. By Lemma 15.4, $\Gamma(M/K)$ is abelian. Theorem 12.2 applied to $L/K$ yields

$$\Gamma(M/K) \cong \Gamma(L/K)/\Gamma(L/M)$$

Now Theorem 14.4(3) shows that $\Gamma(L/K)$ is soluble, completing the induction step. □

We can now complete the proof of the main result:

*Proof of Theorem* 15.2. Let $K_0$ be the fixed field of $\Gamma(L/K)$, and let $N/M$ be the normal closure of $M/K_0$. Then

$$K \subseteq K_0 \subseteq L \subseteq M \subseteq N$$

Since $M/K_0$ is radical, Lemma 15.3 implies that $N/K_0$ is a normal radical extension. By Lemma 15.6, $\Gamma(N/K_0)$ is soluble.

By Theorem 11.14, the extension $L/K_0$ is normal. By Theorem 12.2

$$\Gamma(L/K_0) \cong \Gamma(N/K_0)/\Gamma(N/L)$$

Theorem 14.4(2) implies that $\Gamma(L/K_0)$ is soluble. But $\Gamma(L/K) = \Gamma(L/K_0)$, so $\Gamma(L/K)$ is soluble. □

The idea of this proof is simple: a radical extension is a series of extensions by $n$th roots; such extensions have abelian Galois groups; so the Galois group of a radical extension is made up by fitting together a sequence of abelian groups. Unfortunately there are technical problems in carrying out the proof; we need to throw in roots of unity, and we have to make various extensions normal before the Galois correspondence can be used. These obstacles are similar to those encountered by Abel and overcome by his Theorem on Natural Irrationalities in Section 8.8.

Now we translate back from fields to polynomials, and in doing so revert to Galois's original viewpoint.

**Definition 15.7.** Let $f$ be a polynomial over a subfield $K$ of $\mathbb{C}$, with splitting field $\Sigma$ over $K$. The *Galois group* of $f$ over $K$ is the Galois group $\Gamma(\Sigma/K)$.

Let $G$ be the Galois group of a polynomial $f$ over $K$ and let $\partial f = n$. If $\alpha \in \Sigma$ is a zero of $f$, then $f(\alpha) = 0$, so for any $g \in G$

$$f(g(\alpha)) = g(f(\alpha)) = 0$$

Hence each element $g \in G$ induces a permutation $g'$ of the set of zeros of $f$ in $\Sigma$. Distinct elements of $G$ induce distinct permutations, since $\Sigma$ is generated by the zeros of $f$. It follows easily that the map $g \mapsto g'$ is a group monomorphism of $G$ into the group $\mathbb{S}_n$ of all permutations of the zeros of $f$. By identifying $G$ with its image in $\mathbb{S}_n$ we can think of $G$ as a group of permutations on the zeros of $f$. This, in effect, was how Galois thought of the Galois group, and for many years afterwards the only groups considered by mathematicians were permutation groups and groups of transformations of variables. Arthur Cayley was the first to propose a definition for an abstract group, although it seems that the earliest satisfactory axiom system for groups was given by Leopold Kronecker in 1870 (Huntingdon 1905).

We may restate Theorem 15.2 as:

**Theorem 15.8.** *Let $f$ be a polynomial over a subfield $K$ of $\mathbb{C}$. If $f$ is soluble by radicals, then the Galois group of $f$ over $K$ is soluble.* □

The converse also holds: see Theorem 18.22.

Thus to find a polynomial not soluble by radicals it suffices to find one whose Galois group is not soluble. There are two main ways of doing this. One is to look at the general polynomial of degree $n$, which we introduced in Chapter 8 Section 8.7, but this approach has the disadvantage that it does not show that there are specific polynomials with rational coefficients that are insoluble by radicals. The alternative approach, which we now pursue, is to exhibit a specific polynomial with rational coefficients whose Galois group is not soluble. Since Galois groups are hard to calculate, a little low cunning is necessary, together with some knowledge of the symmetric group.

---

## 15.2 An Insoluble Quintic

Watch carefully; there is nothing up my sleeve...

**Lemma 15.9.** *Let $p$ be a prime, and let $f$ be an irreducible polynomial of degree $p$ over $\mathbb{Q}$. Suppose that $f$ has precisely two non-real zeros in $\mathbb{C}$. Then the Galois group of $f$ over $\mathbb{Q}$ is isomorphic to the symmetric group $\mathbb{S}_p$.*

*Proof.* By the Fundamental Theorem of Algebra, Theorem 2.4, $\mathbb{C}$ contains the splitting field $\Sigma$ of $f$. Let $G$ be the Galois group of $f$ over $\mathbb{Q}$, considered as a permutation group on the zeros of $f$. These are distinct by Proposition 9.14, so $G$ is (isomorphic to) a subgroup of $\mathbb{S}_p$. When we construct the splitting field of $f$ we first adjoin an element of degree $p$, so $[\Sigma : \mathbb{Q}]$ is divisible by $p$. By Theorem 12.2(1), $p$ divides the order of $G$. By Cauchy's Theorem 14.15, $G$ has an element of order $p$. But the only elements of $\mathbb{S}_p$ having order $p$ are the $p$-cycles. Therefore $G$ contains a $p$-cycle.

Complex conjugation is a $\mathbb{Q}$-automorphism of $\mathbb{C}$, and therefore induces a $\mathbb{Q}$-automorphism of $\Sigma$. This leaves the $p-2$ real zeros of $f$ fixed, while transposing the two non-real zeros. Therefore $G$ contains a 2-cycle.

By choice of notation for the zeros, and if necessary taking a power of the $p$-cycle, we may assume that $G$ contains the 2-cycle $(1\,2)$ and the $p$-cycle $(1\,2\ldots p)$. We claim that these generate the whole of $\mathbb{S}_p$, which will complete the proof. To prove the claim, let $c = (1\,2\ldots p), t = (12)$, and let $H$ be the group generated by $c$ and $t$. Then $H$ contains $c^{-1}tc = (2\,3)$, hence $c^{-1}(2\,3)c = (3\,4), \ldots$ and hence all transpositions $(m\ m+1)$. Then $H$ contains

$$(1\,2)(2\,3)(1\,2) = (1\,3) \qquad (1\,3)(3\,4)(1\,3) = (1\,4)$$

and so on, and therefore contains all transpositions $(1\,m)$. Finally, $H$ contains all products $(1\,m)(1\,r)(1\,m) = (m\,r)$ with $1 < m < r$. But every element of $\mathbb{S}_n$ is a product of transpositions, so $H = \mathbb{S}_p$. Finally, $H \subseteq G \subseteq \mathbb{S}_p$, so $G = \mathbb{S}_p$. □

We can now exhibit a specific quintic polynomial over $\mathbb{Q}$ that is not soluble by radicals.

**Theorem 15.10.** *The polynomial $t^5 - 6t + 3$ over $\mathbb{Q}$ is not soluble by radicals.*

*Proof.* Let $f(t) = t^5 - 6t + 3$. By Eisenstein's Criterion, $f$ is irreducible over $\mathbb{Q}$. We shall show that $f$ has precisely three real zeros, each with multiplicity 1, and hence has two non-real zeros. Since 5 is prime, by Lemma 15.9 the Galois group of $f$ over $\mathbb{Q}$ is $\mathbb{S}_5$. By Corollary 14.8, $\mathbb{S}_5$ is not soluble. By Theorem 15.8, $f(t) = 0$ is not soluble by radicals.

It remains to show that $f$ has exactly three real zeros, each of multiplicity 1. Now $f(-2) = -17$, $f(-1) = 8$, $f(0) = 3$, $f(1) = -2$, and $f(2) = 23$. A rough sketch of the graph of $y = f(x)$ looks like Figure 24. This certainly appears to give only three real zeros, but we must be rigorous. By Rolle's theorem, the zeros of $f$ are separated by zeros of $Df$. Moreover, $Df = 5t^4 - 6$, which has two zeros at $\pm\sqrt[4]{6/5}$. Clearly $f$ and $Df$ are coprime, so $f$ has no repeated zeros (this also follows by irreducibility) so $f$ has at most three real zeros. But certainly $f$ has at least three real zeros, since a continuous function defined on the real line cannot change sign except by passing through 0. Therefore $f$ has precisely three real zeros, and the result follows. □

**Remark 15.11.** It might seem strange that such an unremarkable polynomial has such a large Galois group—any of the 120 permutations of the zeros. After all, the Galois group is the symmetry group. Why is this rather ordinary polynomial so symmetric? The answer is that $\mathbb{S}_n$ symmetry is the default for any 'typical' polynomial of degree $n$. For any polynomial, the elementary

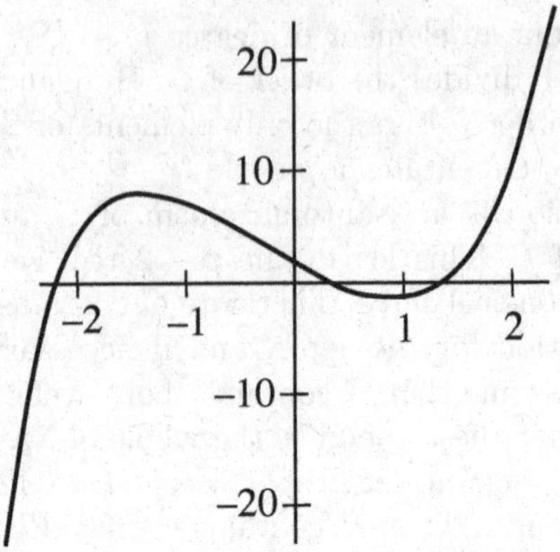


FIGURE 24: A quintic with three real zeros.

symmetric polynomials in its zeros give ± the coefficients. The Galois group is the subgroup that preserves any additional algebraic relations among the zeros. The more closely the zeros are related, the smaller the Galois group becomes.

---

## 15.3 Other Methods

Of course this is not the end of the story. There are more ways of killing a quintic than choking it with radicals. Having established the inadequacy of radicals for solving the problem, it is natural to look further afield.

First, some quintics *are* soluble by radicals. See Chapter 1 Section 1.4 and Berndt, Spearman and Williams (2002). What of the others, though?

On a mundane level, numerical methods can be used to find the zeros (real or complex) to any required degree of accuracy. In 1303 (see Joseph 2000) the Chinese mathematician Zhu Shijie wrote about what was later called Horner's method in the West; there it was long credited to the otherwise unremarkable William George Horner, who discovered it in 1819. For hand calculations it is a useful practical method, but there are many others. The mathematical theory of such numerical methods can be far from mundane—but from the algebraic point of view it is unilluminating.

Another way of solving the problem is to say, in effect, 'What's so special about radicals?' In 1786 the Swedish mathematician Erland Bring proved that the general quintic can be reduced to the form $x^5 + px + q$ without solving any equations of degree higher than cubic, but this theorem went unnoticed. Suppose for any real number $a$ we define the *ultraradical* of $a$ to be the unique real zero of $t^5 + t + a$. In his 3-volume treatise *Mathematical Researches* of

1832–1835, George Jerrard rediscovered Bring's result, presumably independently, and observed that it implies that the quintic equation can be solved using radicals and ultraradicals. See Kollros (1949) page 19 and King (1996); details are in Adamchik and Jeffrey (2003).

Instead of inventing new tools we can refashion existing ones. Charles Hermite made the remarkable discovery that the quintic equation can be solved in terms of 'elliptic modular functions', special functions of classical mathematics which arose in a quite different context, the integration of algebraic functions. The method is analogous to the trigonometric solution of the cubic equation, Exercise 1.8. In a triumph of mathematical unification, Klein (1913) succeeded in connecting together the quintic equation, elliptic functions, and the rotation group of the regular icosahedron. The latter is isomorphic to the alternating group $\mathbb{A}_5$, which we have seen plays a key part in the theory of the quintic. Klein's work helped to explain the unexpected appearance of elliptic functions in the theory of polynomial equations; these ideas were subsequently generalised by Henri Poincaré to cover polynomials of arbitrary degree.

---

## EXERCISES

15.1 Find radical extensions of $\mathbb{Q}$ containing the following elements of $\mathbb{C}$, by exhibiting suitable radical sequences (See Definition 8.12):

(a) $(\sqrt{11} - \sqrt[7]{23})/\sqrt[4]{5}$

(b) $(\sqrt{6} + 2\sqrt[3]{5})^4$

(c) $(2\sqrt[5]{5} - 4)/\sqrt{1 + \sqrt{99}}$

15.2 What is the Galois group of $t^p - 1$ over $\mathbb{Q}$ for prime $p$?

15.3 Show that the polynomials $t^5 - 4t + 2$, $t^5 - 4t^2 + 2$, $t^5 - 6t^2 + 3$, and $t^7 - 10t^5 + 15t + 5$ over $\mathbb{Q}$ are not soluble by radicals.

15.4 Solve the sextic equation

$$t^6 - t^5 + t^4 - t^3 + t^2 - t + 1 = 0$$

satisfied by a primitive 14th root of unity, in terms of radicals (*Hint:* Put $u = t + 1/t$.)

15.5 Solve the sextic equation

$$t^6 + 2t^5 - 5t^4 + 9t^3 - 5t^2 + 2t + 1 = 0$$

by radicals (*Hint:* Put $u = t + 1/t$.)

15.6* If $L/K$ is a radical extension in $\mathbb{C}$ and $M$ is an intermediate field, show that $M/K$ need not be radical.

15.7 If $p$ is an irreducible polynomial over $K \subseteq \mathbb{C}$ and at least one zero of $p$ is expressible by radicals, prove that every zero of $p$ is expressible by radicals.

15.8* If $K \subseteq \mathbb{C}$ and $\alpha^2 = a \in K$, $\beta^2 = b \in K$, and none of $a$, $b$, $ab$ are squares in $K$, prove that $K(\alpha, \beta)/K$ has Galois group $\mathbb{Z}_2 \times \mathbb{Z}_2$.

15.9* Show that if $N$ is an integer such that $|N| > 1$, and $p$ is prime, then the quintic equation

$$x^5 - Npx + p = 0$$

cannot be solved by radicals.

15.10* Suppose that a quintic equation $f(t) = 0$ over $\mathbb{Q}$ is irreducible, and has one real root and two complex conjugate pairs. Does an argument similar to that of Lemma 15.9 prove that the Galois group contains $\mathbb{A}_5$? If so, why? If not, why not?

15.11* Prove the Theorem on Natural Irrationalities using the Galois correspondence.

15.12 Bring and Jerrard proved that any quintic polynomial over $\mathbb{C}$ can be transformed to the form $f(t) = t^5 + pt + q$ using radical expressions. Show that the zeros of such a polynomial $f$ can be expressed using ultraradicals, and deduce that any quintic over $\mathbb{C}$ can be solved by a formula invoving radicals and ultraradicals.

15.13 Mark the following true or false.

(a) Every quartic equation over a subfield of $\mathbb{C}$ can be solved by radicals.

(b) Every radical extension is finite.

(c) Every finite extension is radical.

(d) The order of the Galois group of a polynomial of degree $n$ divides $n!$

(e) Any reducible quintic polynomial can be solved by radicals.

(f) There exist quartics with Galois group $\mathbb{S}_4$.

(g) An irreducible polynomial of degree 11 with exactly two non-real zeros has Galois group $\mathbb{S}_{11}$.

(h) The normal closure of a radical extension is radical.

(i) $\mathbb{A}_5$ has 120 elements.

# Chapter 16

## Abstract Rings and Fields

Having seen how Galois Theory works in the context assumed by its inventor, we can generalise everything to a much broader context. Instead of subfields of $\mathbb{C}$, we can consider arbitrary fields. This step goes back to Weber in 1895, but first achieved prominence in the work of Emil Artin in lectures of 1926, later published as Artin (1948). With the increased generality, new phenomena arise, and these must be dealt with.

One such phenomenon relates to the Fundamental Theorem of Algebra, which does not hold in an arbitrary field. We can get round this by constructing an analogue, the 'algebraic closure' of a field, in which *every* polynomial splits into linear factors. However, the construction of an algebraic closure requires the same machinery that we need to develop Galois theory for polynomial equations—that is, field extensions of finite degree. So we concentrate on developing that machinery, which centres on the abstract properties of field extensions, especially finite ones. For completeness we construct the algebraic closure of a field in Section 17.9.

A more significant problem is that a general field $K$ need not contain $\mathbb{Q}$ as a subfield. The reason is that sums $1 + 1 + \cdots + 1$ can behave in novel ways. In particular, such a sum may be zero. If it is, then the smallest number of 1s involved must be a prime $p$, and $K$ contains a subfield isomorphic to $\mathbb{Z}_p$, the integers modulo $p$. Such fields are said to have 'characteristic' $p$, and they introduce significant complications into the theory. The most important complication is that irreducible polynomials need not be separable; that is, they may have multiple zeros. Separability is automatic for subfields of $\mathbb{C}$, so it has not been seen to play a major role up to this point. However, behind the scenes at has been one of the two significant constraints that make Galois theory work, the other being normality. From now on, separability has to be taken a lot more seriously, and it has a substantial effect.

Rethinking the old results in the new context provides good revision and reinforcement, and it explains where the general concepts come from. Nonetheless, if you seriously work through the material and do not just accept that everything works, you will come to appreciate that the abstract approach has the advantage of generality and (in its own peculiar way) simplicity.

DOI: 10.1201/9781003213949-16

## 16.1 Rings and Fields

Today's concepts of 'ring' and 'field' are the brainchildren of Dedekind, who introduced them as a way of systematising algebraic number theory; their influence spread, reinforced by the growth of abstract algebra under the influence of Weber, Hilbert, Emmy Noether, and Bartel Leenert van der Waerden. These concepts are motivated by the observation that the classical number systems $\mathbb{Z}$, $\mathbb{Q}$, $\mathbb{R}$, and $\mathbb{C}$ enjoy a long list of useful algebraic properties. Specifically, $\mathbb{Z}$ is a 'ring' and the others are 'fields'.

The formal definition of a ring is:

**Definition 16.1.** A *ring* $R$ is a set, equipped with two operations of addition (denoted $a+b$) and multiplication (denoted $ab$), satisfying the following axioms:

(A1) $a+b=b+a$ for all $a,b\in R$.

(A2) $(a+b)+c=a+(b+c)$ for all $a,b,c\in R$.

(A3) There exists $0\in R$ such that $0+a=a$ for all $a\in R$.

(A4) Given $a\in R$, there exists $-a\in R$ such that $a+(-a)=0$.

(M1) $ab=ba$ for all $a,b\in R$.

(M2) $(ab)c=a(bc)$ for all $a,b,c\in R$.

(M3) There exists $1\in R$ such that $1a=a$ for all $a\in R$.

(D) $a(b+c)=ab+ac$ for all $a,b,c\in R$.

When we say that addition and multiplication are 'operations' on $R$, we automatically imply that if $a,b\in R$ then $a+b, ab\in R$, so $R$ is 'closed' under each of these operations. Some axiom systems for rings include these conditions as explicit axioms.

Axioms (A1) and (M1) are the *commutative laws* for addition and multiplication, respectively. Axioms (A2) and (M2) are the *associative laws* for addition and multiplication, respectively. Axiom (D) is the *distributive law.* The element 0 is called the *additive identity* or *zero element*; the element 1 is called the *multiplicative identity* or *unity element.* The element $-a$ is the *additive inverse* or *negative* of $a$. The word 'the' is justified here because 0 is unique, and for any given $a\in F$ the inverse $-a$ is unique. The condition $1\neq 0$ in (M3) excludes the trivial ring with one element.

The modern convention is that axioms (M1) and (M3) are optional for rings. Any ring that satisfies (M1) is said to be *commutative*, and any ring that satisfies (M3) is a *ring with* 1. However, in this book the phrase 'commutative ring with 1' is shortened to 'ring', because we do not require greater generality.

**Examples 16.2.** (1) The classical number systems $\mathbb{Z}$, $\mathbb{Q}$, $\mathbb{R}$, $\mathbb{C}$ are all rings.
(2) The set of natural numbers $\mathbb{N}$ is not a ring, because axiom (A4) fails.
(3) The set $\mathbb{Z}[\mathrm{i}]$ of all complex numbers of the form $a + bi$, with $a, b \in \mathbb{Z}$, is a ring.
(4) The set of polynomials $\mathbb{Z}[t]$ over $\mathbb{Z}$ is a ring, as the usual name 'ring of polynomials' indicates.
(5) The set of polynomials $\mathbb{Z}[t_1, \ldots, t_n]$ in $n$ indeterminates over $\mathbb{Z}$ is a ring.
(6) If $n$ is any integer, the set $\mathbb{Z}_n$ of integers modulo $n$ is a ring.

If $R$ is a ring, then we can define subtraction by

$$a - b = a + (-b) \qquad a, b \in R$$

The axioms ensure that all of the usual algebraic rules of manipulation, except those for division, hold in any ring.

Two extra axioms are required for a field:

**Definition 16.3.** A *field* is a ring $F$ satisfying the extra axioms

(M4) Given $a \in F$, with $a \neq 0$, there exists $a^{-1} \in F$ such that $aa^{-1} = 1$.

(M5) $1 \neq 0$.

Without condition (M5) the set $\{0\}$ would be a field with one element: this causes problems and is usually avoided.

We call $a^{-1}$ the *multiplicative inverse* of $a$. This inverse is also unique. If $F$ is a field, then we can define division by

$$a/b = ab^{-1} \qquad a, b \in F, b \neq 0$$

The axioms ensure that all the usual algebraic rules of manipulation, including those for division, hold in any field.

**Examples 16.4.** (1) The classical number systems $\mathbb{Q}$, $\mathbb{R}$, $\mathbb{C}$ are all fields.
(2) The set of integers $\mathbb{Z}$ is not a field, because axiom (M4) fails.
(3) The set $\mathbb{Q}[\mathrm{i}]$ of all complex numbers of the form $a + bi$, with $a, b \in \mathbb{Q}$, is a field.
(4) The set of polynomials $\mathbb{Q}[t]$ over $\mathbb{Q}$ is not a field, because axiom (M4) fails.
(5) The set of rational expressions $\mathbb{Q}(t)$ over $\mathbb{Q}$ is a field.
(6) The set of rational expressions $\mathbb{Q}(t_1, \ldots, t_n)$ in $n$ indeterminates over $\mathbb{Q}$ is a field.
(7) The set $\mathbb{Z}_2$ of integers modulo 2 is a field. The multiplicative inverse of the only nonzero element 1 is $1^{-1} = 1$. In this field, $1 + 1 = 0$. So $1 + 1 \neq 0$ does not count as one of the 'usual laws of algebra'. Note that it involves an inequality; the statement $1 + 1 = 2$ is true in $\mathbb{Z}_2$. What is not true is that $2 \neq 0$.
(8) The set $\mathbb{Z}_6$ of integers modulo 6 is not a field, because axiom (M4) fails. In fact, the elements $2, 3, 4$ do not have multiplicative inverses. Indeed, $2.3 = 0$

but $2, 3 \neq 0$, a phenomenon that cannot occur in a field: if $F$ is a field, and $a, b \neq 0$ in $F$ but $ab = 0$, then $a = abb^{-1} = 0b^{-1} = 0$, a contradiction.
(9) The set $\mathbb{Z}_5$ of integers modulo 5 is a field. The multiplicative inverses of the nonzero elements are $1^{-1} = 1, 2^{-1} = 3, 3^{-1} = 2, 4^{-1} = 4$. In this field, $1+1+1+1+1 = 0$, because $1+1 = 2$, $2+1 = 3$, $3+1 = 4$, and $4+1 = 0$.
(10) The set $\mathbb{Z}_1$ of integers modulo 1 is *not* a field. It consists of the single element 0, and so violates (M5) which states that $1 \neq 0$. This is a sensible convention since 1 is not prime.

The fields $\mathbb{Z}_2$ and $\mathbb{Z}_5$, or more generally $\mathbb{Z}_p$ where $p$ is prime (see Theorem 16.7 below), are prototypes for an entirely new kind of field, with unusual properties. For example, the formula for solving quadratic equations fails spectacularly over $\mathbb{Z}_2$. Suppose that we want to solve

$$t^2 + at + b = 0$$

where $a, b \in \mathbb{Z}_2$. Completing the square involves rewriting the equation in terms of $(t + a/2)$. But $a/2 = a/0$, which makes no sense. The standard quadratic formula involves division by 2 and also makes no sense. Nevertheless, many choices of $a, b$ here lead to soluble equations:

$$\begin{aligned} t^2 &= 0 \quad \text{has solution } t = 0 \\ t^2 + 1 &= 0 \quad \text{has solution } t = 1 \\ t^2 + t &= 0 \quad \text{has solutions } t = 0, 1 \\ t^2 + t + 1 &= 0 \quad \text{has no solution} \end{aligned}$$

---

## 16.2 General Properties of Rings and Fields

We briefly develop some of the basic properties of rings and fields, with emphasis on structural features that let us construct examples of fields. Among these features are the presence or absence of 'divisors of zero' (like $2, 3 \in \mathbb{Z}_6$), leading to the concept of an integral domain, and the notion of an ideal in a ring, leading to quotient rings and a general construction for interesting fields. Most readers will have encountered these ideas before; if not, it may be a good idea to find an introductory textbook and work through the first two or three chapters. For example, Fraleigh (1989) and Sharpe (1987) cover the relevant material.

**Definition 16.5.** (1) A *subring* of a ring $R$ is a non-empty subset $S$ of $R$ such that $1 \in S$ and if $a, b \in S$ then $a + b \in S$, $a - b \in S$, and $ab \in S$.
(2) A *subfield* of a field $F$ is a subset $S$ of $F$ containing the elements 0 and 1, such that if $a, b \in S$ then $a + b, a - b, ab \in S$, and further if $a \neq 0$ then $a^{-1} \in S$.

(3) An *ideal* of a ring $R$ is a subset $I \subseteq R$ such that if $a, b \in I$ then $-a$ and $a + b$ are in $I$, and if $i \in I$ and $r \in R$ then $ir$ and $ri$ lie in $I$.

Thus $\mathbb{Z}$ is a subring of $\mathbb{Q}$, and $\mathbb{R}$ is a subfield of $\mathbb{C}$, while the set $2\mathbb{Z}$ of even integers is an ideal of $\mathbb{Z}$. Because we assume (M3), an ideal is generally not a subring. It satisfies all ring axioms except (M3).

If $R, S$ are rings, then a *ring homomorphism* $\phi : R \to S$ is a map that satifies three conditions:

$$\phi(1) = 1 \quad \phi(r_1+r_2) = \phi(r_1)+\phi(r_2) \quad \phi(r_1r_2) = \phi(r_1)\phi(r_2) \quad \text{for all } r_1, r_2 \in R$$

The *kernel* $\ker \phi$ of $\phi$ is $\{r : \phi(r) = 0\}$. It is an ideal of $R$. An *isomorphism* is a homomorphism that is one-to-one and onto; a *monomorphism* is a homomorphism that is one-to-one. A homomorphism is a monomorphism if and only if its kernel is zero.

The most important property of an ideal is the possibility of working modulo that ideal, or, more abstractly, constructing the 'quotient ring' by that ideal. Specifically, if $I$ is an ideal of the ring $R$, then the *quotient ring* $R/I$ consists of the cosets $I + s$ of $I$ in $R$ (considering $R$ as a group under addition) The operations in the quotient ring are:

$$\begin{aligned}(I + r) + (I + s) &= I + (r + s) \\ (I + r)(I + s) &= I + (rs)\end{aligned}$$

where $r$, $s \in R$ and $I + r$ is the coset $\{i + r : i \in I\}$. The zero element is $I + 0 = I$, and the unity element is $I + 1$.

**Examples 16.6.** (1) Let $n\mathbb{Z}$ be the set of integers divisible by a fixed integer $n$. This is an ideal of $\mathbb{Z}$, and the quotient ring $\mathbb{Z}_n = \mathbb{Z}/n\mathbb{Z}$ is the ring of integers modulo $n$, that is, $\mathbb{Z}_n$.
(2) Let $R = K[t]$ where $K$ is a subfield of $\mathbb{C}$, and let $m(t)$ be an irreducible polynomial over $K$. Define $I = \langle m(t) \rangle$ to be the set of all multiples of $m(t)$. Then $I$ is an ideal, and $R/I$ is what we previously denoted by $K[t]/\langle m \rangle$ in Chapter 5. This quotient is a field.
(3) We can perform the same construction as in Example 2, without taking $m$ to be irreducible. We still get a quotient ring, but if $m$ is reducible the quotient is no longer a field.

When $I$ is an ideal of $R$, there is a natural ring homomorphism $\phi : R \to R/I$, defined by $\phi(r) = I + r$. Its kernel is $I$.

We shall need the following property of $\mathbb{Z}_n$, which explains the differences we found among $\mathbb{Z}_2$, $\mathbb{Z}_5$, and $\mathbb{Z}_6$.

**Theorem 16.7.** *The ring $\mathbb{Z}_n$ is a field if and only if $n$ is a prime number.*

*Proof.* First suppose that $n$ is not prime. If $n = 1$, then $\mathbb{Z}_n = \mathbb{Z}/\mathbb{Z}$, which has only one element and so cannot be a field. If $n > 1$ then $n = rs$ where $r$ and $s$ are integers less than $n$. Putting $I = n\mathbb{Z}$,

$$(I + r)(I + s) = I + rs = I$$

But $I$ is the zero element of $\mathbb{Z}/I$, while $I+r$ and $I+s$ are non-zero. Since in a field the product of two non-zero elements is non-zero, $\mathbb{Z}/I$ cannot be a field.

Now suppose that $n$ is prime. Let $I+r$ be a non-zero element of $\mathbb{Z}/I$. Then $r$ and $n$ are coprime, so by standard properties of $\mathbb{Z}$ there exist integers $a$ and $b$ such that $ar+bn=1$. Therefore

$$(I+a)(I+r) = (I+1) - (I+n)(I+b) = I+1$$

and similarly

$$(I+r)(I+a) = I+1$$

Since $I+1$ is the identity element of $\mathbb{Z}/I$, we have found a multiplicative inverse for the given element $I+r$. Thus every non-zero element of $\mathbb{Z}/I$ has an inverse, so that $\mathbb{Z}_n = \mathbb{Z}/I$ is a field. □

From now on, when dealing with $\mathbb{Z}_n$, we revert to the usual convention and write the elements as $0, 1, 2, \ldots, n-1$ rather than $I, I+1, I+2, \ldots, I+n-1$.

---

## 16.3 Polynomials Over General Rings

We now introduce polynomials with coefficients in a given ring. The main point to bear in mind is that identifying polynomials with functions, as we cheerfully did in Chapter 2 for coefficients in $\mathbb{C}$, is no longer a good idea, because Proposition 2.3, which states that polynomials defining the same function are equal, need not be true when the coefficients belong to a general ring.

Indeed, consider the ring $\mathbb{Z}_2$. Suppose that $f(t) = t^2+1$ and $g(t) = t^4+t^3+t+1$. There are numerous reasons to want these to be different polynomials, the most obvious being that they have different coefficients and different degrees. But if we interpret them as functions from $\mathbb{Z}_2$ to itself, we find that $f(0) = 1 = g(0)$ and $f(1) = 0 = g(1)$. As functions, $f$ and $g$ are equal.

It turns out that this problem arises because the ring is finite. Since finite rings (especially finite fields) are important, we need a definition of 'polynomial' that does not rely on interpreting it as a function. We did this in Section 2.1 for polynomials over $\mathbb{C}$, and the same idea works for any ring.

To be specific, let $R$ be a ring. We define a *polynomial over $R$ in the indeterminate $t$* to be an expression

$$r_0 + r_1 t + \cdots + r_n t^n$$

where $r_0, \ldots, r_n \in R$, $0 \leq n \in \mathbb{Z}$, and $t$ is undefined. Again, for set-theoretic purity we can replace such an expression by the sequence $(r_0, \ldots, r_n)$, as in Exercise 2.2. The elements $r_0, \ldots, r_n$ are the *coefficients* of the polynomial.

Two polynomials are defined to be equal if and only if the corresponding coefficients are equal (with the understanding that powers of $t$ not occurring in the polynomial may be taken to have zero coefficient). The sum and the product of two polynomials are defined using the same formulas (2.1, 2.2) as in Section 2.1, but now the $r_i$ belong to a general ring. It is straightforward to check that the set of all polynomials over $R$ in the indeterminate $t$ is a ring—the *ring of polynomials over $R$ in the indeterminate $t$*. As before, we denote this by the symbol $R[t]$. We can also define polynomials in several indeterminates $t_1, t_2, \ldots$ and obtain the polynomial ring $R[t_1, t_2, \ldots]$. Again, each polynomial $f \in R[t]$ defines a function from $R$ to $R$. We use the same symbols $f$, to denote this function. If $f(t) = \sum r_i t^i$ then $f(\alpha) = \sum r_i \alpha^i$, for $\alpha \in R$. We reiterate that two distinct polynomials over $R$ may give rise to the same function on $R$.

Proposition 2.3 is still true when $R = \mathbb{R}, \mathbb{Q}$, or $\mathbb{Z}$, with the same proof. The definition of 'degree' also applies to these rings, as does the proof of Proposition 2.2.

## 16.4 The Characteristic of a Field

In Proposition 4.4 we observed that every subfield of $\mathbb{C}$ must contain $\mathbb{Q}$. The main step in the proof was that the subfield contains all elements $1+1+\cdots+1$, that is, it contains $\mathbb{N}$, hence $\mathbb{Z}$, hence $\mathbb{Q}$.

The same idea *nearly* works for any field. However, a finite field such as $\mathbb{Z}_5$ cannot contain $\mathbb{Q}$, or even anything isomorphic to $\mathbb{Q}$, because $\mathbb{Q}$ is infinite. How does the proof fail? As we have already seen in Example 16.4(9), the equation $1+1+1+1+1 = 0$ holds in $\mathbb{Z}_5$. So we can build up a unique smallest subfield just as before—but now it need not be isomorphic to $\mathbb{Q}$.

Pursuing this line of thought leads to:

**Definition 16.8.** The *prime subfield* of a field $K$ is the intersection of all subfields of $K$.

It is easy to see that the intersection of any collection of subfields of $K$ is a subfield (the intersection is not empty since every subfield contains 0 and 1), and therefore the prime subfield of $K$ is the *unique* smallest subfield of $K$. The fields $\mathbb{Q}$ and $\mathbb{Z}_p$ ($p$ prime) have no proper subfields, so are equal to their prime subfields. The next theorem shows that these are the only fields that can occur as prime subfields.

**Theorem 16.9.** *For every field $K$, the prime subfield of $K$ is isomorphic either to the field $\mathbb{Q}$ of rationals or the field $\mathbb{Z}_p$ of integers modulo a prime number $p$.*

*Proof.* Let $K$ be a field, $P$ its prime subfield. Then $P$ contains 0 and 1, and therefore contains the elements $n^*(n \in \mathbb{Z})$ defined by

$$n^* = \begin{cases} 1+1+\ldots+1 \ (n \text{ times}) & \text{if } n > 0 \\ 0 & \text{if } n = 0 \\ -(-n)^* & \text{if } n < 0 \end{cases}$$

A short calculation using the distributive law (D) and induction shows that the map $^* : \mathbb{Z} \to P$ so defined is a ring homomorphism. Two distinct cases arise.

(1) $n^* = 0$ for some $n \neq 0$. Since also $(-n)^* = 0$, there exists a smallest positive integer $p$ such that $p^* = 0$. If $p$ is composite, say $p = rs$ where $r$ and $s$ are smaller positive integers, then $r^*s^* = p^* = 0$, so either $r^* = 0$ or $s^* = 0$, contrary to the definition of $p$. Therefore $p$ is prime. The elements $n^*$ form a ring isomorphic to $\mathbb{Z}_p$, which is a field by Theorem 16.7. This must be the whole of $P$, since $P$ is the smallest subfield of $K$.

(2) $n^* \neq 0$ if $n \neq 0$. Then $P$ must contain all the elements $m^*/n^*$ where $m$, $n$ are integers and $n \neq 0$. These form a subfield isomorphic to $\mathbb{Q}$ (by the map which sends $m^*/n^*$ to $m/n$), which is necessarily the whole of $P$. □

The distinction among possible prime subfields is summed up by:

**Definition 16.10.** The *characteristic* of a field $K$ is 0 if the prime subfield of $K$ is isomorphic to $\mathbb{Q}$, and $p$ if the prime subfield of $K$ is isomorphic to $\mathbb{Z}_p$.

For example, the fields $\mathbb{Q}$, $\mathbb{R}$, $\mathbb{C}$ all have characteristic zero, since in each case the prime subfield is $\mathbb{Q}$. The field $\mathbb{Z}_p$ ($p$ prime) has characteristic $p$. We shall see later that there are other fields of characteristic $p$: for an example, see Exercise 16.7.

The elements $n^*$ defined in the proof of Theorem 16.9 are of considerable importance in what follows. It is conventional to omit the asterisk and write $n$ instead of $n^*$. This abuse of notation will cause no confusion as long as it is understood that $n$ may be zero in the field without being zero as an integer. Thus in $\mathbb{Z}_5$ we have $10 = 0$ and $2 = 7 = -3$. This difficulty does not arise in fields of characteristic zero.

With this convention, a product $nk$ ($n \in \mathbb{Z}$, $k \in K$) makes sense, and

$$nk = \pm(k + \cdots + k)$$

where $k$ appears $|n|$ times and the $\pm$ sign is the sign of $n$.

**Lemma 16.11.** *If $K$ is a subfield of $L$, then $K$ and $L$ have the same characteristic.*

*Proof.* In fact, $K$ and $L$ have the same prime subfield. □

**Lemma 16.12.** *If $k$ is a non-zero element of the field $K$, and if $n$ is an integer such that $nk = 0$, then $n$ is a multiple of the characteristic of $K$.*

*Proof.* We must have $n = 0$ in $K$, that is, in old notation, $n^* = 0$. If the characteristic is 0, this implies that $n = 0$ (as an integer). If the characteristic is $p > 0$, it implies that $n$ is a multiple of $p$. □

---

## 16.5 Integral Domains

The ring $\mathbb{Z}$ has an important property, which is shared by many of the other rings that we shall be studying: if $mn = 0$ where $m, n$ are integers, then $m = 0$ or $n = 0$. We abstract this property as:

**Definition 16.13.** A ring $R$ is an *integral domain* if $rs = 0$, for $r, s \in R$, implies that $r = 0$ or $s = 0$.

We often express this condition as '$D$ has no zero-divisors', where a *zero-divisor* is a non-zero element $a \in D$ for which there exists a non-zero element $b \in D$ such that $ab = 0$.

**Examples 16.14.** (1) The integers $\mathbb{Z}$ form an integral domain.
(2) Any field is an integral domain. For suppose $K$ is a field and $rs = 0$. Then either $s = 0$, or $r = rss^{-1} = 0s^{-1} = 0$.
(3) The ring $\mathbb{Z}_6$ is not an integral domain. As observed earlier, in this ring $2.3 = 0$ but $2, 3 \neq 0$.
(4) The polynomial ring $\mathbb{Z}[t]$ is an integral domain. One way to prove this is to consider the corresponding functions. Recall that the zero polynomial is denoted by $\mathbf{0}$. If $f(t)g(t) = \mathbf{0}$ as polynomials, but $f(t), g(t) \neq \mathbf{0}$, then we can find an element $x \in \mathbb{Z}$ such that $f(x) \neq 0, g(x) \neq 0$. (Just choose $x$ different from the finite set of zeros of $f$ together with zeros of $g$.) But then $f(x)g(x) \neq \mathbf{0}$, a contradiction.

A more elegant and more general proof for Example 16.14(4) leads to Lemma 16.17 below.

It turns out that a ring is an integral domain if and only if it is (isomorphic to) a subring of some field. To understand how this comes about, we analyse when it is possible to *embed* a ring $R$ in a field—that is, find a field containing a subring isomorphic to $R$. Thus $\mathbb{Z}$ can be embedded in $\mathbb{Q}$. This particular example has the property that every element of $\mathbb{Q}$ is a fraction whose numerator and denominator lie in $\mathbb{Z}$. We wish to generalise this situation.

**Definition 16.15.** A *field of fractions* of the ring $R$ is a field $K$ containing a subring $R'$ isomorphic to $R$, such that every element of $K$ can be expressed in the form $r/s$ for $r$, $s \in R'$, where $s \neq 0$.

To see how to construct a field of fractions for $R$, we analyse how $\mathbb{Z}$ is embedded in $\mathbb{Q}$. We can think of a rational number, written as a fraction

$r/s$, as an ordered pair $(r, s)$ of integers. However, the same rational number corresponds to many distinct fractions: for instance $\frac{2}{3} = \frac{4}{6} = \frac{10}{15}$ and so on. Therefore the pairs (2, 3), (4, 6), and (10, 15) must be treated as if they are 'the same'. The way to achieve this is to define an equivalence relation that makes them equivalent to each other. In general $(r, s)$ represents the same rational as $(t, u)$ if and only if $r/s = t/u$, that is, $ru = st$. In this form the condition involves only the arithmetic of $\mathbb{Z}$. By generalising these ideas we obtain:

**Theorem 16.16.** *Every integral domain possesses a field of fractions.*

*Proof.* Let $R$ be an integral domain, and let $S$ be the set of all ordered pairs $(r, s)$ where $r$ and $s$ lie in $R$ and $s \neq 0$. Define a relation $\sim$ on $S$ by

$$(r, s) \sim (t, u) \iff ru = st$$

It is easy to verify that $\sim$ is an equivalence relation; we denote the equivalence class of $(r, s)$ by $[r, s]$. The set $F$ of equivalence classes provides the required field of fractions. First we define the operations on $F$ by

$$\begin{aligned} [r, s] + [t, u] &= [ru + ts, su] \\ [r, s][t, u] &= [rt, su] \end{aligned}$$

Then we perform a long series of computations to show that $F$ has all the required properties. Since these computations are routine we shall not perform them here, but if you have never seen them, you should check them for yourself, see Exercise 16.8. What you have to prove is:

(1) The operations are well defined. That is to say, if $(r, s) \sim (r', s')$ and $(t, u) \sim (t', u')$, then

$$\begin{aligned} [r, s] + [t, u] &= [r', s'] + [t', u'] \\ [r, s][t, u] &= [r', s'][t', u'] \end{aligned}$$

(2) They are operations on $F$ (this is where we need to know that $R$ is an integral domain).

(3) $F$ is a field.

(4) The map $R \to F$ which sends $r \to [r, 1]$ is a monomorphism.

(5) $[r, s] = [r, 1]/[s, 1]$.

□

It can be shown (Exercise 16.9) that for a given integral domain $R$, all fields of fractions are isomorphic. We can therefore refer to the field constructed above as *the* field of fractions of $R$. It is customary to identify an element $r \in R$ with its image $[r, 1]$ in $F$, whereupon $[r, s] = r/s$.

A short calculation reveals a useful property:

**Lemma 16.17.** *If $R$ is an integral domain and $t$ is an indeterminate, then $R[t]$ is an integral domain.*

*Proof.* Suppose that

$$f = f_n t^n + \cdots + f_1 t + f_1 \qquad g = g_m t^m + \cdots + g_1 t + g_0$$

where $f_n \neq 0 \neq g_m$ and all the coefficients lie in $R$. The coefficient of $t^{m+n}$ in $fg$ is $f_n g_m$, which is non-zero since $R$ is an integral domain. Thus if $f$, $g$ are non-zero then $fg$ is non-zero. This implies that $R[t]$ is an integral domain, as claimed. □

**Corollary 16.18.** *If $F$ is a field, then the polynomial ring $F[t_1, \ldots, t_n]$ in $n$ indeterminates is an integral domain for any $n$.*

*Proof.* Write $F[t_1, \ldots, t_n] = F[t_1][t_2, \ldots, t_n]$ and use induction. □

Proposition 2.2, about the degrees of $f+g$ and $fg$, applies to polynomials over any integral domain.

Theorem 16.16 implies that when $R$ is an integral domain, $R[t]$ has a field of fractions. We call this the *field of rational expressions in $t$ over $R$* and denote by $R(t)$. Its elements are of the form $p(t)/q(t)$ where $p$ and $q$ are polynomials and $q$ is not the zero polynomial. Similarly $R[t_1, \ldots, t_n]$ has a field of fractions $R(t_1, \ldots, t_n)$. Rational expressions can be considered as fractions $p(t)/q(t)$, where $p, q \in R[t]$ and $q$ is not the zero polynomial $\mathbf{0}$. If we add two such fractions together, or multiply them, the result is another such fraction. In fact, by the usual rules of algebra,

$$\frac{p(t)}{q(t)}\frac{r(t)}{s(t)} = \frac{p(t)r(t)}{q(t)s(t)}$$
$$\frac{p(t)}{q(t)} + \frac{r(t)}{s(t)} = \frac{p(t)s(t) + q(t)r(t)}{q(t)s(t)}$$

We can also divide and subtract such expressions:

$$\frac{p(t)}{q(t)} \bigg/ \frac{r(t)}{s(t)} = \frac{p(t)s(t)}{q(t)r(t)}$$
$$\frac{p(t)}{q(t)} - \frac{r(t)}{s(t)} = \frac{p(t)s(t) - q(t)r(t)}{q(t)s(t)}$$

where in the first equation we assume $r(t)$ is not the zero polynomial.

The Division Algorithm and the Euclidean Algorithm work for polynomials over any field, without change. Therefore the entire theory of factorisation of polynomials, including irreducibles, works for polynomials in $K[t]$ whose coefficients lie in any field $K$.

## EXERCISES

16.1 If $I$ is an ideal of a ring $R$ and $1 \in I$, prove that $I = R$. Thus proper ideals are not subrings according to Definition 16.5.

16.2 Is it possible for a proper ideal of a ring $R$ to be a ring, with its own unity element, under the same operations as those in $R$?

16.3 Write out addition and multiplication tables for $\mathbb{Z}_6$, $\mathbb{Z}_7$, and $\mathbb{Z}_8$. Which of these rings are integral domains? Which are fields?

16.4 Define a *prime field* to be a field with no proper subfields. Show that the prime fields (up to isomorphism) are precisely $\mathbb{Q}$ and $\mathbb{Z}_p$ ($p$ prime).

16.5 Find the prime subfield of $\mathbb{Q}$, $\mathbb{R}$, $\mathbb{C}$, $\mathbb{Q}(t)$, $\mathbb{R}(t)$, $\mathbb{C}(t)$, $\mathbb{Z}_5(t)$, $\mathbb{Z}_{17}(t_1, t_2)$.

16.6 Show that the following tables define a field.

| $+$ | $0$ | $1$ | $\alpha$ | $\beta$ |
|---|---|---|---|---|
| $0$ | $0$ | $1$ | $\alpha$ | $\beta$ |
| $1$ | $1$ | $0$ | $\beta$ | $\alpha$ |
| $\alpha$ | $\alpha$ | $\beta$ | $0$ | $1$ |
| $\beta$ | $\beta$ | $\alpha$ | $1$ | $0$ |

| $\cdot$ | $0$ | $1$ | $\alpha$ | $\beta$ |
|---|---|---|---|---|
| $0$ | $0$ | $0$ | $0$ | $0$ |
| $1$ | $0$ | $1$ | $\alpha$ | $\beta$ |
| $\alpha$ | $0$ | $\alpha$ | $\beta$ | $1$ |
| $\beta$ | $0$ | $\beta$ | $1$ | $\alpha$ |

Find its prime subfield $P$.

16.7 Prove properties (1–5) listed in the construction of the field of fractions of an integral domain in Theorem 16.16.

16.8 Let $D$ be an integral domain with a field of fractions $F$. Let $K$ be any field. Prove that any monomorphism $\phi : D \to K$ has a unique extension to a monomorphism $\psi : F \to K$ defined by

$$\psi(a/b) = \phi(a)/\phi(b)$$

for $a, b \in D$. By considering another field of fractions $F'$ for $D$, with $\phi$ being the inclusion map, show that fields of fractions are unique up to isomorphism.

16.9 Let $K = \mathbb{Z}_2$. Describe the subfields of $K(t)$ of the form:

(a) $K(t^2)$
(b) $K(t+1)$
(c) $K(t^5)$
(d) $K(t^2+1)$

16.10 Does the condition $\partial(f+g) \leq \max(\partial f, \partial g)$ hold for polynomials $f, g$ over a general ring?

By considering the polynomials $3t$ and $2t$ over $\mathbb{Z}_6$ show that the equality $\partial(fg) = \partial f + \partial g$ fails for polynomials over a general ring $R$. What if $R$ is an integral domain?

16.11 Mark the following true or false:

(a) Every integral domain is a field.

(b) Every field is an integral domain.

(c) If $F$ is a field, then $F[t]$ is a field.

(d) If $F$ is a field, then $F(t)$ is a field.

(e) $\mathbb{Z}(t)$ is a field.

# Chapter 17

# Abstract Field Extensions and Galois Groups

Having defined rings and fields, and equipped ourselves with several methods for constructing them, we are now in a position to attack the general structure of an abstract field extension. Our previous work with subfields of $\mathbb{C}$ paves the way, and most of the effort goes into making minor changes to terminology and checking carefully that the underlying ideas generalise in the obvious manner.

We begin by extending the classification of simple extensions to general fields. Having done that, we assure ourselves that the theory of normal extensions, including their relation to splitting fields, carries over to the general case. A new issue, separability, comes into play when the characteristic of the field is not zero. The main result is that the Galois correspondence can be set up for any finite separable normal extension, and it then has exactly the same properties that we have already proved over $\mathbb{C}$.

***Convention on Generalisations.*** Much of this chapter consists of routine verification that theorems previously stated and proved for subfields or subrings of $\mathbb{C}$ remain valid for general rings and fields—and have essentially the same proofs. As a standing convention, when we refer to 'Lemma X.Y' with X $\leq$ 15, or make similar references to theorems, propositions, and so on, we mean the generalisation of Lemma X.Y to an arbitrary ring or field. Usually we do not restate Lemma X.Y in its new form. In cases where the proof requires a new method, or extra hypotheses, we are more specific. Moreover, some of the most important theorems are restated explicitly.

## 17.1 Minimal Polynomials

**Definition 17.1.** A *field extension* is a monomorphism $\iota : K \to L$, where $K, L$ are fields.

Usually we identify $K$ with its image $\iota(K)$, and in this case $K$ becomes a subfield of $L$.

We write $L/K$ for an extension where $K$ is a subfield of $L$. In this case, $\iota$ is the inclusion map.

DOI: 10.1201/9781003213949-17

We define the *degree* $[L : K]$ of an extension $L/K$ exactly as in Chapter 6. Namely, consider $L$ as a vector space over $K$ and take its dimension. The Tower Law remains valid and has exactly the same proof.

In Chapter 16 we observed that all of the usual properties of factorisation of polynomials over $\mathbb{C}$ carry over, without change, to general polynomials. (Even Gauss's Lemma and Eisenstein's Criterion can be generalised to polynomials over suitable rings, but we do not discuss such generalisations here.) Specifically, the definitions of reducible and irreducible polynomials, uniqueness of factorisation into irreducibles, and the concept of a greatest common divisor, or gcd, carry over to the general case. Moreover, if $K$ is a field and $h \in K[t]$ is a gcd of $f, g \in K[t]$, then there exist $a, b \in K[t]$ such that $h = af + bg$. As before, a polynomial is *monic* if its term of highest degree has coefficient 1, and a monic gcd is unique.

If $L/K$ is a field extension and $\alpha \in L$, the same dichotomy arises: either $\alpha$ is a zero of some polynomial $f \in K[t]$, or it is not. In the first case $\alpha$ is *algebraic* over $K$; in the second case $\alpha$ is *transcendental* over $K$.

An element $\alpha \in L$ that is algebraic over $K$ has a well-defined *minimal polynomial* $m(t) \in K[t]$; this is the unique monic polynomial over $K$ of smallest degree such that $m(\alpha) = 0$.

## 17.2 Simple Algebraic Extensions

As before, we can define the subfield of $L$ *generated by* a subset $X \subseteq L$, together with some subfield $K$, and we employ the same notation $K(X)$ for this field. We say that it is obtained by *adjoining* $X$ to $K$. The concept of a simple extension generalises in the obvious way.

We mimic the classification of simple extensions in $\mathbb{C}$ of Chapter 5. Simple transcendental extensions are easy to analyse, and we obtain the same result: every simple transcendental extension $K(\alpha)$ of $K$ is isomorphic to $K(t)/K$, the field of rational expressions in one indeterminate $t$. Moreover, there is an isomorphism that carries $t$ to $\alpha$.

The algebraic case is slightly trickier: again the key is irreducible polynomials. The result that opens up the whole area is:

**Theorem 17.2.** *Let $K$ be a field and suppose that $m \in K[t]$ is irreducible and monic. Let $I$ be the ideal of $K[t]$ consisting of all multiples of $m$. Then $K[t]/I$ is a field, and there is a natural monomorphism $\iota : K \to K[t]/I$ such that $\iota(k) = I + k$. Moreover, $I + t$ is a zero of $m$, which is its minimal polynomial.*

*Proof.* First, observe that $I$ really is an ideal (Exercise 17.1). We know on general nonsense grounds that $K[t]/I$ is a ring. So suppose that $I + f \in K[t]/I$ is not the zero element, which in this case means that $f \notin I$. Then $f$ is not a multiple of $m$, and since $m$ is irreducible, the gcd of $f$ and $m$ is 1. Therefore there exist $a, b \in K[t]$ such that $af + bm = 1$. We claim that the multiplicative inverse of $I + f$ is $I + a$. To prove this, compute:

$$(I + f)(I + a) = I + fa = I + (1 - bm) = I + 1$$

since $bm \in I$ by definition. But $I + 1$ is the multiplicative identity of $K[t]/I$. Therefore $K[t]/I$ is a field.

Define $\iota : K \to K[t]/I$ by $\iota(k) = I + k$. It is easy to check that $\iota$ is a homomorphism. We show that it is one-to-one. If $a \neq b \in K$ then clearly $a - b \notin \langle m \rangle$, so $\iota(a) \neq \iota(b)$. Therefore $\iota$ is a monomorphism.

It is easy to see that the minimal polynomial of $I + t \in K[t]/I$ over $K$ is $m(t)$. Indeed, $m(I + t) = I + m(t) = I + 0$. (This is the only place we use the fact that $m$ is monic. But if $m$ is irreducible and not monic, then some multiple $km$, with $k \in K$, is irreducible and monic; moreover, $m$ and $km$ determine the same ideal $I$.) □

This proof can be made more elegant and more general: see Exercise 17.2. We can (and do) identify $K$ with its image $\iota(K)$, so we can assume without loss of generality that $K \subseteq K[t]/I$. We now prove a classification theorem for simple algebraic extensions. (Here the symbol / has two meanings: 'quotient' if it refers to an ideal, 'over' when it refers to a field extension.)

**Theorem 17.3.** *Let $K(\alpha)/K$ be a simple algebraic extension, where $\alpha$ has minimal polynomial $m$ over $K$. Then $K(\alpha)/K$ is isomorphic to $K[t]/I/K$, where $I$ is the ideal of $K[t]$ consisting of all multiples of $m$. Moreover, there is a natural isomorphism in which $\alpha \mapsto$ the coset $I + t$.*

*Proof.* Define a map $\phi : K[t] \to K(\alpha)$ by $\phi(f(t)) = f(\alpha)$. This is clearly a ring homomorphism. Its image is the whole of $K(\alpha)$, and its kernel consists of all multiples of $m(t)$ by Lemma 5.5. Now $K(\alpha) = \text{im}(\phi) \cong K[t]/\ker(\phi) = K[t]/I$, as required. □

We can now prove a preliminary version of the result that $K$ and $m$ between them determine the extension $K(\alpha)$.

**Theorem 17.4.** *Suppose $K(\alpha)/K$ and $K(\beta)/K$ are simple algebraic extensions, such that $\alpha$ and $\beta$ have the same minimal polynomial $m$ over $K$. Then the two extensions are isomorphic, and the isomorphism of the large fields can be taken to map $\alpha$ to $\beta$.*

*Proof.* This is an immediate corollary of Theorem 17.3. □

---

## 17.3 Splitting Fields

In Chapter 9 we defined the term 'splitting field': a polynomial $f \in K[t]$ splits in $L$ if it can be expressed as a product of linear factors over $L$, and the splitting field $\Sigma$ of $f$ is the smallest such $L$. There, we appealed to the Fundamental Theorem of Algebra to construct the splitting field for any given complex polynomial. In the general case, the Fundamental Theorem of Algebra is not available to us. (There is a version of it, in Section 17.9, but in order to prove that version we must be able to construct splitting fields *without* appealing to that version of the Fundamental Theorem of Algebra.) And there is no longer a unique splitting field—though splitting fields are unique up to isomorphism.

We start by generalising Definitions 9.1 and 9.3.

**Definition 17.5.** If $K$ is a field, $L$ is an extension of $K$, and $f$ is a nonzero polynomial over $K$, then $f$ *splits* over $L$ if it can be expressed as a product of linear factors

$$f(t) = k(t - \alpha_1) \dots (t - \alpha_n)$$

where $k \in K$ and $\alpha_1, \dots, \alpha_n \in L$.

**Definition 17.6.** Let $K$ be a field and let $\Sigma$ be an extension of $K$. Then $\Sigma$ is a *splitting field* for the polynomial $f$ over $K$ if

(1) $f$ splits over $\Sigma$.

(2) If $K \subseteq \Sigma' \subseteq \Sigma$ and $f$ splits over $\Sigma'$ then $\Sigma' = \Sigma$.

Our aim is to show that for any field $K$, any polynomial over $K$ has a splitting field $\Sigma$, and this splitting field is unique up to isomorphism of extensions.

The work that we have already done allows us to construct, in the abstract, any *simple* extension of a field $K$. Specifically, any simple transcendental extension $K(\alpha)$ of $K$ is isomorphic to the field $K(t)$ of rational expressions in $t$ over $K$. And if $m \in K[t]$ is irreducible and monic, and $I$ is the ideal of $K[t]$ consisting of all multiples of $m$, then $K[t]/I$ is a simple algebraic extension $K(\alpha)$ of $K$ where $\alpha = I + t$ has minimal polynomial $m$ over $K$. Moreover, all simple algebraic extensions of $K$ arise (up to isomorphism) by this construction.

**Definition 17.7.** We refer to these constructions as *adjoining* $\alpha$ to $K$.

When we were working with subfields $K$ of $\mathbb{C}$, we could assume that the element(s) being adjoined were in $\mathbb{C}$, so all we had to do was take the field they generate, together with $K$. Now we do not have a big field in which to work, so we have to create the fields along with the elements we need.

We construct a splitting field by adjoining to $K$ elements that are to be thought of as the zeros of $f$. We already know how to do this for irreducible polynomials, see Theorem 17.2, so we split $f$ into irreducible factors and work on these separately.

**Theorem 17.8.** *If $K$ is any field and $f$ is any nonzero polynomial over $K$, then there exists a splitting field for $f$ over $K$.*

*Proof.* Use induction on the degree $\partial f$. If $\partial f = 1$ there is nothing to prove, for $f$ splits over $K$. If $f$ does not split over $K$ then it has an irreducible factor $f_1$ of degree $> 1$. Using Theorem 5.6 we adjoin $\sigma_1$ to $K$, where $f_1(\sigma_1) = 0$. Then in $K(\sigma_1)[t]$ we have $f = (t - \sigma_1)g$ where $\partial g = \partial f - 1$. By induction, there is a splitting field $\Sigma$ for $g$ over $K(\sigma_1)$. But then $\Sigma$ is clearly a splitting field for $f$ over $K$. □

It would appear at first sight that we might construct different splitting fields for $f$ by varying the choice of irreducible factors. In fact splitting fields (for given $f$ and $K$) are unique up to isomorphism. The statements and proofs are exactly as in Lemma 9.5 and Theorem 9.6, and we do not repeat them here.

## 17.4 Normality

As before, the key properties that drive the Galois correspondence are normality and separability. We discuss normality in this section, and separability in the next.

Because we suppressed explicit use of '$\subseteq \mathbb{C}$' from our earlier definition, it remains seemingly unchanged:

**Definition 17.9.** A field extension $L/K$ is *normal* if every irreducible polynomial $f$ over $K$ that has at least one zero in $L$ splits in $L$.

So does the proof of the main result about normality and splitting fields:

**Theorem 17.10.** *A field extension $L/K$ is normal and finite if and only if $L$ is a splitting field for some polynomial over $K$.*

*Proof.* The same as for Theorem 9.9, except that 'the splitting field' becomes 'a splitting field'. □

Finally we need to discuss the concept of a normal closure in the abstract context. For subfields of $\mathbb{C}$ the normal closure of an extension $L/K$ is an extension $N$ of $L$ such that $N/K$ is normal, and $N$ is as small as possible subject to this condition. We proved existence by taking a suitable splitting field, yielding a normal extension of $K$ containing $L$, and then finding the unique smallest subfield with those two properties.

For abstract fields, we have to proceed in a similar but technically different manner. The proof of Theorem 11.6 still constructs a normal closure, because this is defined there using a splitting field, which we construct using Theorem 17.8. The only difference is that the normal closure is now unique *up to isomorphism.* That is, if $N_1/K$ and $N_2/K$ are normal closures of $L/K$, then the extensions $N_1/L$ and $N_2/L$ are isomorphic. This follows because splitting fields are unique up to isomorphism, as remarked immediately after Theorem 17.8.

---

## 17.5 Separability

We generalise Definition 9.10:

**Definition 17.11.** An irreducible polynomial $f$ over a field $K$ is *separable* over $K$ if it has no multiple zeros in a splitting field.

Since a splitting field is unique up to isomorphism, it is irrelevant which splitting field we use to check this property.

**Example 17.12.** Consider $f(t) = t^2 + t + 1$ over $\mathbb{Z}_2$. This time we cannot use $\mathbb{C}$, so we must go back to the basic construction for a splitting field. The field $\mathbb{Z}_2$ has two elements, 0 and 1. It is easy to see that $f$ is irreducible, since $f(0) = f(1) = 1 \neq 0$, so we may adjoin an element $\zeta$ such that $\zeta$ has minimal polynomial $f$ over $\mathbb{Z}_2$. Then $\zeta^2 + \zeta + 1 = 0$ so that $\zeta^2 = 1 + \zeta$ (remember, the characteristic is 2) and the elements $0, 1, \zeta, 1+\zeta$ form a field. This follows from Theorem 5.10. It can also be verified directly by working out addition and multiplication tables:

| $+$ | $0$ | $1$ | $\zeta$ | $1+\zeta$ |
|---|---|---|---|---|
| $0$ | $0$ | $1$ | $\zeta$ | $1+\zeta$ |
| $1$ | $1$ | $0$ | $1+\zeta$ | $\zeta$ |
| $\zeta$ | $\zeta$ | $1+\zeta$ | $0$ | $1$ |
| $1+\zeta$ | $1+\zeta$ | $\zeta$ | $1$ | $0$ |

| $\cdot$ | $0$ | $1$ | $\zeta$ | $1+\zeta$ |
|---|---|---|---|---|
| $0$ | $0$ | $0$ | $0$ | $0$ |
| $1$ | $0$ | $1$ | $\zeta$ | $1+\zeta$ |
| $\zeta$ | $0$ | $\zeta$ | $1+\zeta$ | $1$ |
| $1+\zeta$ | $0$ | $1+\zeta$ | $1$ | $\zeta$ |

A typical calculation for the second table runs like this:

$$\zeta(1+\zeta) = \zeta + \zeta^2 = \zeta + \zeta + 1 = 1$$

Therefore $\mathbb{Z}_2(\zeta)$ is a field with four elements. Now $f$ splits over $\mathbb{Z}_2(\zeta)$:

$$t^2 + t + 1 = (t - \zeta)(t - 1 - \zeta)$$

but over no smaller field. Hence $\mathbb{Z}_2(\zeta)$ is a splitting field for $f$ over $\mathbb{Z}_2$.

We have now reached the point at which the theory of fields of prime characteristic $p$ starts to differ markedly from that for characteristic zero. A major difference is that separability (see Definition 9.10) can, and often does, fail. To investigate this phenomenon, we introduce a new term:

**Definition 17.13.** An irreducible polynomial over a field $K$ is *inseparable* over $K$ if it is not separable over $K$.

We are now ready to prove the existence of a very useful map.

**Lemma 17.14.** *Let $K$ be a field of characteristic $p > 0$. Then the map $\phi : K \to K$ defined by $\phi(k) = k^p$ $(k \in K)$ is a field monomorphism. If $K$ is finite, $\phi$ is an automorphism.*

*Proof.* Let $x, y \in K$. Then

$$\phi(xy) = (xy)^p = x^p y^p = \phi(x)\phi(y)$$

By the binomial theorem,

$$\phi(x+y) = (x+y)^p = x^p + px^{p-1}y + \binom{p}{2}x^{p-2}y^2 + \cdots + pxy^{p-1} + y^p \quad (17.1)$$

Since the characteristic is $p$, Lemma 3.23 implies that the sum in (17.1) reduces to its first and last terms, and

$$\phi(x+y) = x^p + y^p = \phi(x) + \phi(y)$$

We have now proved that $\phi$ is a homomorphism.

To show that $\phi$ is one-to-one, suppose that $\phi(x) = \phi(y)$. Then $\phi(x-y) = 0$. So $(x-y)^p = 0$, so $x = y$. Therefore $\phi$ is a monomorphism.

If $K$ is finite, then any monomorphism $K \to K$ is automatically onto by counting elements, so $\phi$ is an automorphism in this case. □

**Definition 17.15.** If $K$ is a field of characteristic $p > 0$, the map $\phi : K \to K$ defined by $\phi(k) = k^p$ $(k \in K)$ is the *Frobenius monomorphism* or *Frobenius map* of $K$; indeed, sometimes it is referred to as *the Frobenius.* When $K$ is finite, $\phi$ is called the *Frobenius automorphism* of $K$.

If you try this on the field $\mathbb{Z}_5$, it turns out that $\phi$ is the identity map, which is not very inspiring. The same goes for $\mathbb{Z}_p$ for any prime $p$. But for the field of Example 17.12 we have $\phi(0) = 0$, $\phi(1) = 1$, $\phi(\zeta) = 1 + \zeta$, $\phi(1+\zeta) = \zeta$, so that $\phi$ is not always the identity.

**Example 17.16.** We use the Frobenius map to give an example of an inseparable polynomial. Let $K_0 = \mathbb{Z}_p$ for prime $p$. Let $K = K_0(u)$ where $u$ is transcendental over $K_0$, and let

$$f(t) = t^p - u \in K[t]$$

Let $\Sigma$ be a splitting field for $f$ over $K$, and let $\tau$ be a zero of $f$ in $\Sigma$. Then $\tau^p = u$. Now use the Frobenius map:

$$(t - \tau)^p = t^p - \tau^p = t^p - u = f(t)$$

Thus if $\sigma^p - u = 0$ then $(\sigma - \tau)^p = 0$ so that $\sigma = \tau$; all the zeros of $f$ in $\Sigma$ are *equal*.

It remains to show that $f$ is irreducible over $K$. Suppose that $f = gh$ where $g, h \in K[t]$, and $g$ and $h$ are monic and have lower degree than $f$. We must have $g(t) = (t - \tau)^s$ where $0 < s < p$ by uniqueness of factorisation. Hence the constant coefficient $(-\tau)^s$ of $g$ lies in $K$. This implies that $\tau \in K$, for there exist integers $a$ and $b$ such that $as + bp = 1$, and since $\tau^{as+bp} \in K$ it follows that $\tau \in K$. Then $\tau = v(u)/w(u)$ where $v, w \in K_0[u]$, so

$$v(u)^p - u(w(u))^p = 0$$

But the terms of highest degree cannot cancel. Hence $f$ is irreducible.

The formal derivative $Df$ of a polynomial $f$ can be defined for any underlying field $K$:

**Definition 17.17.** Suppose that $K$ is a field, and let

$$f(t) = a_0 + a_1 t + \cdots + a_n t^n \in K[t]$$

Then the *formal derivative* of $f$ is the polynomial

$$Df = a_1 + 2a_2 t + \cdots + n a_n t^{n-1}$$

Here the elements $2, \ldots, n$ belong to $K$, not $\mathbb{Z}$. In fact they are what we briefly wrote as $2^*, \ldots, n^*$ in the proof of Theorem 16.9.

Lemma 9.13 states that a polynomial $f \neq 0$ has a multiple zero in a splitting field if and only if $f$ and $Df$ have a common factor of degree $\geq 1$. This lemma remains valid over any field, and has the same proof. Using the formal derivative, we can characterise inseparable irreducible polynomials:

**Proposition 17.18.** *If $K$ is a field of characteristic 0, then every irreducible polynomial over $K$ is separable over $K$.*

*If $K$ has characteristic $p > 0$, then an irreducible polynomial $f$ over $K$ is inseparable if and only if*

$$f(t) = k_0 + k_1 t^p + \cdots + k_r t^{rp}$$

*where $k_0, \ldots, k_r \in K$.*

*Proof.* By Lemma 9.13, an irreducible polynomial $f$ over $K$ is inseparable if and only if $f$ and $Df$ have a common factor of degree $\geq 1$. If so, then since $f$ is irreducible and $Df$ has smaller degree than $f$, we must have $Df = 0$. Thus if

$$f(t) = a_0 + \cdots + a_m t^m$$

then $n a_n = 0$ for all integers $n > 0$. For characteristic 0 this is equivalent to $a_n = 0$ for all $n$. For characteristic $p > 0$ it is equivalent to $a_n = 0$ if $p$ does not divide $n$. Let $k_i = a_{ip}$, and the result follows. ☐

The condition on $f$ for inseparability over fields of characteristic $p$ can be expressed by saying that only powers of $t$ that are multiples of $p$ occur. That is $f(t) = g(t^p)$ for some polynomial $g$ over $K$.

We now define two more uses of the word 'separable'.

**Definition 17.19.** If $L/K$ is an extension then an algebraic element $\alpha \in L$ is *separable* over $K$ if its minimal polynomial over $K$ is separable over $K$.

An algebraic extension $L/K$ is a *separable extension* if every $\alpha \in L$ is separable over $K$.

For algebraic extensions, separability carries over to intermediate fields.

**Lemma 17.20.** *Let $L/K$ be a separable algebraic extension and let $M$ be an intermediate field. Then $M/K$ and $L/M$ are separable.*

*Proof.* Clearly $M/K$ is separable. Let $\alpha \in L$, and let $m_K$ and $m_M$ be its minimal polynomials over $K, M$ respectively. Now $m_M | m_K$ in $M[t]$. But $\alpha$ is separable over $K$ so $m_K$ is separable over $K$, hence $m_M$ is separable over $M$. Therefore $L/M$ is a separable extension. □

We end this section by proving that an extension generated by the zeros of a separable polynomial is separable. To prove this, we first prove:

**Lemma 17.21.** *Let $L/K$ be a field extension where the fields have characteristic $p$, and let $\alpha \in L$ be algebraic over $K$. Then $\alpha$ is separable over $K$ if and only if $K(\alpha^p) = K(\alpha)$.*

*Proof.* Since $\alpha$ is a zero of $t^p - \alpha^p \in K(\alpha^p)[t]$, which equals $(t-\alpha)^p$ by the Frobenius map, the minimal polynomial of $\alpha$ over $K(\alpha^p)$ must divide $(t-\alpha)^p$ and hence be $(t-\alpha)^s$ for some $s \leq p$.

If $\alpha$ is separable over $K$ then it is separable over $K(\alpha^p)$. Therefore $(t-\alpha)^s$ has simple zeros, so $s = 1$. Therefore $\alpha \in K(\alpha^p)$, so $K(\alpha^p) = K(\alpha)$.

For the converse, suppose that $\alpha$ is inseparable over $K$. Then its minimal polynomial over $K$ has the form $g(t^p)$ for some $g \in K[t]$. Thus $\alpha$ has degree $p\,\partial g$ over $K$. In contrast, $\alpha^p$ is a zero of $g$, which has smaller degree $\partial g$. Thus $K(\alpha^p)$ and $K(\alpha)$ have different degrees over $K$, so cannot be equal. □

**Theorem 17.22.** *If $L/K$ is a field extension such that $L$ is generated over $K$ by a set of separable algebraic elements, then $L/K$ is separable.*

*Proof.* We may assume that $K$ has characteristic $p$. It is sufficient to prove that the set of elements of $L$ that are separable over $K$ is closed under addition, subtraction, multiplication, and division. (Indeed, subtraction and division are enough.) We give the proof for addition: the other cases are similar.

Suppose that $\alpha, \beta \in L$ are separable over $K$. Observe that

$$K(\alpha + \beta, \beta) = K(\alpha, \beta) = K(\alpha^p, \beta^p) = K(\alpha^p + \beta^p, \beta^p) \qquad (17.2)$$

using Lemma 17.21 for the middle equality. Now consider the towers

$$K \subseteq K(\alpha + \beta) \subseteq K(\alpha + \beta, \beta)$$
$$K \subseteq K(\alpha^p + \beta^p) \subseteq K(\alpha^p + \beta^p, \beta^p)$$

and consider the corresponding degrees. Apply the Frobenius map to minimal polynomials to see that

$$[K(\alpha^p + \beta^p, \beta^p) : K(\alpha^p + \beta^p)] \leq [K(\alpha + \beta, \beta) : K(\alpha + \beta)]$$

and

$$[K(\alpha^p + \beta^p) : K] \geq [K(\alpha + \beta) : K]$$

However,

$$[K(\alpha^p + \beta^p, \beta^p) : K] = [K(\alpha + \beta, \beta) : K]$$

by (17.2). Now the Tower Law implies that the above inequalities of degrees must actually be equalities. □

## 17.6 Galois Theory for Abstract Fields

Finally, we can set up the Galois correspondence as in Chapter 12. Everything works, provided that we work with a normal separable field extension rather than just a normal one. As we remarked in that context, separability is automatic for subfields of $\mathbb{C}$. So there should be no difficulty in reworking the theory in the more general context. In particular, Theorem 11.14 requires separability for fields of prime characteristic.

We record one key definition in the general setting:

**Definition 17.23.** Let $L/K$ be a normal separable extension. The *Galois group* $\Gamma(L/K)$ is the group of all $K$-automorphisms of $L$ under composition.

Because of its importance, we also restate the Fundamental Theorem of Galois Theory:

**Theorem 17.24 (Fundamental Theorem of Galois Theory, General Case).** *If $L/K$ is a finite separable normal field extension, with Galois group $G$, and if $\mathcal{F}, \mathcal{G}, ^*, ^\dagger$ are defined as before, then:*

(1) *The Galois group $G$ has order $[L : K]$.*

(2) *The maps $^*$ and $^\dagger$ are mutual inverses, and set up an order-reversing one-to-one correspondence between $\mathcal{F}$ and $\mathcal{G}$.*

(3) *If $M$ is an intermediate field, then*

$$[L : M] = |M^*| \qquad [M : K] = |G|/|M^*|$$

(4) *An intermediate field $M$ is a normal extension of $K$ if and only if $M^*$ is a normal subgroup of $G$.*

(5) *If an intermediate field $M$ is a normal extension of $K$, then the Galois group of $M/K$ is isomorphic to the quotient group $G/M^*$.*

*Proof.* Mimic the proof of Theorem 12.2 and look out for steps that require separability. □

Another thing to look out for is the uniqueness of the splitting field of a polynomial: now it is unique only up to isomorphism. For example, we defined the Galois group of a polynomial $f$ over $K$ to be the Galois group of $\Sigma/K$, where $\Sigma$ is the splitting field of $f$. When $K$ is a subfield of $\mathbb{C}$, the subfield $\Sigma$ is unique. In general it is unique up to isomorphism, so the Galois group of $f$ is unique up to isomorphism. That suits us fine.

What about radical extensions? In characteristic $p$, inseparability raises its ugly head, and its effect is serious. For example, $t^p - 1 = (t-1)^p$, by the Frobenius map, so the only $p$th root of unity is 1. The definition of 'radical extension' has to be changed in characteristic $p$, and we shall not go into the details. However, everything carries through unchanged to fields with characteristic 0.

## 17.7 Conjugates and Minimal Polynomials

Suppose that $\alpha \in \mathbb{C}$ is algebraic over a subfield $K \subseteq \mathbb{C}$, with minimal polynomial $m(t)$. Recall Definition 5.7: the *conjugates* of $\alpha$ are the zeros of $m(t)$. To generalise this concept to an arbitrary field $K$ we replace $\mathbb{C}$ by a splitting field $\Sigma$ for $m(t)$. Corollary 5.13, generalised to this setting, implies that the conjugates of $\alpha$ are its images under elements of the Galois group. Moreover, if $\Sigma/K$ is separable, the zeros of $m(t)$ are distinct by Definitions 17.11 and 17.19.

These results suggest considering the polynomial

$$p(t) = \prod_{g \in \Gamma} (t - g(\alpha)) \tag{17.3}$$

whose zeros are the images of $\alpha$ under $\Gamma$. Observe that $p$ depends on $\alpha$ but this is not made explicit in the notation.

**Proposition 17.25.** *Let $L/K$ be a normal separable field extension with Galois group $\Gamma$, and let $\alpha \in L$. Then the polynomial $p(t)$ in (17.3) belongs to $K[t]$.*

*Proof.* The group $\Gamma$ acts on $L[t]$ as follows. If

$$f(t) = a_n t^n + \cdots + a_0 \in L[t]$$

and $g \in \Gamma$, define

$$f^g(t) = g(a_n)t^n + \cdots + g(a_0) \in L[t]$$

Then $f^g(t) = f(t)$ if and only if all coefficients $a_j$ are fixed by $g$; that is, $g(a_j) = a_j$. Since the fixed field of $\Gamma$ is $K$, we deduce that:

$$f^g(t) = f(t) \text{ for all } g \in \Gamma \iff f(t) \in K[t] \tag{17.4}$$

We claim that $p^g(t) = p(t)$ for all $g \in \Gamma$. Indeed, for each $g$,

$$p^g(t) = \prod_{h \in \Gamma} (t - gh(\alpha))$$

By Lemma 10.4, as $h$ runs through $\Gamma$, so does $gh$; therefore $p^g(t) = p(t)$. Now (17.4) implies that $p(t) \in K[t]$. □

This result implies that in $K[t]$, the polynomial $p(t)$ is divisible by the minimal polynomial $m(t)$ of $\alpha$ over $K$. In fact, more is true. As motivation, we consider an example.

**Example 17.26.** Recall Example 13.3, the polynomial $f(t) = t^4 - 5t^2 + 6$ over $\mathbb{Q}$. It factorises as $(t^2 - 2)(t^2 - 3)$, so its zeros are $\pm\sqrt{2}, \pm\sqrt{3}$ and its splitting field over $\mathbb{Q}$ is $K = \mathbb{Q}(\sqrt{2}, \sqrt{3})$.

The Galois group $\Gamma$ of $f$ is isomorphic to $\mathbb{Z}_2 \times \mathbb{Z}_2$, given by maps $\sqrt{2} \mapsto \pm\sqrt{2}$ and $\sqrt{3} \mapsto \pm\sqrt{3}$ with all combinations of signs.

This polynomial depends on $\alpha$. For example,

$$\begin{aligned}
p(\sqrt{2}) &= (t - \sqrt{2})(t + \sqrt{2})(t - \sqrt{2})(t + \sqrt{2}) = (t^2 - 2)^2 \\
p(\sqrt{3}) &= (t - \sqrt{3})(t - \sqrt{3})(t + \sqrt{3})(t + \sqrt{3}) = (t^2 - 3)^2 \\
p(\sqrt{6}) &= (t - \sqrt{6})(t + \sqrt{6})(t + \sqrt{6})(t - \sqrt{6}) = (t^2 - 6)^2 \\
p(\sqrt{2} + \sqrt{3}) &= (t - \sqrt{2} - \sqrt{3})(t + \sqrt{2} - \sqrt{3})(t - \sqrt{2} + \sqrt{3})(t + \sqrt{2} + \sqrt{3}) \\
&= t^4 - 10t^2 + 1
\end{aligned}$$

In the first three cases $p(t)$ is the square of the minimal polynomial of $\alpha$. In the fourth case it is the minimal polynomial of $\alpha$.

These results are no coincidence. The next result shows that $p(t)$ is always a power of the minimal polynomial of $\alpha$. In particular, if $L = K(\alpha)$ then $p(t)$ is equal to the minimal polynomial of $\alpha$.

**Proposition 17.27.** (1) *Let $L/K$ be a normal separable field extension with Galois group $\Gamma$, and let $\alpha \in L$. Let the minimal polynomial of $\alpha$ over $K$ be $m(t) \in K[t]$, and let $p(t)$ be the polynomial defined by* (17.3). *Let $H = K(\alpha)^* \subseteq \Gamma$. Then $p(t) = m(t)^d$ where $d = |H|$.*

(2) *If $L = K(\alpha)$ then $p(t)$ is the minimal polynomial of $\alpha$ over $K$. In particular, it is irreducible over $K$.*

*Proof.* (1) The main point is that for some $\alpha$, distinct elements of $\Gamma$ may have the same effect on $\alpha$, as in Example 17.26. To see how this happens, let $g_1, g_2 \in \Gamma$. Then

$$\begin{aligned} g_1(\alpha) = g_2(\alpha) &\iff g_2^{-1}g_1(\alpha) = \alpha \\ &\iff g_2^{-1}g_1 \in K(\alpha)^* = H \\ &\iff g_1 = g_2 h \text{ for some } h \in H \end{aligned}$$

Decompose $\Gamma$ into left cosets $gH$. Let $g_1, \ldots, g_k$ be a set if representatives, one from each such coset. Then $k = |\Gamma|/|H| = |\Gamma|/d$. By the above calculation, the $g_i$ are distinct. Now

$$\Gamma = g_1H \mathbin{\dot\cup} g_2H \mathbin{\dot\cup} \cdots \mathbin{\dot\cup} g_kH$$

where $\dot\cup$ indicates disjoint union. Therefore

$$\begin{aligned} p(t) &= \prod_{i=1}^{k} \prod_{h \in H} (t - g_ih(\alpha)) \\ &= \prod_{i=1}^{k} \prod_{h \in H} (t - g_i(\alpha)) \quad \text{since } h \in H \\ &= \left( \prod_{i=1}^{k} (t - g_i(\alpha)) \right)^d = q(t)^d \end{aligned}$$

where

$$q(t) = \prod_{i=1}^{k} (t - g_i(\alpha))$$

and $d = |\Gamma|/k = |H|$.

We claim that $q(t)$ is the minimal polynomial of $\alpha$. Proposition 17.25 tells us that $p(t) \in K[t]$. We claim that $q(t) \in K[t]$. We know that $p(t) = q(t)^d$ as polynomials over $L$. Apply an arbitrary element $g \in \Gamma$ to deduce that $q^g(t)^d = p^g(t) = p(t)$. Therefore $q^g(t)^d = q(t)^d$ as elements of $L[t]$. Both $q^g(t)$ and $q(t)$ are monic. Factorising into irreducibles of $L[t]$ we deduce that $q^g(t) = q(t)$. Since $g \in \Gamma$ is arbitrary, $q(t)$ is fixed by $\Gamma$, so $q(t) \in K[t]$.

Certainly $q(\alpha) = 0$, so the minimal polynomial $m(t)$ of $\alpha$ over $K$ divides $q(t)$. But $q(t)$ is divisible by all distinct images $g(\alpha)$ of $\alpha$, each of which is a zero of $m(t)$. Moreover, all zeros of $m(t)$ arise in this manner. Since $m(t)$ is irreducible and $L/K$ is separable, the zeros of $m(t)$ are simple. Therefore $q(t) = m(t)$ as required.

(2) Follows immediately because $L = K(\alpha)$ implies that $H = K(\alpha)^* = \mathbf{1}$, so $d = 1$. Therefore $p(t) = m(t)$, which is irreducible over $K$ since it is a minimal polynomial. □

Finally, we remark that the power $d$ can also be expressed as $d = [L : K(\alpha)]$.

## 17.8 The Primitive Element Theorem

Using Galois theory we can give a simple proof of a generalisation of the classical Primitive Element Theorem 6.13. This can be found, for example, in Weintraub (2021), together with some implications for normal subgroups of Galois groups. First, we need a simple but subtle lemma about vector spaces.

**Lemma 17.28.** *Let $V$ be a vector space over an infinite field $K$. Then $V$ is not the union of finitely many proper subspaces.*

*Proof.* Suppose for a contradiction that

$$V = W_1 \cup \cdots \cup W_k$$

where the $W_i$ are proper subspaces. The result is obvious for $k = 1$, so we may assume that $k \geq 2$.

Among all such expressions, make $k$ as small as possible. Then $V$ is not the union of a proper subset of the $W_j$. Define

$$\widehat{W}_i = \left( \bigcup_{j \neq i} W_j \right)$$

For $1 \leq i \leq k$, choose a vector

$$v_i \in W_i \setminus \widehat{W}_i$$

so that $v_i$ lies in $W_i$ but no other $W_j$. The set

$$S = \{v_1 + \alpha v_2 : \alpha \in K\} \subseteq V$$

is infinite, so at least two distinct members of $S$ lie in the same $W_j$. This implies that $v_1, v_2 \in W_j$. But $v_i$ lies in $W_i$ but no other $W_j$, so $j = 1$ and $j = 2$ both hold, a contradiction. □

This result fails for vector spaces over finite fields: see Exercise 17.3.

**Theorem 17.29 (Abstract Primitive Element Theorem).** *Let $K$ be an infinite field and let $L/K$ be a finite separable extension. Then there exists $\theta \in L$ such that $L = K(\theta)$.*

*Proof.* If $L = K$ there is nothing to prove, so we may assume $L \neq K$. The Galois group $\Gamma(L/K)$ is finite since $L/K$ is finite. Therefore $\Gamma(L/K)$ has only a finite number of subgroups. By the Galois correspondence, there are only finitely many intermediate fields $M$ with $K \subsetneq M \subsetneq L$. As vector spaces over $K$, each $M$ is a vector subspace of $L$ of smaller dimension than $L$. Therefore the union $U$ of these subspaces is a proper subset of $L$ by Lemma 17.28. If $\theta \notin U$ then the subfield generated by $\theta$ is not $K$ or any intermediate $M$, so it must equal $L$. Therefore $L = K(\theta)$. □

Example 19.11 below shows that the condition of separability cannot be omitted in the statement of this theorem. However, some inseparable extensions have a primitive element. It can be proved that a necessary and sufficient condition for a primitive element to exist is that there are only finitely many intermediate fields.

Theorem 17.29 is also true if $K$ is a finite field. This is a simple consequence of Corollary 19.9, proved later.

We have now reworked the entire theory established in previous chapters, generalising from subfields of $\mathbb{C}$ to arbitrary fields. The main technical condition that does not apply to all fields is separability, so the theory often assumes this as an extra condition.

## 17.9 Algebraic Closure of a Field

We end this chapter by proving existence of the algebraic closure of any field. This is a *minimal* algebraically closed field containing $K$. (There can be larger ones—for instance, let $\overline{K}$ be any algebraically closed field containing $K$, form the larger field $\overline{K}(t)$, and then take the algebraic closure of *that*.)

We can safely say 'the' algebraic closure, because it is unique up to isomorphism, but we omit the proof. The main technical ideas are set-theoretic.

**Theorem 17.30.** *For any field $K$, there exists a field extension $\overline{K}/K$ such that $\overline{K}$ is algebraically closed, and no proper subfield of $\overline{K}$ is algebraically closed.*

The proof uses Zorn's Lemma. An equivalent approach employs transfinite induction, which in turn is based on the concept of an (infinite) *ordinal*. We give a brief introduction for readers unfamiliar with these ideas. The results of this section are not used elsewhere, and it can safely be skipped.

First, we state Zorn's Lemma (but call it a theorem because it's too important to be a lemma.)

**Theorem 17.31 (Zorn's Lemma).** *Let $S$ be a partially ordered set in which every totally ordered subset has an upper bound. Then $S$ contains a maximal element.* □

Proofs depend on suitable set-theoretic axioms, and it would take us too far afield to go over this topic in detail, so we assume Zorn's Lemma without proof. It is equivalent to the axiom of choice (given any set $S$ of sets we can choose a single element from each member of $S$) and the well-ordering principle (any set $S$ can be given a total ordering in which every descending chain of elements stops—that is, there is a bijection from $S$ to some ordinal). See Krantz (2002).

What we would *like* to do is let $\mathcal{S}$ be the set of all algebraic extensions of $K$ and define a partial order on $\mathcal{S}$ by $L \leq M$ if and only if $L \subseteq M$. Given any totally ordered subset, its union is an upper bound, and lies in $\mathcal{S}$. By Zorn's Lemma there is a maximal element $H$. This must be algebraically closed. If not, there is some polynomial $f(t)$ that does not split in $H$. Extend $H$ to a splitting field $\Sigma$. Then $\Sigma \supsetneq H$ and $\Sigma/H$ is algebraic, so $\Sigma/K$ is algebraic. This contradicts maximality of $H$.

Unfortunately there are set-theoretic problems with this proof. The main one is that $\mathcal{S}$ is too large to be a set. It is a 'class'. There are set-theoretic paradoxes involving classes: the best known is Russell's Paradox. (Consider the set of all sets that do not contain themselves. If this set contains itself, it does not; if it does not, it does.) We must therefore proceed in a manner that is consistent with axiomatic set theory, which leads to:

*Proof of Theorem 17.30*:

Let $S$ be the set of all monic irreducible polynomials in $K[t]$. (This is a set because $K[t]$ is.) For each $p_\lambda \in S$ introduce new indeterminates $u_{\lambda 1}, \ldots, u_{\lambda d_\lambda}$ where $d_\lambda = \partial f_\lambda$. Let $R$ be the polynomial ring over $K$ in infinitely many indeterminates $u_{\lambda j}$ for $1 \leq j \leq d_\lambda$ and all $\lambda$. Write

$$f_\lambda - \prod_{j=1}^{d_\lambda}(t - u_{\lambda j}) = \sum_{k=0}^{d_\lambda} r_{\lambda k} t^k \in R[t]$$

where the coefficients $r_{\lambda k} \in R$. Let $I$ be the ideal of $R$ generated by all $r_{\lambda k}$. Clearly $I \neq R$. Let $\mathcal{S}$ be the set of all ideals of $R$ that contain $I$. By Zorn's Lemma there is a maximal ideal in $\mathcal{S}$; that is, a maximal ideal $M$ of $R$ such that $I \subseteq M$. We claim that every polynomial over $K$ splits in $K_1 = R/M$. To prove this, observe that since $M$ is a maximal ideal, $K_1$ is a field. There is a monomorphism $K \to K_1$ mapping each $x \in K$ to the corresponding constant polynomial over $R$, modulo $M$. Again working modulo $M$, every polynomial $f_\lambda \in K[t]$ splits as the product

$$f_\lambda(t) = \prod_{j=1}^{d_\lambda}(t - u_{\lambda j}) \qquad (\text{mod } M)$$

in $K_1$.

We have not yet constructed an algebraic closure of $K$, however: although every polynomial over $K$ splits in $K_1[t]$, there might be polynomials over $K_1$ that do not split in $K_1[t]$. To get round this obstacle we construct a (countably infinite) tower

$$K = K_0 \subseteq K_1 \subseteq K_2 \subseteq \cdots$$

where $K_{j+1}$ is obtained from $K_j$ using the same construction that gives $K_1$ from $K$. Let

$$\overline{K} = \bigcup_{j=0}^{\infty} K_j$$

We claim that $\overline{K}$ is an algebraic closure of $K$.

First, $\overline{K}$ is a field, since it is the union of a tower of fields. In more detail, suppose $x, y$ in $\overline{K}$. Then $x \in K_j$ and $y \in K_k$ for some $j, k$. Let $l = \max(j, k)$; then $x, y \in K_l$. Therefore $x+y, x-y, xy$ are in $K_l$, hence in $\overline{K}$. The same goes for $y^{-1}$ if $y \neq 0$. Each field axiom involves only finitely many elements—at most three—so for each choice of these elements, they lie in some common $K_l$. But the axioms hold in $K_l$. Therefore they hold in $\overline{K}$.

There is a monomorphism $K \to \overline{K}$ (the monomorphism from $K \to K_1$ composed with the inclusion map of $K_1$ in $\overline{K}$). Let $p(t) \in \overline{K}[t]$. Then $p(t)$ has finitely many nonzero coefficients, which lie in various $K_j$. If $l$ is the maximum of these $j$, the polynomial $p(t)$ lies in $K_l$. Therefore it splits in $K_{l+1}$ by construction. But this means that it splits in $\overline{K}$. Therefore $\overline{K}$ is algebraically closed. □

Using similar methods, it can be proved that the extension $\overline{K}/K$ is unique up to isomorphism. The main point is that any normal algebraic extension of $K$ embeds naturally in $\overline{K}$. We omit the details.

---

## EXERCISES

17.1 Let $K$ be a field, and let $f(t) \in K[t]$. Prove that the set of all multiples of $f$ is an ideal of $K[t]$.

17.2 Let $\phi : K \to R$ be a ring homomorphism, where $K$ is a field and $R$ is a ring. Prove that $\phi$ is one-to-one. (Recall that in this book rings have identity elements 1 and homomorphisms preserve such elements.)

17.3 Show that Lemma 17.28 is false for a finite field.

17.4* Let $L = K(t)$ be the field of rational expressions over a field $K$.

Show that the six maps

$$\begin{aligned}
\iota &: f(t) \mapsto f(t)\\
\alpha &: f(t) \mapsto f(1-t)\\
\beta &: f(t) \mapsto f\left(\frac{1}{t}\right)\\
\gamma &: f(t) \mapsto f\left(1-\frac{1}{t}\right)\\
\delta &: f(t) \mapsto f\left(\frac{1}{1-t}\right)\\
\varepsilon &: f(t) \mapsto f\left(\frac{t}{t-1}\right)
\end{aligned}$$

are automorphisms of $L$. Prove that they form a group $G$ of order 6 under composition, and that $G$ is isomorphic to $\mathbb{D}_3 \cong \mathbb{S}_3$.

Let $F$ be the fixed field of $G$ acting on $L$. Prove that the rational expression

$$\phi(t) = \frac{(t^2 - t + 1)^3}{t^2(t-1)^2}$$

belongs to $F$. Deduce that $F = K(\phi)$ and $[L : F] = 6$. Use the Galois correspondence to find all intermediate fields between $F$ and $L$.

17.5* Prove that algebraic closures are unique up to isomorphism. More precisely, if $K$ is any field and $A, B$ are algebraic closures of $K$, show that the extensions $A/K$ and $B/K$ are isomorphic.

17.6 Let $\mathbb{A}$ denote the set of all complex numbers that are algebraic over $\mathbb{Q}$. The elements of $\mathbb{A}$ are called *algebraic numbers.* Show that $\mathbb{A}$ is a field, as follows.

(a) Prove that a complex number $\alpha \in \mathbb{A}$ if and only if $[\mathbb{Q}(\alpha) : \mathbb{Q}] < \infty$.

(b) Let $\alpha, \beta \in \mathbb{A}$. Use the Tower Law to show that $[\mathbb{Q}(\alpha, \beta) : \mathbb{Q}] < \infty$.

(c) Use the Tower Law to show that $[\mathbb{Q}(\alpha + \beta) : \mathbb{Q}] < \infty$, $[\mathbb{Q}(-\alpha) : \mathbb{Q}] < \infty$, $[\mathbb{Q}(\alpha\beta) : \mathbb{Q}] < \infty$, and if $\alpha \neq 0$ then $[\mathbb{Q}(\alpha^{-1}) : \mathbb{Q}] < \infty$.

(d) Therefore $\mathbb{A}$ is a field.

17.7 Prove that $\mathbb{R}[t]/\langle t^2 + 1\rangle$ is isomorphic to $\mathbb{C}$.

17.8 Find the minimal polynomials over the small field of the following elements in the following extensions:

(a) $\alpha$ in $K/P$ where $K$ is the field of Exercise 16.6 and $P$ is its prime subfield.

(b) $\alpha$ in $\mathbb{Z}_3(t)(\alpha)/\mathbb{Z}_3(t)$ where $t$ is indeterminate and $\alpha^2 = t + 1$.

17.9 For which of the following values of $m(t)$ do there exist extensions $K(\alpha)$ of $K$ for which $\alpha$ has minimal polynomial $m(t)$?

(a) $m(t) = t^2 + 1, K = \mathbb{Z}_3$

(b) $m(t) = t^2 + 1, K = \mathbb{Z}_5$

(c) $m(t) = t^7 - 3t^6 + 4t^3 - t - 1, K = \mathbb{R}$

17.10 Show that for fields for characteristic 2 there may exist quadratic equations that cannot be solved by adjoining square roots of elements in the field. (*Hint:* Try $\mathbb{Z}_2$.)

17.11 Show that we can solve quadratic equations over a field of characteristic 2 if as well as square roots we adjoin elements $\sqrt[*]{k}$ defined to be solutions of the equation

$$(\sqrt[*]{k})^2 + \sqrt[*]{k} = k.$$

17.12 Show that the two zeros of $t^2 + t - k = 0$ in the previous question are $\sqrt[*]{k}$ and $1 + \sqrt[*]{k}$.

17.13 Let $K = \mathbb{Z}_3$. Find all irreducible quadratics over $K$, and construct all possible extensions of $K$ by an element with quadratic minimal polynomial. Into how many isomorphism classes do these extensions fall? How many elements do they have?

17.14 Mark the following true or false.

(a) The minimal polynomial over a field $K$ of any element of an algebraic extension of $K$ is irreducible over $K$.

(b) Every monic irreducible polynomial over a field $K$ can be the minimal polynomial of some element $\alpha$ in a simple algebraic extension of $K$.

(c) A transcendental element does not have a mimimum polynomial.

(d) Any field has infinitely many non-isomorphic simple transcendental extensions.

(e) Splitting fields for a given polynomial are unique.

(f) Splitting fields for a given polynomial are unique up to isomorphism.

(g) The polynomial $t^6 - t^3 + 1$ is separable over $\mathbb{Z}_3$.

# Chapter 18

# The General Polynomial Equation

As we saw in Chapter 8, the so-called 'general' polynomial is in fact very special. It is a polynomial whose coefficients do not satisfy any algebraic relations. This property makes it in some respects simpler to work with than, say, a polynomial over $\mathbb{Q}$, and in particular it is easier to calculate its Galois group. As a result, we can show that the general quintic polynomial is not soluble by radicals without assuming as much group theory as we did in Chapter 15, and without having to prove the Theorem on Natural Irrationalities, Theorem 8.15.

Chapter 15 makes it clear that the Galois group of the general polynomial of degree $n$ should be the whole symmetric group $\mathbb{S}_n$, and we will show that this contention is correct. This immediately leads to the insolubility of the general quintic. Moreover, our knowledge of the structure of $\mathbb{S}_2$, $\mathbb{S}_3$, and $\mathbb{S}_4$ can be used to find a unified method to solve the general quadratic, cubic, and quartic equations. Further work, not described here, leads to a method for solving any quintic that *is* soluble by radicals, and finding out whether this is the case: see Berndt, Spearman and Williams (2002).

## 18.1 Transcendence Degree

Previously, we have avoided transcendental extensions. Indeed the assumption that extensions are finite has been central to the theory. We now need to consider a wider class of extensions, which still have a flavour of finiteness. This class was mentioned in passing in Definition 4.3, but it now takes centre stage.

**Definition 18.1.** An extension $L/K$ is *finitely generated* if $L = K(\alpha_1, \ldots, \alpha_n)$ where $n$ is finite. The elements $\alpha_i$ are *generators* of $L$ over $K$.

Here the $\alpha_j$ may be either algebraic or transcendental over $K$.

**Proposition 18.2.** *If $L/K$ is finitely generated and algebraic, it is a finite extension.*

*Proof.* Use the Tower Law and induction on the number $n$ of generators. □

DOI: 10.1201/9781003213949-18

**Definition 18.3.** If $\alpha_1, \ldots, \alpha_n$ are transcendental elements over a field $K$, all lying inside some extension $L$ of $K$, then they are *independent* if there is no non-trivial polynomial $p \in K[t_1, \ldots, t_n]$ such that $p(\alpha_1, \ldots, \alpha_n) = 0$ in $L$.

Thus, for example, if $t$ is transcendental over $K$ and $u$ is transcendental over $K(t)$, then $K(t, u)$ is a finitely generated extension of $K$, and $t, u$ are independent. On the other hand, $t$ and $u = t^2 + 1$ are both transcendental over $K$, but are connected by the polynomial equation $t^2 + 1 - u = 0$, so are not independent.

We now prove a condition for a set to consist of independent transcendental elements.

**Lemma 18.4.** *Let $K \subseteq M$ be fields, $\alpha_1, \ldots, \alpha_r \in M$, and suppose that $\alpha_1, \ldots, \alpha_{r-1}$ are independent transcendental elements over $K$. Then the following conditions are equivalent:*

(1) *$\alpha_r$ is transcendental over $\alpha_1, \ldots, \alpha_{r-1}$*

(2) *$\alpha_1, \ldots, \alpha_r$ are independent transcendental elements over $K$.*

*Proof.* We show that (1) is false if and only if (2) is false, which is equivalent to the above statement.

Suppose (2) is false. Let $p(t_1, \ldots, t_r) \in K[t_1, \ldots, t_r]$ be a nonzero polynomial such that $p(\alpha_1, \ldots, \alpha_r) = 0$. Write $p = \sum_{j=1}^{n} p_j t_r^j$ where each $p_j \in K[t_1, \ldots, t_{r-1}]$. That is, think of $p$ as a polynomial in $t_r$ with coefficients not evolving $t_r$. Since $p$ is nonzero, some $p_j$ must be nonzero. Because $\alpha_1, \ldots, \alpha_{r-1}$ are independent transcendental elements over $K$, the polynomial $p_j$ remains nonzero when we substitute $\alpha_i$ for $t_i$, with $1 \leq i \leq r-1$. This substitution turns $p$ into a nonzero polynomial over $K(\alpha_1, \ldots, \alpha_{r-1})$ satisifed by $\alpha_r$, so (1) fails.

The converse uses essentially the same idea. If (1) fails, then $\alpha_r$ satisfies a polynomial in $t_r$ with coefficients in $K(\alpha_1, \ldots, \alpha_{r-1})$. Multiplying by the denominators of the coefficients we may assume the coefficients lie in $K[\alpha_1, \ldots, \alpha_{r-1}]$. But now we have constructed a nonzero polynomial in $K[t_1, \ldots, t_r]$ satisfied by the $\alpha_j$, so (2) fails. □

The next result describes the structure of a finitely generated extension. The main point is that we can adjoin a number of independent transcendental elements first, with algebraic ones coming afterwards.

**Lemma 18.5.** *If $L/K$ is finitely generated, then there exists an intermediate field $M$ such that*

(1) *$M = K(\alpha_1, \ldots, \alpha_r)$ where the $\alpha_i$ are independent transcendental elements over $K$.*

(2) *$L/M$ is a finite extension.*

*Proof.* We know that $L = K(\beta_1, \ldots, \beta_n)$. If all the $\beta_j$ are algebraic over $K$, then $L/K$ is finite by Lemma 6.11 and we may take $M = K$. Otherwise some $\beta_i$ is transcendental over $K$. Call this $\alpha_1$. If $L/K(\alpha_1)$ is not finite, there exists some $\beta_k$ transcendental over $K(\alpha_1)$. Call this $\alpha_2$. Since $n$ is finite, this process eventually stops when $M = K(\alpha_1, \ldots, \alpha_r)$ and $L/M$ is finite. By Lemma 18.4, the $\alpha_j$ are independent transcendental elements over $K$. □

A result due to Ernst Steinitz says that the integer $r$ that gives the number of independent transcendental elements does not depend on the choice of $M$.

**Lemma 18.6 (Steinitz Exchange Lemma).** *With the notation of Lemma* 18.5, *if there is another intermediate field* $N = K(\beta_1, \ldots, \beta_s)$ *such that* $\beta_1, \ldots, \beta_s$ *are independent transcendental elements over* $K$ *and* $L/N$ *is finite, then* $r = s$.

*Proof.* The idea of the proof is that if there is a nontrivial polynomial relation involving $\alpha_i$ and $\beta_j$, then we can swap them, leaving the field concerned the same except for some finite extension. Inductively, we replace successive $\alpha_i$ by $\beta_j$ until all $\beta_j$ have been used, proving that $s \leq r$. By symmetry, $r \leq s$ and we are finished.

The details require some care. We claim inductively on $m$, that:

If $0 \leq m \leq s$, then renumbering the $\alpha_j$ if necessary,

(1) $L/K(\beta_1, \ldots, \beta_m, \alpha_{m+1}, \ldots, \alpha_r)$ is finite.

(2) $\beta_1, \ldots, \beta_m, \alpha_{m+1}, \ldots, \alpha_r$ are independent transcendental elements over $K$.

The renumbering simplifies the notation and is also carried out inductively. No $\alpha_j$ is renumbered more than once.

Claims (1, 2) are true when $m = 0$; in this case, no $\beta_i$ occurs, and the conditions are the same as those in Lemma 18.5.

Assuming (1, 2), we must prove the corresponding claims for $m+1$. To be explicit, these are:

(1′) $L/K(\beta_1, \ldots, \beta_{m+1}, \alpha_{m+2}, \ldots, \alpha_r)$ is finite.

(2′) $\beta_1, \ldots, \beta_{m+1}, \alpha_{m+2}, \ldots, \alpha_r$ are independent transcendental elements over $K$.

We have $m + 1 \leq s$, so $\beta_{m+1}$ exists. It is algebraic over $K(\beta_1, \ldots, \beta_m, \alpha_{m+1}, \ldots, \alpha_r)$ by (1). Therefore there is some polynomial equation

$$p(\beta_1 \ldots, \beta_{m+1}, \alpha_{m+1}, \ldots, \alpha_r) = 0 \tag{18.1}$$

in which both $\beta_{m+1}$ and some $\alpha_j$ actually occur. (That is, each appears in some term with a nonzero coefficient.) Renumbering if necessary, we can assume that this $\alpha_j$ is $\alpha_{m+1}$. Define four fields:

$$\begin{aligned}
K_0 &= K(\beta_1 \ldots, \beta_{m+1}, \alpha_{m+1}, \ldots, \alpha_r) \\
K_1 &= K(\beta_1 \ldots, \beta_m, \alpha_{m+1}, \ldots, \alpha_r) \\
K_2 &= K(\beta_1 \ldots, \beta_{m+1}, \alpha_{m+2}, \ldots, \alpha_r) \\
K_3 &= K(\beta_1 \ldots, \beta_m, \alpha_{m+2}, \ldots, \alpha_r)
\end{aligned}$$

Then $K_3 \subseteq K_1$, $K_3 \subseteq K_2$, $K_1 \subseteq K_0$, $K_2 \subseteq K_0$.

To prove (1′), observe that $K_0 \supseteq K$, and $L/K_1$ is finite by (2), so $L/K_0$ is finite. But $K_0/K_2$ is finite by (18.1). By the Tower Law, $L/K_2$ is finite. This is (2′).

To prove (2′), suppose it is false. Then there is a polynomial equation

$$p(\beta_1 \ldots, \beta_{m+1}, \alpha_{m+2}, \ldots, \alpha_r) = 0$$

The element $\beta_{m+1}$ actually occurs in some nonzero term, otherwise (2) is false. Therefore $\beta_{m+1}$ is algebraic over $K_3$, so $K_2/K_3$ is finite, so $L/K_3$ is finite by (1′) which we have already proved. Therefore $K_1/K_3$ is finite, but this contradicts (1).

This completes the induction. Continuing up to $m = s$ we deduce that $s \leq r$. Similarly $r \leq s$, so $r = s$. □

**Definition 18.7.** The integer $r$ defined in Lemma 18.5 is the *transcendence degree* of $L/K$. By Lemma 18.6, the value of $r$ is well-defined.

For example consider $K(t, \alpha, u)/K$, where $t$ is transcendental over $K$, $\alpha^2 = t$, and $u$ is transcendental over $K(t, \alpha)$. Then $M = K(t, u)$ where $t$ and $u$ are independent transcendental elements over $K$, and

$$[K(t, \alpha, u) : M] = [M(\alpha) : M]$$

is finite. The transcendence degree is 2.

The degree $[L : M]$ of the algebraic part is not an invariant, see Exercise 18.3.

Recall that $K(t_1, \ldots, t_r)$ denotes the field of rational expressions in $r$ indeterminates $t_i$. It is straightforward to show by induction on $r$ that an extension $K(\alpha_1, \ldots, \alpha_r)/K$ by independent transcendental elements $\alpha_i$ is isomorphic to $K(t_1, \ldots, t_r)/K$. In consequence:

**Proposition 18.8.** *A finitely generated extension $L/K$ has transcendence degree $r$ if and only if there is an intermediate field $M$ such that $L$ is a finite extension of $M$ and $M/K$ is isomorphic to $K(t_1, \ldots, t_r)/K$.*

**Corollary 18.9.** *if $L/K$ is a finitely generated extension, and $E$ is a finite extension of $L$, then the transcendence degrees of $E$ and $L$ over $K$ are equal.*

---

## 18.2 Elementary Symmetric Polynomials

Usually we are given a polynomial and wish to find its zeros. But it is also possible to work in the opposite direction: given the zeros and their multiplicities, reconstruct the polynomial. This is a far easier problem which has a complete general solution, as we saw in Section 8.7 for complex polynomials. We recap the main ideas.

Consider a monic polynomial of degree $n$ having its full quota of $n$ zeros (counting multiplicities). It is therefore a product of $n$ linear factors

$$f(t) = (t - \alpha_1) \dots (t - \alpha_n)$$

where the $\alpha_j$ are the zeros in $K$ (not necessarily distinct). Suppose that

$$f(t) = t^n + a_{n-1}t^{n-1} + \dots + a_1 t + a_0$$

If we expand the first product and equate coefficients with the second expression, we get the expected result:

$$\begin{aligned}
a_{n-1} &= -(\alpha_1 + \dots + \alpha_n) \\
a_{n-2} &= (\alpha_1\alpha_2 + \alpha_1\alpha_3 + \dots + \alpha_{n-1}\alpha_n) \\
&\dots \\
a_0 &= (-1)^n \alpha_1\alpha_2 \dots \alpha_n
\end{aligned}$$

The expressions in $\alpha_1, \dots, \alpha_n$ on the right are the elementary symmetric polynomials of Chapter 8, but now they are more generally interpreted as elements of $K[t_1, \dots, t_n]$ and evaluated at $t_j = \alpha_j$, for $1 \leq j \leq n$.

The elementary symmetric polynomials are symmetric in the sense that they are unchanged by permuting the indeterminates $t_j$. This property suggests:

**Definition 18.10.** A polynomial $q \in K[t_1, \dots, t_n]$ is *symmetric* if

$$q(t_{\sigma(1)}, \dots, t_{\sigma(n)}) = q(t_1, \dots, t_n)$$

for all permutations $\sigma \in \mathbb{S}_n$.

There are other symmetric polynomials apart from the elementary ones, for example $t_1^2 + \dots + t_n^2$, but they can all be expressed in terms of elementary symmetric polynomials:

**Theorem 18.11.** *Over a field $K$, any symmetric polynomial in $t_1, \dots, t_n$ can be expressed as a polynomial of smaller or equal degree in the elementary symmetric polynomials $s_r(t_1, \dots, t_n)$, for $r = 0, \dots, n$.*

*Proof.* See Exercise 8.4 (generalised to any field). □

A slightly weaker version of this result is proved in Corollary 18.13. We need Theorem 18.11 to prove that $\pi$ is transcendental (Chapter 24). The quickest proof of Theorem 18.11 is by induction, and full details can be found in any of the older algebra texts such as Salmon (1885) page 57 or Van der Waerden (1953) page 81.

## 18.3 The General Polynomial

Let $K$ be any field, and let $t_1, \ldots, t_n$ be independent transcendental elements over $K$. The symmetric group $\mathbb{S}_n$ can be made to act as a group of $K$-automorphisms of $K(t_1, \ldots, t_n)$ by defining

$$\sigma(t_i) = t_{\sigma(i)}$$

for all $\sigma \in \mathbb{S}_n$, and extending any rational expressions $\phi$ by defining

$$\sigma(\phi(t_1, \ldots, t_n)) = \phi(t_{\sigma(1)}, \ldots, t_{\sigma(n)})$$

It is easy to prove that $\sigma$, extended in this way, is a $K$-automorphism.

For example, if $n = 4$ and $\sigma$ is the permutation

$$\begin{pmatrix} 1234 \\ 2431 \end{pmatrix}$$

then $\sigma(t_1) = t_2$, $\sigma(t_2) = t_4$, $\sigma(t_3) = t_3$, and $\sigma(t_4) = t_1$. Moreover, as a typical case,

$$\sigma\left(\frac{t_1^5 t_4}{t_2^4 - 7t_3}\right) = \frac{t_2^5 t_1}{t_4^4 - 7t_3}$$

Clearly distinct elements of $\mathbb{S}_n$ give rise to distinct $K$-automorphisms.

The fixed field $F$ of $\mathbb{S}_n$ obviously contains all the symmetric polynomials in the $t_i$, and in particular the elementary symmetric polynomials $s_r = s_r(t_1, \ldots, t_n)$. We show that these generate $F$.

**Lemma 18.12.** *With the above notation, $F = K(s_1, \ldots, s_n)$. Moreover,*

$$[K(t_1, \ldots, t_n) : K(s_1, \ldots, s_n)] = n! \tag{18.2}$$

*Proof.* Clearly $L = K(t_1, \ldots, t_n)$ is a splitting field of $f(t)$ over both $K(s_1, \ldots, s_n)$ and the possibly larger field $F$. Since $\mathbb{S}_n$ fixes both of these fields, the Galois group of each extension contains $\mathbb{S}_n$, so must equal $\mathbb{S}_n$. Therefore the fields $F$ and $K(s_1, \ldots, s_n)$ are equal. Equation (18.2) follows by the Galois correspondence. □

**Corollary 18.13.** *Every symmetric polynomial in $t_1, \ldots, t_n$ over $K$ can be written as a rational expression in $s_1, \ldots, s_n$.*

*Proof.* By definition, symmetric polynomials are precisely those that lie inside the fixed field $F$ of $\mathbb{S}_n$. By Lemma 18.12, $F = K(s_1, \ldots, s_n)$. □

Compare this result with Theorem 18.11.

**Lemma 18.14.** *With the above notation, $s_1, \ldots, s_n$ are independent transcendental elements over $K$.*

*Proof.* By (18.2), $K(t_1, \ldots, t_n)$ is a finite extension of $K(s_1, \ldots, s_n)$. By Corollary 18.9 they both have the same transcendence degree over $K$, namely $n$. Therefore the $s_j$ are independent, for otherwise the transcendence degree of $K(s_1, \ldots, s_n)/K$ would be smaller than $n$. □

**Definition 18.15.** Let $K$ be a field and let $s_1, \ldots, s_n$ be independent transcendental elements over $K$. The *general polynomial of degree* $n$ 'over' $K$ is the polynomial

$$t^n - s_1 t^{n-1} + s_2 t^{n-2} - \cdots + (-1)^n s_n$$

over the field $K(s_1, \ldots, s_n)$.

The quotation marks are used because technically the polynomial is over the field $K(s_1, \ldots, s_n)$, not over $K$.

**Theorem 18.16.** *For any field $K$ let $g$ be the general polynomial of degree $n$ 'over' $K$, and let $\Sigma$ be a splitting field for $g$ over $K(s_1, \ldots, s_n)$. Then the zeros $t_1, \ldots, t_n$ of $g$ in $\Sigma$ are independent transcendental elements over $K$, and the Galois group of $\Sigma/K(s_1, \ldots, s_n)$ is the symmetric group $\mathbb{S}_n$.*

*Proof.* The extension $\Sigma/K(s_1, \ldots, s_n)$ is finite by Theorem 9.9, so the transcendence degree of $\Sigma/K$ is equal to that of $K(s_1, \ldots, s_n)/K$, namely $n$. Since $\Sigma = K(t_1, \ldots, t_n)$, the $t_j$ are independent transcendental elements over $K$, since any algebraic relation between them would lower the transcendence degree. The $s_j$ are now the elementary symmetric polynomials in $t_1, \ldots, t_n$ by Theorem 18.11. As above, $\mathbb{S}_n$ acts as a group of automorphisms of $\Sigma = K(t_1, \ldots, t_n)$, and by Lemma 18.12 the fixed field is $K(s_1, \ldots, s_n)$. By Theorem 11.14, $\Sigma/K(s_1, \ldots, s_n)$ is separable and normal (normality also follows from the definition of $\Sigma$ as a splitting field), and by Theorem 10.5 its degree is $|\mathbb{S}_n| = n!$. Then by Theorem 17.24(1) the Galois group has order $n!$, and contains $\mathbb{S}_n$, so it equals $\mathbb{S}_n$. □

Theorem 15.8 and Corollary 14.8 imply:

**Theorem 18.17.** *If $K$ is a field of characteristic zero and $n \geq 5$, the general polynomial of degree $n$ 'over' $K$ is not soluble by radicals.* □

---

## 18.4 Cyclic Extensions

Theorem 18.17 does not imply that any particular polynomial over $K$ of degree $n \geq 5$ is not soluble by radicals, because the general polynomial 'over' $K$ is actually a polynomial over the extension field $K(s_1, \ldots, s_n)$, with $n$ independent transcendental elements $s_j$. For example, the theorem does not rule out the possibility that every quintic over $K$ might be soluble by radicals, but that the formula involved varies so much from case to case that no general formula holds.

However, when the general polynomial of degree $n$ 'over' $K$ can be solved by radicals, it is easy to deduce a solution by radicals of *any* polynomial of degree $n$ over $K$, by substituting elements of $K$ for $s_1, \dots, s_n$ in that solution. This is the source of the generality of the general polynomial. From Theorem 18.17, the best that we can hope for using radicals is a solution of polynomials of degree $\leq 4$. We fulfil this hope by analysing the structure of $\mathbb{S}_n$ for $n \leq 4$, and appealing to a converse to Theorem 15.8. This converse is proved by showing that 'cyclic extensions'—extensions with cyclic Galois group—are closely linked to radicals.

**Definition 18.18.** Let $L/K$ be a finite normal extension with Galois group $G$. The *norm* of an element $a \in L$ is

$$N(a) = \tau_1(a)\tau_2(a)\dots\tau_n(a)$$

where $\tau_1, \dots, \tau_n$ are the elements of $G$.

Clearly $N(a)$ lies in the fixed field of $G$ (use Lemma 10.4) so if the extension is also separable, then $N(a) \in K$.

The next result is traditionally referred to as Hilbert's Theorem 90 from its appearance in his 1893 report on algebraic numbers.

**Theorem 18.19 (Hilbert's Theorem 90).** *Let $L/K$ be a finite normal extension with cyclic Galois group $G$ generated by an element $\tau$. Then $a \in L$ has norm $N(a) = 1$ if and only if*

$$a = b/\tau(b)$$

*for some $b \in L$, where $b \neq 0$.*

*Proof.* Let $|G| = n$. If $a = b/\tau(b)$ and $b \neq 0$ then

$$\begin{aligned} N(a) &= a\tau(a)\tau^2(a)\dots\tau^{n-1}(a) \\ &= \frac{b}{\tau(b)}\frac{\tau(b)}{\tau^2(b)}\frac{\tau^2(b)}{\tau^3(b)}\cdots\frac{\tau^{n-1}(b)}{\tau^n(b)} \\ &= 1 \end{aligned}$$

since $\tau^n = 1$.

Conversely, suppose that $N(a) = 1$. Let $c \in L$, and define

$$\begin{aligned} d_0 &= ac \\ d_1 &= (a\tau(a))\tau(c) \\ &\dots \\ d_j &= (a\tau(a)\dots\tau^i(a))\tau^i(c) \end{aligned}$$

for $0 \leq j \leq n-1$. Then

$$d_{n-1} = N(a)\tau^{n-1}(c) = \tau^{n-1}(c)$$

Further,

$$d_{j+1} = a\tau(d_j) \qquad (0 \leq j \leq n-2)$$

Define

$$b = d_0 + d_1 + \cdots + d_{n-1}$$

We choose $c$ to make $b \neq 0$. Suppose on the contrary that $b = 0$ for all choices of $c$. Then for any $c \in L$

$$\lambda_0\tau^0(c) + \lambda_1\tau(c) + \cdots + \lambda_{n-1}\tau^{n-1}(c) = 0$$

where

$$\lambda_j = a\tau(a)\ldots\tau^j(a)$$

belongs to $L$. Hence the distinct automorphisms $\tau^j$ are linearly dependent over $L$, contrary to Lemma 10.1.

Therefore we can choose $c$ so that $b \neq 0$. But now

$$\begin{aligned}\tau(b) &= \tau(d_0) + \cdots + \tau(d_{n-1}) \\ &= \tfrac{1}{a}(d_1 + \cdots + d_{n-1}) + \tau^n(c) \\ &= \tfrac{1}{a}(d_0 + \cdots + d_{n-1}) \\ &= b/a\end{aligned}$$

Thus $a = b/\tau(b)$ as claimed. □

**Theorem 18.20.** *Suppose that $L/K$ is a finite separable normal extension whose Galois group $G$ is cyclic of prime order $p$, generated by $\tau$. Assume that the characteristic of $K$ is 0 or is prime to $p$, and that $t^p - 1$ splits in $K$. Then $L = K(\alpha)$, where $\alpha$ is a zero of an irreducible polynomial $t^p - a$ over $K$ for some $a \in K$.*

*Proof.* The $p$ zeros of $t^p - 1$ form a group of order $p$, which must therefore be cyclic, since any group of prime order is cyclic. Because a cyclic group consists of powers of a single element, the zeros of $t^p - 1$ are the powers of some $\varepsilon \in K$ where $\varepsilon^p = 1$. But then

$$N(\varepsilon) = \varepsilon \ldots \varepsilon = 1$$

since $\varepsilon \in K$, so $\tau^i(\varepsilon) = \varepsilon$ for all $i$. By Theorem 18.19, $\varepsilon = \alpha/\tau(\alpha)$ for some $\alpha \in L$. Therefore

$$\tau(\alpha) = \varepsilon^{-1}\alpha \qquad \tau^2(\alpha) = \varepsilon^{-2}\alpha \qquad \ldots \qquad \tau^j(\alpha) = \varepsilon^{-j}\alpha$$

and $a = \alpha^p$ is fixed by $G$, so lies in $K$. Now $K(\alpha)$ is a splitting field for $t^p - a$ over $K$. The $K$-automorphisms $1, \tau, \ldots, \tau^{p-1}$ map $\alpha$ to distinct elements, so they give $p$ distinct $K$-automorphisms of $K(\alpha)$. By Theorem 17.24(1) the degree $[K(\alpha) : K] \geq p$. But $[L : K] = |G| = p$, so $L = K(\alpha)$. Hence $t^p - a$ is the minimal polynomial of $\alpha$ over $K$, otherwise we would have $[K(\alpha) : K] < p$. Being a minimal polynomial, $t^p - a$ is irreducible over $K$. □

We can now prove the promised converse to Theorem 15.8. Compare with Lemma 8.17(2).

**Theorem 18.21.** *Let $K$ be a field of characteristic 0 and let $L/K$ be a finite normal extension with soluble Galois group $G$. Then there exists an extension $R$ of $L$ such that $R/K$ is radical.*

*Proof.* All extensions are separable since the characteristic is 0. Use induction on $|G|$. The result is clear when $|G| = 1$. If $|G| \neq 1$, consider a maximal proper normal subgroup $H$ of $G$, which exists since $G$ is a finite group. Then $G/H$ is simple, since $H$ is maximal and is also soluble by Theorem 14.4(2). By Theorem 14.6, $G/H$ is cyclic of prime order $p$. Let $N$ be a splitting field over $L$ of $t^p - 1$. Then $N/K$ is normal, for by Theorem 9.9 $L$ is a splitting field over $K$ of some polynomial $f$, so $N$ is a splitting field over $L$ of $(t^p - 1)f$, which implies that $N/K$ is normal by Theorem 9.9.

The Galois group of $N/L$ is abelian by Lemma 15.6, and by Theorem 17.24(5) $\Gamma(L/K)$ is isomorphic to $\Gamma(N/K)/\Gamma(N/L)$. By Theorem 14.4(3), $\Gamma(N/K)$ is soluble. Let $M$ be the subfield of $N$ generated by $K$ and the zeros of $t^p - 1$. Then $N/M$ is normal. Now $M/K$ is clearly radical, and since $L \subseteq N$ the desired result will follow provided we can find an extension $R$ of $N$ such that $R/M$ is radical.

We claim that the Galois group of $N/M$ is isomorphic to a subgroup of $G$. Let us map any $M$-automorphism $\tau$ of $N$ into its restriction $\tau|_L$. Since $L/K$ is normal, $\tau|_L$ is a $K$-automorphism of $L$, and there is a group homomorphism

$$\phi : \Gamma(N/M) \to \Gamma(L/K).$$

If $\tau \in \ker(\phi)$ then $\tau$ fixes all elements of $M$ and $L$, which generate $N$. Therefore $\tau = 1$, so $\phi$ is a monomorphism, which implies that $\Gamma(N/M)$ is isomorphic to a subgroup $J$ of $\Gamma(L/K)$.

If $J = \phi(\Gamma(N/M))$ is a proper subgroup of $G$, then by induction there is an extension $R$ of $N$ such that $R/M$ is radical.

The remaining possibility is that $J = G$. Then we can find a subgroup $H \lhd \Gamma(N/M)$ of index $p$, namely $H = \phi^{-1}(H)$. Let $P$ be the fixed field $H^\dagger$. Then $[P : M] = p$ by Theorem 17.24(3), $P/M$ is normal by Theorem 17.24(4), and $t^p - 1$ splits in $M$. By Theorem 18.20, $P = M(\alpha)$ where $\alpha^p = a \in M$. But $N/P$ is a normal extension with soluble Galois group of order smaller than $|G|$, so by induction there exists an extension $R$ of $N$ such that $R/P$ is radical. But then $R/M$ is radical, and the theorem is proved. □

To extend this result to fields of characteristic $p > 0$, radical extensions must be defined differently. As well as adjoining elements $\alpha$ such that $\alpha^n$ lies in the given field, we must also allow adjunction of elements $\alpha$ such that $\alpha^p - \alpha$ lies in the given field (where $p$ is the same as the characteristic). It is then true that a polynomial is soluble by radicals if and only if its Galois group is soluble. The proof is different because we have to consider extensions of degree $p$ over fields of characteristic $p$. Then Theorem 18.20 breaks down, and second type above come in. If we do not modify the definition radicals then although every soluble polynomial has soluble erse need not hold—indeed, some quadratic polynomials with roup are not soluble by radicals, see Exercises 18.12 and 18.13. ting field is always a normal extension, we have:

**2.** *Over a field of characteristic zero, a polynomial is soluble d only if it has a soluble Galois group.*

orems 15.8 and 18.21.

□

---

## ng Equations of Degree Four or Less

polynomial of degree $n$ has Galois group $\mathbb{S}_n$, and we know his is soluble (Chapter 14). Theorem 18.22 therefore implies $K$ of characteristic zero, the general polynomial of degree ed by radicals. We already know this from the classical tricks ut now we can use the structure of the symmetric group to fied way, why those tricks work.

### ions

linear polynomial is

$$t - s_1$$

is a zero.

group here is trivial, and adds little to the discussion except the zero must lie in $K$.

### uations

quadratic polynomial is

$$t^2 - s_1t + s_2$$

$t_1$ and $t_2$. The Galois group $\mathbb{S}_2$ consists of the identity and a ng $t_1$ and $t_2$. By Hilbert's Theorem 90, Theorem 18.19, there ement which, when acted on by the nontrivial element of $\mathbb{S}_2$, a primitive square root of 1; that is, by $-1$. Obviously $t_1 - t_2$ y. Therefore $(t_1 - t_2)^2$ is fixed by $\mathbb{S}_2$, so lies in $K(s_1, s_2)$. By on

$$(t_1 - t_2)^2 = s_1^2 - 4s_2$$

$$
\begin{aligned}
t_1 - t_2 &= \pm\sqrt{s_1^2 - 4s_2} \\
t_1 + t_2 &= s_1
\end{aligned}
$$

ve the familiar formula

$$t_1, t_2 = \frac{s_1 \pm \sqrt{s_1^2 - 4s_2}}{2}$$

## quations

eneral cubic polynomial is

$$t^3 - s_1 t^2 + s_2 t - s_3$$

eros be $t_1, t_2, t_3$. The Galois group $\mathbb{S}_3$ has a series

$$1 \lhd \mathbb{A}_3 \lhd \mathbb{S}_3$$

ian quotients.

ated once more by Hilbert's Theorem 90, Theorem 18.19, we adjoin at $\omega \neq 1$ such that $\omega^3 = 1$. Consider

$$y = t_1 + \omega t_2 + \omega^2 t_3$$

ents of $\mathbb{A}_3$ permute $t_1$, $t_2$, and $t_3$ cyclically, and therefore multiply $y$ er of $\omega$. Hence $y^3$ is fixed by $\mathbb{A}_3$. Similarly if

$$z = t_1 + \omega^2 t_2 + \omega t_3$$

fixed by $\mathbb{A}_3$. Now any odd permutation in $\mathbb{S}_3$ interchanges $y^3$ and $z^3$, $+ z^3$ and $y^3 z^3$ are fixed by the whole of $\mathbb{S}_3$, hence lie in $K(s_1, s_2, s_3)$. formulas are given in the final section of this chapter.) Hence $y^3$ and ros of a quadratic over $K(s_1, s_2, s_3)$ which can be solved as in part ng cube roots we know $y$ and $z$. But since

$$s_1 = t_1 + t_2 + t_3$$

that

$$
\begin{aligned}
t_1 &= \tfrac{1}{3}(s_1 + y + z) \\
t_2 &= \tfrac{1}{3}(s_1 + \omega^2 y + \omega z) \\
t_3 &= \tfrac{1}{3}(s_1 + \omega y + \omega^2 z)
\end{aligned}
$$

## ations

quartic polynomial is

$$t^4 - s_1t^3 + s_2t^2 - s_3t + s_4$$

$t_1, t_2, t_3, t_4$. The Galois group $\mathbb{S}_4$ has a series

$$1 \lhd \mathbb{V} \lhd \mathbb{A}_4 \lhd \mathbb{S}_4$$

otients, where

$$\mathbb{V} = \{1, (1\,2)(3\,4), (1\,3)(2\,4), (1\,4)(2\,3)\}$$

-group. It is therefore natural to consider the three expressions

$$\begin{aligned} y_1 &= (t_1 + t_2)(t_3 + t_4) \\ y_2 &= (t_1 + t_3)(t_2 + t_4) \\ y_3 &= (t_1 + t_4)(t_2 + t_3) \end{aligned}$$

uted among themselves by any permutation in $\mathbb{S}_4$, so that ry symmetric polynomials in $y_1, y_2, y_3$ lie in $K(s_1, s_2, s_3, s_4)$. las are indicated below). Then $y_1, y_2, y_3$ are the zeros of a lynomial over $K(s_1, s_2, s_3, s_4)$ called the *resolvent cubic*. Since

$$t_1 + t_2 + t_3 + t_4 = s_1$$

e quadratic polynomials whose zeros are $t_1 + t_2$ and $t_3 + t_4$, $t_4$, $t_1 + t_4$ and $t_2 + t_3$. From these it is easy to find $t_1, t_2, t_3, t_4$.

---

## cit Formulas

eness, we now state, for degrees 3 and 4, the explicit formulas is alluded to above. For details of the calculations, see Van 953, pages 177–182). Compare with Section 1.4.

irnhaus transformation

$$u = t - \tfrac{1}{3}s_1$$

eral cubic polynomial to

$$u^3 + pu + q$$

e zeros of this it is an easy matter to deduce their values for c. The above procedure for this polynomial leads to

$$\begin{aligned} y^3 + z^3 &= -27q \\ y^3z^3 &= -27p^3 \end{aligned}$$

that $y^3$ and $z^3$ are the zeros of the quadratic polynomial

$$t^2 + 27qt - 27p^3$$

is Cardano's formula (1.8). See Figure 25.

25: Cardano, the first person to publish solutions of cubic and quartic
.

The Tschirnhaus transformation

$$u = t - \tfrac{1}{4}s_1$$

he quartic to the form

$$t^4 + pt^2 + qt + r$$

ove procedure,

$$\begin{aligned} y_1 + y_2 + y_3 &= 2p \\ y_1y_2 + y_1y_3 + y_2y_3 &= p^2 - 4r \\ y_1y_2y_3 &= -q^2 \end{aligned}$$

vent cubic is

$$t^3 - 2pt^2 + (p^2 - 4r)t + q^2 \tag{18.3}$$

sed form of (1.12) with $t = -2u$). Its zeros are $y_1, y_2, y_3$, and

$$\begin{aligned} t_1 &= \tfrac{1}{2}(\sqrt{-y_1} + \sqrt{-y_2} + \sqrt{-y_3}) \\ t_2 &= \tfrac{1}{2}(\sqrt{-y_1} - \sqrt{-y_2} - \sqrt{-y_3}) \\ t_3 &= \tfrac{1}{2}(-\sqrt{-y_1} + \sqrt{-y_2} - \sqrt{-y_3}) \\ t_4 &= \tfrac{1}{2}(-\sqrt{-y_1} - \sqrt{-y_2} + \sqrt{-y_3}) \end{aligned}$$

f the square roots must be chosen so that

$$\sqrt{-y_1}\sqrt{-y_2}\sqrt{-y_3} = -q$$

---

## S

ountable field and $L/K$ is finitely generated, show that $L$ is Hence show that $\mathbb{R}/\mathbb{Q}$ and $\mathbb{C}/\mathbb{Q}$ are not finitely generated.

the transcendence degrees of the following extensions:

$u, v, w)/\mathbb{Q}$ where $t, u, v, w$ are independent transcendental eles over $\mathbb{Q}$.

$u, v, w)/\mathbb{Q}$ where $t^2 = 2, u$ is transcendental over $\mathbb{Q}(t)$, $t + 5$, and $w$ is transcendental over $\mathbb{Q}(t, u, v)$.

$u, v)/\mathbb{Q}$ where $t^2 = u^3 = v^4 = 7$.

in Lemma 18.5 the degree $[L : M]$ is not independent of the $M$. (*Hint:* Consider $K(t^2)$ as a subfield of $K(t)$.)

hat $K \subseteq L \subseteq M$, and each of $M/K$, $L/K$ is finitely generated. $M/K$ and $L/K$ have the same transcendence degree if and $/L$ is finite.

eld $K$ show that $t^3 - 3t^2 + 3$ is either irreducible or splits in By Exercise 11.2(e), if $\alpha$ is a zero then so are $\beta = \frac{2\alpha-3}{\alpha-1}$ and Investigate when these expressions are defined and when they ent zeros.)

a field of characteristic zero. Suppose that $L/K$ is finite, noreparable with Galois group $G$. For any $a \in L$ define the *trace*

$$T(a) = \tau_1(a) + \cdots + \tau_n(a)$$

$\ldots, \tau_n$ are the distinct elements of $G$. Show that $T(a) \in K$ $T$ is a surjective map $L \to K$.

the previous exercise $G$ is cyclic with generator $\tau$, show that $T(a) =$ and only if $a = b - \tau(b)$ for some $b \in L$.

e by radicals the following polynomial equations over $\mathbb{Q}$:

$t^3 - 7t + 5 = 0$

$t^3 - 7t + 6 = 0$

$t^4 + 5t^3 - 2t - 1 = 0$

$t^4 + 4t + 2 = 0$

w that a finitely generated algebraic extension is finite, and hence an algebraic extension that is not finitely generated.

$\theta$ have minimal polynomial

$$t^3 + at^2 + bt + c$$

$\mathbb{Q}$. Find necessary and sufficient conditions in terms of $a, b, c$ such $\theta = \phi^2$ where $\phi \in \mathbb{Q}(\theta)$. (*Hint:* Consider the minimal polynomial .) Hence or otherwise express $\sqrt[3]{28} - 3$ as a square in $\mathbb{Q}(\sqrt[3]{28})$, and $- \sqrt[3]{4}$ as a square in $\mathbb{Q}(\sqrt[3]{5}, \sqrt[3]{2})$. (See Ramanujan 1962 page 329.)

$\Gamma$ be a finite group of automorphisms of $K$ with fixed field $K_0$. Let transcendental over $K$. For each $\sigma \in \Gamma$ show there is a unique omorphism $\sigma'$ of $K(t)$ such that

$$\sigma'(k) = \sigma(k) \quad (k \in K)$$
$$\sigma'(t) = t$$

w that the $\sigma'$ form a group $\Gamma'$ isomorphic to $\Gamma$, with fixed field $K_0(t)$.

$K$ be a field of characteristic $p$. Suppose that $f(t) = t^p - t - \alpha \in$ ]. If $\beta$ is a zero of $f$, show that the zeros of $f$ are $\beta + k$ where $0, 1, \ldots, p-1$. Deduce that either $f$ is irreducible over $K$ or $f$ splits $K$.

in exercise 18.12 is irreducible over $K$, show that the Galois group of cyclic. State and prove a characterisation of finite normal separable nsions with soluble Galois group in characteristic $p$.

k the following true or false.

Every finite extension is finitely generated.

Every finitely generated extension is algebraic.

The transcendence degree of a finitely generated extension is invariant under isomorphism.

(d) If $t_1, \ldots, t_n$ are independent transcendental elements, then their elementary symmetric polynomials are also independent transcendental elements.

(e) The Galois group of the general polynomial of degree $n$ is soluble for all $n$.

(f) The general quintic polynomial is soluble by radicals.

(g) The only proper subgroups of $\mathbb{S}_3$ are 1 and $\mathbb{A}_3$.

(h) The transcendence degree of $\mathbb{Q}(t)/\mathbb{Q}$ is 1.

(i) The transcendence degree of $\mathbb{Q}(t^2)/\mathbb{Q}$ is 2.

# Chapter 19

# Finite Fields

Fields that have finitely many elements are important in many branches of mathematics, including number theory, group theory, and projective geometry. They also have practical applications, especially to the coding of digital communications, see Lidl and Niederreiter (1986), and, especially for the history, Thompson (1983).

The most familiar examples of such fields are the fields $\mathbb{Z}_p$ of integers modulo a prime $p$, but these are not all. In this chapter we give a complete classification of all finite fields. It turns out that a finite field is uniquely determined up to isomorphism by the number of elements that it contains, that this number must be a power of a prime, and that for every prime $p$ and integer $n > 0$ there exists a field with $p^n$ elements. All these facts were discovered by Galois, though not in this terminology.

## 19.1 Structure of Finite Fields

We begin by proving the second of these three statements.

**Theorem 19.1.** *If $F$ is a finite field, then $F$ has characteristic $p > 0$, and the number of elements of $F$ is $p^n$ where $n$ is the degree of $F$ over its prime subfield.*

*Proof.* Let $P$ be the prime subfield of $F$. By Theorem 16.9, $P$ is isomorphic either to $\mathbb{Q}$ or to $\mathbb{Z}_p$ for prime $p$. Since $\mathbb{Q}$ is infinite, $P \cong \mathbb{Z}_p$. Therefore $F$ has characteristic $p$. By Theorem 6.1, $F$ is a vector space over $P$. This vector space has finitely many elements, so $[F : P] = n$ is finite. Let $x_1, \ldots, x_n$ be a basis for $F$ over $P$. Every element of $F$ is uniquely expressible in the form

$$\lambda_1 x_1 + \cdots + \lambda_n x_n$$

where $\lambda_1, \ldots, \lambda_n \in P$. Each $\lambda_j$ may be chosen in $p$ ways since $|P| = p$, hence there are $p^n$ such expressions. Therefore $|F| = p^n$.

$\square$

Thus there do not exist fields with $6, 10, 12, 14, 18, 20, \ldots$ elements. Notice the contrast with group theory, where there exist groups of any given

DOI: 10.1201/9781003213949-19

order. However, there exist non-isomorphic groups with equal orders. To show that this cannot happen for finite fields, we recall the Frobenius map, Definition 17.15, which maps $x$ to $x^p$ and is an automorphism when the field is finite by Lemma 17.14. We use the Frobenius automorphism to establish a basic uniqueness theorem for finite fields:

**Theorem 19.2.** *Let $p$ be any prime number and let $q = p^n$ where $n$ is any integer $> 0$. A field $F$ has $q$ elements if and only if it is a splitting field for $f(t) = t^q - t$ over the prime subfield $P \cong \mathbb{Z}_p$ of $F$.*

*Proof.* Suppose that $|F| = q$. The set $F \backslash \{0\}$ forms a group under multiplication, of order $q - 1$, so if $0 \neq x \in F$ then $x^{q-1} = 1$. Hence $x^q - x = 0$. Since $0^q - 0 = 0$, every element of $F$ is a zero of $t^q - t$, so $f(t)$ splits in $F$. Since the zeros of $f$ exhaust $F$, they certainly generate it, so $F$ is a splitting field for $f$ over $P$.

Conversely, let $K$ be a splitting field for $f$ over $\mathbb{Z}_p$. Since $Df = -1$, which is prime to $f$, all the zeros of $f$ in $K$ are distinct, so $f$ has exactly $q$ zeros. The set of zeros is precisely the set of elements fixed by $\phi^n$, that is, its fixed field. So the zeros form a field, which must therefore be the whole splitting field $K$. Therefore $|K| = q$.

□

Since splitting fields exist and are unique up to isomorphism, we deduce a complete classification of finite fields:

**Theorem 19.3.** *A finite field has $q = p^n$ elements where $p$ is a prime number and $n$ is a positive integer. For each such $q$ there exists, up to isomorphism, precisely one field with $q$ elements, which can be constructed as a splitting field for $t^q - t$ over $\mathbb{Z}_p$.*

**Definition 19.4.** The *Galois Field* $\mathbb{F}_q$ is the unique field with $q$ elements.

(The alternative notation $\mathbf{F}_q$ is also used, and $\mathbb{GF}(q)$ or $\mathbf{GF}(q)$ are common in older textbooks.)

---

## 19.2 The Multiplicative Group

The above classification of finite fields, although a useful result in itself, does not give any detailed information on their deeper structure. There are many questions we might ask—what are the subfields? How many are there? What are the Galois groups? We content ourselves with proving one important theorem, which gives the structure of the multiplicative group $F \backslash \{0\}$ of any finite field $F$. First we need to know a little more about abelian groups.

**Definition 19.5.** The *exponent* $e(G)$ of a finite group $G$ is the least common multiple of the orders of the elements of $G$.

The order of any element of $G$ divides the order $|G|$, so $e(G)$ divides $|G|$. In general, $G$ need not possess an element of order $e(G)$. For example if $G = \mathbb{S}_3$ then $e(G) = 6$, but $G$ has no element of order 6. Abelian groups are better behaved in this respect:

**Lemma 19.6.** *Any finite abelian group $G$ contains an element of order $e(G)$.*

*Proof.* Let $e = e(G) = p_1^{\alpha_1} \dots p_n^{\alpha_n}$ where the $p_j$ are distinct primes and $\alpha_j \geq 1$. The definition of $e(G)$ implies that for each $j$, the group $G$ must possess an element $g_j$ whose order is divisible by $p_j^{\alpha_j}$. Then a suitable power $a_j$ of $g_j$ has order $p_j^{\alpha_j}$. Define

$$g = a_1 a_2 \dots a_n \tag{19.1}$$

Suppose that $g^m = 1$ where $m \geq 1$. Then

$$a_j^m = a_1^{-m} \dots a_{j-1}^{-m} a_{j+1}^{-m} \dots a_n^{-m}$$

So if

$$q = p_1^{\alpha_1} \dots p_{j-1}^{\alpha_{j-1}} p_{j+1}^{\alpha_{j+1}} \dots p_n^{\alpha_n}$$

then $a_j^{mq} = 1$. But $q$ is prime to the order of $a_j$, so $p_j^{\alpha_j}$ divides $m$. Hence $e$ divides $m$. But clearly $g^e = 1$. Hence $g$ has order $e$, which is what we want. □

**Corollary 19.7.** *If $G$ is a finite abelian group such that $e(G) = |G|$, then $G$ is cyclic.*

*Proof.* The element $g$ in (19.1) generates $G$. □

We can apply this corollary immediately.

**Theorem 19.8.** *If $G$ is a finite subgroup of the multiplicative group $K \backslash \{0\}$ of a field $K$, then $G$ is cyclic.*

*Proof.* Since multiplication in $K$ is commutative, $G$ is an abelian group. Let $e = e(G)$. For any $x \in G$ we have $x^e = 1$, so that $x$ is a zero of the polynomial $t^e - 1$ over $K$. By Theorem 3.30 there are at most $e$ zeros of this polynomial, so $|G| \leq e$. But $e \leq |G|$, hence $e = |G|$; by Corollary 19.7, $G$ is cyclic. □

**Corollary 19.9.** *The multiplicative group of a finite field is cyclic.* □

Therefore for any finite field $F$ there is at least one element $x$ such that every non-zero element of $F$ is a power of $x$. We give two examples.

**Examples 19.10.** (1) The field $\mathbb{F}_{11}$. The powers of 2, in order, are

$$1, 2, 4, 8, 5, 10, 9, 7, 3, 6, 1$$

so 2 generates the multiplicative group. On the other hand, the powers of 4 are

$$1, 4, 5, 9, 3, 1$$

so 4 does not generate the group.

(2) The field $\mathbb{F}_{25}$. This can be constructed as a splitting field for $t^2 - 2$ over $\mathbb{Z}_5$, since $t^2 - 2$ is irreducible and of degree 2. We can therefore represent the elements of $\mathbb{F}_{25}$ in the form $a + b\alpha$ where $\alpha^2 = 2$. There is no harm in writing $\alpha = \sqrt{2}$. By trial and error we are led to consider the element $2 + \sqrt{2}$. Successive powers of this are

$$\begin{array}{ccccccc} 1 & 2+\sqrt{2} & 1+4\sqrt{2} & 4\sqrt{2} & 3+3\sqrt{2} & 2+4\sqrt{2} & 2 \\ 4+2\sqrt{2} & 2+3\sqrt{2} & 3\sqrt{2} & 1+\sqrt{2} & 4+3\sqrt{2} & 4 & \\ 3+4\sqrt{2} & 4+\sqrt{2} & \sqrt{2} & 2+2\sqrt{2} & 3+\sqrt{2} & 3 & \\ 1+3\sqrt{2} & 3+2\sqrt{2} & 2\sqrt{2} & 4+4\sqrt{2} & 1+2\sqrt{2} & 1 & \end{array}$$

Hence $2 + \sqrt{2}$ generates the multiplicative group.

Aside from computer algorithms, there is no known procedure for finding a generator other than enlightened trial and error. Fortunately the existence of a generator often provides sufficient information.

## 19.3 Counterexample to the Primitive Element Theorem

The theory of finite fields leads to a counterexample to the Primitive Element Theorem 17.29 in the case that $L/K$ is not separable.

**Example 19.11.** We given an example of a field extension such that Theorem 17.29 is false. This shows that some extra condition, such as separability, is necessary.

Let $p$ be prime. Let $K = \mathbb{F}_p(t, u)$, the field of rational expressions in two independent indeterminates $t, u$ over $\mathbb{F}_p$. Let $L = K(\alpha, \beta)$ where $\alpha^p = t, \beta^p = u$.

The degree $[L : K]$ is $p^2$, because

$$[L : K] = [K(\alpha, \beta) : K] = [K(\alpha, \beta)K(\alpha)][K(\alpha) : K] = p.p = p^2$$

Consider any intermediate field $K(\theta)$. By Lemma 17.14 the Frobenius map $\phi(x) = x^p$ is an automorphism. There is some polynomial $h(x, y) \in K[x, y]$ such that $\theta = h(\alpha, \beta)$. Therefore

$$\theta^p = h(\alpha, \beta)^p = h(\alpha^p \beta^p) = h(t, u) \in K$$

So $[K(\theta) : K] \leq p$. In particular $K(\theta) \neq L$.

This example does not contradict Theorem 17.29 because Example 17.16 implies that the extension $L/K$ is not separable.

## 19.4 Application to Solitaire

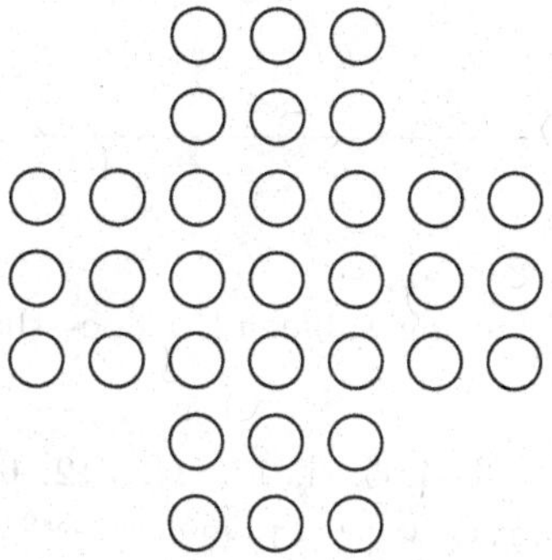

FIGURE 26: The Solitaire board.

Finite fields have an unexpected and elegant application to the recreational pastime of (peg) solitaire, see de Bruijn (1972). Solitaire is played on a board with holes arranged like Figure 26. A peg is placed in each hole, except the centre one, and play proceeds by jumping any peg horizontally or vertically over an adjacent peg into an empty hole; the peg that is jumped over is removed. The player's objective is to remove all pegs except one, which—traditionally—is the peg that occupies the central hole. Can it be another hole? Experiment shows that it can, but suggests that the final peg cannot occupy *any* hole. Which holes are possible?

De Bruijn's idea is to use the field $\mathbb{F}_4$, whose addition and multiplication tables are given in Exercise 16.6, in terms of elements $0, 1, \alpha, \beta$. Consider the holes as a subset of the integer lattice $\mathbb{Z}^2$, with the origin $(0,0)$ at the centre and the axes horizontal and vertical as usual. If $X$ is a set of pegs, define

$$A(X) = \sum_{(x,y)\in X} \alpha^{x+y} \qquad B(X) = \sum_{(x,y)\in X} \alpha^{x-y}$$

Observe that if a legal move changes $X$ to $Y$, then $A(Y) = A(X)$ and $B(Y) = B(X)$. This follows easily from the equation $\alpha^2 + \alpha + 1 = 0$, which in turn follows from the tables. Thus the pair $(A(X), B(X))$ is invariant under any sequence of legal moves.

The starting position $X$ has $A(X) = B(X) = 1$. Therefore any position $Y$ that arises during the game must satisfy $A(Y) = B(Y) = 1$. If the game ends with a single peg on $(x, y)$ then $\alpha^{x+y} = \alpha^{x-y} = 1$. Now $\alpha^3 = 1$, so $\alpha$ has order 3 in the multiplicative group of nonzero elements of $\mathbb{F}_4$. Therefore $x+y, x-y$ are multiples of 3, so $x, y$ are multiples of 3. Thus the only possible end positions are $(-3,0), (0,-3), (0,0), (0,3), (3,0)$. All five of these positions can be obtained by a series of legal moves. The case $(0, 0)$ is the traditional finishing position, and numerous solutions are known. The same final move leads to one of the other positions if we move the other peg, and symmetry yields the rest.

---

## EXERCISES

19.1 For which of the following values of $n$ does there exist a field with $n$ elements?

$$1,\ 2,\ 3,\ 4,\ 5,\ 6,\ 17,\ 24,\ 312,\ 65536,$$
$$65537,\ 83521,\ 103823,\ 2^{82589933} - 1$$

(*Hint:* See 'Mersenne primes' under 'Internet' in the References.)

19.2 Construct fields having 8, 9, and 16 elements.

19.3 Let $\phi$ be the Frobenius automorphism of $\mathbb{F}_{p^n}$. Find the smallest value of $m > 0$ such that $\phi^m$ is the identity map.

19.4 Show that the subfields of $\mathbb{F}_{p^n}$ are isomorphic to $\mathbb{F}_{p^r}$ where $r$ divides $n$, and there exists a unique subfield for each such $r$.

19.5 Show that the Galois group of $\mathbb{F}_{p^n}/\mathbb{F}_p$ is cyclic of order $n$, generated by the Frobenius automorphism $\phi$. Show that for finite fields the Galois correspondence is a bijection, and find the Galois groups of

$$\mathbb{F}_{p^n}/\mathbb{F}_{p^m}$$

whenever $m$ divides $n$.

19.6 Are there any composite numbers $r$ that divide all the binomial coefficients $\binom{r}{s}$ for $1 \leq s \leq r - 1$?

19.7 Find generators for the multiplicative groups of $\mathbb{F}_{p^n}$ when $p^n = 8$, 9, 13, 17, 19, 23, 29, 31, 37, 41, and 49.

19.8 Show that the additive group of $\mathbb{F}_{p^n}$ is a direct product of $n$ cyclic groups of order $p$.

19.9 By considering the field $\mathbb{Z}_2(t)$, show that the Frobenius monomorphism is not always an automorphism.

19.10* For which values of $n$ does $\mathbb{S}_n$ contain an element of order $e(\mathbb{S}_n)$? (*Hint:* Use the cycle decomposition to estimate the maximum order of an element of $\mathbb{S}_n$, and compare this with an estimate of $e(\mathbb{S}_n)$. You may need estimates on the size of the $n$th prime: for example, 'Bertrand's Postulate', which states that the interval $[n, 2n]$ contains a prime for any integer $n \geq 1$.)

19.11* Prove that in a finite field every element is a sum of two squares.

19.12 Mark the following true or false.

(a) There is a finite field with 124 elements.

(b) There is a finite field with 125 elements.

(c) There is a finite field with 126 elements.

(d) There is a finite field with 127 elements.

(e) There is a finite field with 128 elements.

(f) The multiplicative group of $\mathbb{F}_{19}$ contains an element of order 3.

(g) $\mathbb{F}_{2401}$ has a subfield isomorphic to $\mathbb{F}_{49}$.

(h) Any monomorphism from a finite field to itself is an automorphism.

(i) The additive group of a finite field is cyclic.

# Chapter 20

## Regular Polygons

We return with more sophisticated weapons to the time-honoured problem of ruler-and-compass construction, introduced in Chapter 7. We consider the following question: for which values of $n$ can the regular $n$-sided polygon be constructed by ruler and compass?

The ancient Greeks knew constructions for 3-, 5-, and 15-gons; they also knew how to construct a $2n$-gon given an $n$-gon, by the obvious method of bisecting the angles. We describe these constructions in Section 20.1. For about two thousand years little progress was made beyond the Greeks. If you answered Exercises 7.16 or 7.17 you got further than they did. It seemed 'obvious' that the Greeks had found all the constructible regular polygons ... Then, on 30 March 1796, Gauss made the remarkable discovery that the regular 17-gon can be constructed (Figure 27). He was nineteen years old at the time. So pleased was he with this discovery that he resolved to dedicate the rest of his life to mathematics, having until then been unable to decide between that and the study of languages. In his *Disquisitiones Arithmeticae*, reprinted as Gauss (1966), he stated necessary and sufficient conditions for constructibility of the regular $n$-gon, and proved their sufficiency; he claimed to have a proof of necessity although he never published it. Doubtless he did: Gauss knew a proof when he saw one.

### 20.1 What Euclid Knew

Euclid's *Elements* gets down to business straight away. The first regular polygon constructed there is the equilateral triangle, in Book 1 Proposition 1. Figure 28 (left) makes the construction fairly clear.

The square also makes its appearance in Book 1:

**Proposition 46 (Euclid)** *On a given straight line to describe a square.*

In the proof, which we give in detail to illustrate Euclid's style, notation such as [1,31] refers to Proposition 31 of Book 1 of the *Elements*. The proof is taken from Heath (1956), the classic edition of Euclid's *Elements*. Refer to Figure 28 (right) for the lettering.

DOI: 10.1201/9781003213949-20

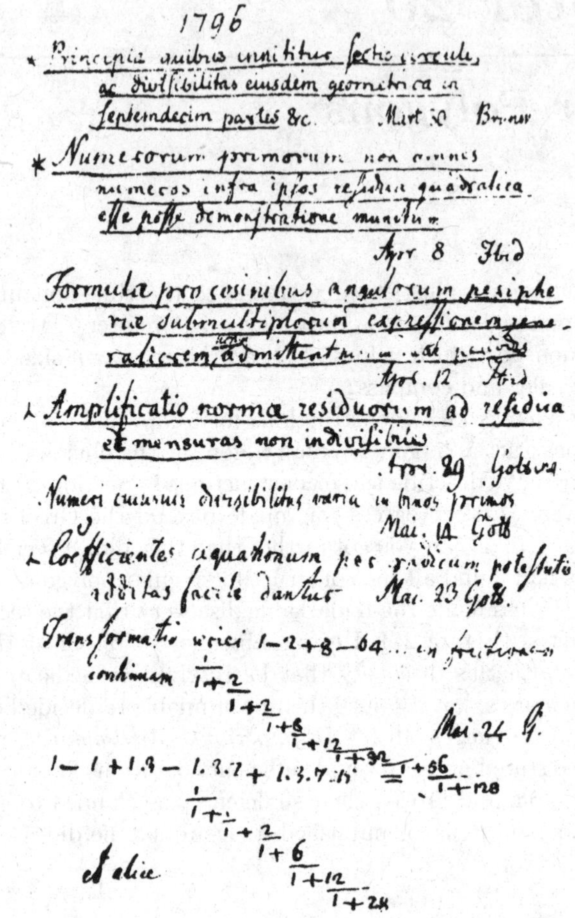

FIGURE 27: The first entry in Gauss's notebook records his discovery that the regular 17-gon can be constructed.

*Proof.* Let AB be the given straight line; thus it is required to describe a square on the straight line AB.

Let AC be drawn at right angles to the straight line AB from the point A on it [1, 11], and let AD be made equal to AB;
through the point D let DE be drawn parallel to AB,
and through the point B let BE be drawn parallel to AD. [1,31]

Therefore ADEB is a parallelogram;

therefore AB is equal to DE, and AD to BE. [1, 34]

But AB is equal to AD;

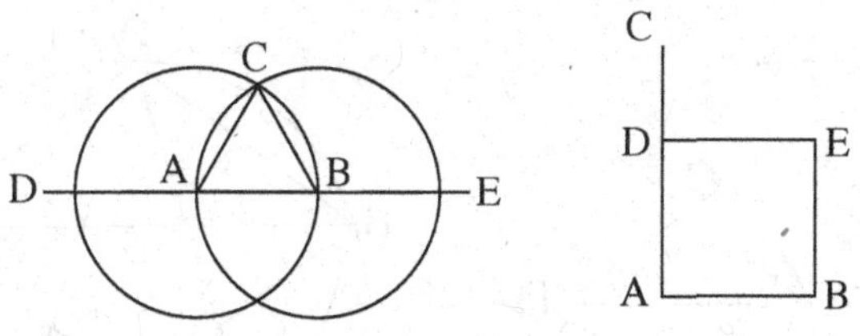


FIGURE 28: *Left*: Euclid's construction of an equilateral triangle. *Right*: Euclid's construction of a square.

therefore the four straight lines BA, AD, DE, EB are equal to one another;
therefore the parallelogram ADEB is equilateral.

I say next that it is also right-angled.

For, since the straight line AD falls upon the parallels AB, DE,
the angles BAD, ADE are equal to two right angles. [1, 29]
But the angle BAD is also right;
therefore the angle ADE is also right.

And in parallelogrammic areas the opposite sides and angles are equal to one another; [1, 34]
therefore each of the opposite angles ABE, BED is also right.
Therefore ADEB is right-angled.

And it was also proved equilateral.

Therefore it is a square; and it is described on the straight line AB. Q.E.F.

□

Here Q.E.F. (quod erat faciendum—that which was to be done) replaces the familiar Q.E.D. (quod erat demonstrandum—that which was to be proved) because this is not a theorem but a construction. In any case, the Latin phrase occurs in later translations: Euclid wrote in Greek. Now imagine you are a Victorian schoolboy—it always *was* a schoolboy in those days—trying to learn Euclid's proof by heart, including the exact choice of letters in the diagrams ...

Euclid's construction of the regular pentagon has to wait until Book 4 Proposition 11, because it depends on some quite sophisticated ideas, notably Proposition 10 of Book 4: *To construct an isosceles triangle having each of the angles at the base double of the remaining one.* In modern terms: construct a triangle with angles $2\pi/5, 2\pi/5, \pi/5$. Euclid's method for doing this is shown in Figure 29. Given AB, find C so that AB×BC = CA$^2$. To do that, see Book 2 Proposition 11, which is itself quite complicated—the construction here is essentially the famous 'golden section', a name that seems to have been introduced in 1835 by Martin Ohm (Herz-Fischler 1998, Livio 2002). Euclid's method is given in Exercise 20.9. Next, draw the circle centre A radius AB, and find D such that BD = AC. Then triangle ABD is the one required.

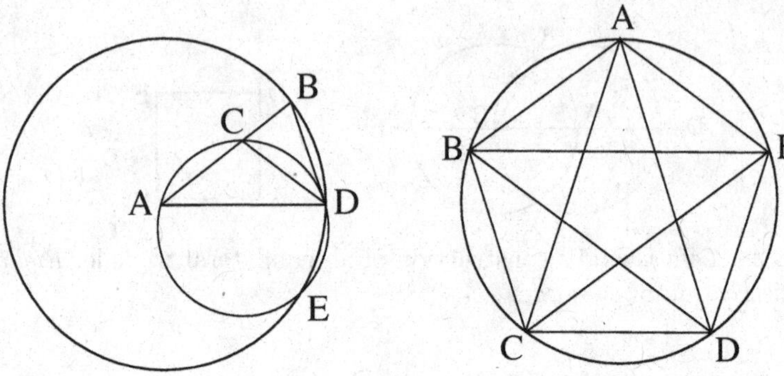


FIGURE 29: *Left*: Euclid's construction of an isosceles triangle with base angles $2\pi/5$. *Right*: Euclid's construction of a regular pentagon. Make ACD similar to triangle ABD in the left-hand Figure and proceed from there.

With this shape of triangle under his belt, Euclid then constructs the regular pentagon: Figure 29 (right) shows his method.

The hexagon occurs in Book 4 Proposition 15, and the 15-gon in Book 4 Proposition 16. Bisection of any angle, Book 1 Proposition 9, effectively completes the Euclidean catalogue of constructible regular polygons.

---

## 20.2 Which Constructions are Possible?

That, however, was not the end of the story.

We derived necessary and sufficient conditions for the existence of a ruler-and-compass construction in Theorem 7.9. We restate it here for convenience in equivalent form as:

**Theorem 20.1.** *Suppose that $K$ is a subfield of $\mathbb{C}$, generated by points in a subset $P \subseteq \mathbb{C}$. Let $\alpha$ lie in an extension $L$ of $K$ such that there exists a finite series of subfields*

$$K = K_0 \subseteq K_1 \subseteq \cdots \subseteq K_r = L$$

*such that $[K_{j+1} : K_j] = 2$ for $j = 0, \dots, r-1$. Then the point $\alpha \in \mathbb{C}$ is constructible from $P$. The converse is also valid.*

There is a more useful, but weaker, version of Theorem 20.1. To prove it, we first need:

**Lemma 20.2.** *If $G$ is a finite group and $|G| = 2^r$ for $r \geq 1$, then its centre $Z(G)$ contains an element of order 2.*

*Proof.* Use the class equation (14.2). We have

$$1 + C_2 + \cdots + C_k = 2^r$$

so some $C_j$ is odd. By Corollary 14.12 this $C_j$ also divides $2^r$, so we must have $|C_j| = 1$. Hence $Z(G) \neq 1$. Now apply Lemma 14.14. □

**Corollary 20.3.** *If $G$ is a finite group and $|G| = 2^r$ then there exists a series*

$$1 = G_0 \subseteq G_1 \subseteq \cdots \subseteq G_r = G$$

*of normal subgroups of $G$, such that $|G_j| = 2^j$ for $0 \leq j \leq r$.*

*Proof.* Use Lemma 20.2 and induction. □

Now we can state and prove the promised modification of Theorem 20.1.

**Proposition 20.4.** *If $K$ is a subfield of $\mathbb{C}$, generated by points in a subset $P \subseteq \mathbb{C}$, and if $\alpha$ lies in a normal extension $L$ of $K$ such that $[L : K] = 2^r$ for some integer $r$, then $\alpha$ is constructible from $P$.*

*Proof.* $L/K$ is separable since the characteristic is zero. Let $G$ be the Galois group of $L/K$. By Theorem 12.2(1) $|G| = 2^r$. By Corollary 20.3, $G$ has a series of normal subgroups

$$1 = G_0 \subseteq G_1 \subseteq \cdots \subseteq G_r = G$$

such that $|G_j| = 2^j$. Let $K_j$ be the fixed field $G^\dagger_{r-j}$. By Theorem 12.2(3) $[K_{j+1} : K_j] = 2$ for all $j$. By Theorem 20.1, $\alpha$ is constructible from $P$. □

---

## 20.3 Regular Polygons

We use a mixture of algebraic and geometric ideas to find those values of $n$ for which the regular $n$-gon is constructible. To save breath, make the following (non-standard) definition:

**Definition 20.5.** The positive integer $n$ is *constructive* if the regular $n$-gon is constructible by ruler and compass.

The first step is to reduce the problem to prime-power values of $n$.

**Lemma 20.6.** *If $n$ is constructive and $m$ divides $n$, then $m$ is constructive. If $m$ and $n$ are coprime and constructive, then $mn$ is constructive.*

*Proof.* If $m$ divides $n$, then we can construct a regular $m$-gon by joining every $d$th vertex of a regular $n$-gon, where $d = n/m$.

If $m$ and $n$ are coprime, then there exist integers $a, b$ such that $am+bn = 1$. Therefore

$$\frac{1}{mn} = a\frac{1}{n} + b\frac{1}{m}$$

Hence from angles $2\pi/m$ and $2\pi/n$ we can construct $2\pi/mn$, and from this we obtain a regular $mn$-gon. □

**Corollary 20.7.** *Suppose that* $n = p_1^{m_1} \dots p_r^{m_r}$ where $p_1, \dots, p_r$ *are distinct primes. Then* $n$ *is constructive if and only if each* $p_j^{m_j}$ *is constructive.* □

Another obvious result:

**Lemma 20.8.** *For any positive integer* $m$*, the number* $2^m$ *is constructive.*

*Proof.* Any angle can be bisected by ruler and compass, and the result follows by induction on $m$. □

This reduces the problem of constructing regular polygons to the case when the number of sides is an odd prime power. Now we bring in the algebra. In the complex plane, the set of $n$th roots of unity forms the vertices of a regular $n$-gon. Further, as we have seen repeatedly, these roots of unity are the zeros in $\mathbb{C}$ of the polynomial

$$t^n - 1 = (t-1)(t^{n-1} + t^{n-2} + \cdots + t + 1)$$

We concentrate on the second factor on the right-hand side: $f(t) = t^{n-1} + t^{n-2} + \cdots + t + 1$. Its zeros are the powers $\zeta^k$ for $1 \leq k \leq n-1$ of a primitive $n$th root of unity

$$\zeta = \mathrm{e}^{2\pi \mathrm{i}/n}$$

**Lemma 20.9.** *Let* $p$ *be a prime such that* $p^n$ *is constructive. Let* $\zeta$ *be a primitive* $p^n$*th root of unity in* $\mathbb{C}$*. Then the degree of the minimal polynomial of* $\zeta$ *over* $\mathbb{Q}$ *is a power of* 2.

*Proof.* Take $\zeta = \mathrm{e}^{2\pi\mathrm{i}/p^n}$. The number $p^n$ is constructive if and only if we can construct $\zeta$ from $\mathbb{Q}$. Hence by Theorem 7.14 $[\mathbb{Q}(\zeta) : \mathbb{Q}]$ is a power of 2. Hence the degree of the minimal polynomial of $\zeta$ over $\mathbb{Q}$ is a power of 2. □

The next step is to calculate the relevant minimal polynomials to find their degrees. It turns out to be sufficient to consider $p$ and $p^2$ only.

**Lemma 20.10.** *If* $p$ *is a prime and* $\zeta$ *is a primitive* $p$*th root of unity in* $\mathbb{C}$*, then the minimal polynomial of* $\zeta$ *over* $\mathbb{Q}$ *is*

$$f(t) = 1 + t + \cdots + t^{p-1}$$

*Proof.* This polynomial is irreducible over $\mathbb{Q}$ by Lemma 3.24. Clearly $\zeta$ is a zero. Therefore it is the minimal polynomial of $\zeta$. □

To prove the case $p^2$, we apply the method of Lemma 3.24.

**Lemma 20.11.** *If $p$ is a prime and $\zeta$ is a primitive $p^2$th root of unity in $\mathbb{C}$, then the minimal polynomial of $\zeta$ over $\mathbb{Q}$ is*

$$g(t) = 1 + t^p + \cdots + t^{p(p-1)}$$

*Proof.* Note that $g(t) = (t^{p^2} - 1)/(t^p - 1)$. Now $\zeta^{p^2} - 1 = 0$ but $\zeta^p - 1 \neq 0$ so $g(\zeta) = 0$. It suffices to show that $g(t)$ is irreducible over $\mathbb{Q}$. As before we make the substitution $t = 1 + u$. Then

$$g(1+u) = \frac{(1+u)^{p^2} - 1}{(1+u)^p - 1}$$

and modulo $p$ this is

$$\frac{(1+u^{p^2}) - 1}{(1+u^p) - 1} = u^{p(p-1)}$$

Therefore $g(1+u) = u^{p(p-1)} + pk(u)$ where $k$ is a polynomial in $u$ over $\mathbb{Z}$. From the alternative expression

$$g(1+u) = 1 + (1+u)^p + \cdots + (1+u)^{p(p-1)}$$

it follows that $k$ has constant term 1. By Eisenstein's Criterion, $g(1+u)$ is irreducible over $\mathbb{Q}$. □

We can now obtain a more specific result than Lemma 15.4 for $p$th roots of unity over $\mathbb{Q}$:

**Theorem 20.12.** *Let $p$ be prime and let $\zeta$ be a primitive $p$th root of unity in $\mathbb{C}$. Then the Galois group of $\mathbb{Q}(\zeta)/\mathbb{Q}$ is cyclic of order $p - 1$.*

*Proof.* This follows the same lines as the proof of Lemma 15.4, but now we can say a little more.

The zeros in $\mathbb{C}$ of $t^p - 1$ are $\zeta^j$, where $0 \leq j \leq p-1$, and these are distinct. These zeros form a group under multiplication, and this group is cyclic, generated by $\zeta$. Therefore any $\mathbb{Q}$-automorphism of $\mathbb{Q}(\zeta)$ is determined by its effect on $\zeta$. Further, $\mathbb{Q}$-automorphisms permute the zeros of $t^p - 1$. Hence any $\mathbb{Q}$-automorphism of $\mathbb{Q}(\zeta)$ has the form

$$\rho_k : \zeta \mapsto \zeta^k$$

and is uniquely determined by this condition.

We claim that every $\rho_k$ is, in fact, a $\mathbb{Q}$-automorphism of $\mathbb{Q}(\zeta)$. The $\zeta^j$ with $j > 0$ are the zeros of $1 + t + \cdots + t^{p-1}$. This polynomial is irreducible over $\mathbb{Q}$ by Lemma 3.24. Therefore it is the minimal polynomial of any of its zeros, namely $\zeta^j$ where $1 \leq j \leq p-1$. By Proposition 11.4, every $\rho_k$ is a $\mathbb{Q}$-automorphism of $\mathbb{Q}(\zeta)$, as claimed.

Clearly $\rho_i\rho_j = \rho_{ij}$, where the product $ij$ is taken modulo $p$. Therefore the Galois group of $\mathbb{Q}(\zeta)/\mathbb{Q}$ is isomorphic to the multiplicative group $\mathbb{Z}_p^*$. This is cyclic by Corollary 19.9. □

We now come to the main result of this chapter.

**Theorem 20.13 (Gauss).** *The regular $n$-gon is constructible by ruler and compass if and only if*

$$n = 2^r p_1 \dots p_s$$

*where $r$ and $s$ are integers $\geq 0$, and $p_1, \dots, p_s$ are distinct odd primes of the form*

$$p_j = 2^{2^{r_j}} + 1$$

*for positive integers $r_j$.*

*Proof.* Let $n$ be constructive. Then $n = 2^r p_1^{m_1} \dots p_s^{m_s}$ where $p_1, \dots, p_s$ are distinct odd primes. By Corollary 20.7, each $p_j^{m_j}$ is constructive. If $m_j \geq 2$ then $p_j^2$ is constructive by Theorem 20.1. Hence the degree of the minimal polynomial of a primitive $p_j^2$th root of unity over $\mathbb{Q}$ is a power of 2 by Lemma 20.9. By Lemma 20.11, $p_j(p_j - 1)$ is a power of 2, which cannot happen since $p_j$ is odd. Therefore $m_j = 1$ for all $j$. Therefore $p_j$ is constructive. By Lemma 3.24

$$p_j - 1 = 2^{s_j}$$

for suitable $s_j$. Suppose that $s_j$ has an odd divisor $a > 1$, so that $s_j = ab$. Then

$$p_j = (2^b)^a + 1$$

which is divisible by $2^b + 1$ since

$$t^a + 1 = (t+1)(t^{a-1} - t^{a-2} + \dots + 1)$$

when $a$ is odd. So $p_j$ cannot be prime. Hence $s_j$ has no odd factors, so

$$s_j = 2^{r_j}$$

for some $r_j > 0$.

This establishes the necessity of the given form of $n$. Now we prove sufficiency. By Corollary 20.7 we need consider only prime-power factors of $n$. By Lemma 20.8, $2^r$ is constructive. We must show that each $p_j$ is constructive. Let $\zeta$ be a primitive $p_j$th root of unity. By Theorem 20.12

$$[\mathbb{Q}(\zeta) : \mathbb{Q}] = p_j - 1 = 2^{s_j}$$

Now $\mathbb{Q}(\zeta)$ is a splitting field for $f(t) = 1 + \dots + t^{p-1}$ over $\mathbb{Q}$, so that $\mathbb{Q}(\zeta)/\mathbb{Q}$ is normal. It is also separable since the characteristic is zero. By Lemma 15.5, the Galois group $\Gamma(\mathbb{Q}(\zeta)/\mathbb{Q})$ is abelian, and by Theorem 20.12 or an appeal to the Galois correspondence it has order $2^{s_j}$. By Proposition 20.4, $\zeta \in \mathbb{C}$ is constructible. □

## 20.4 Fermat Numbers

The problem of finding all constructible regular polygons now reduces to number theory, and there the question has a longer history. In 1640 Pierre de Fermat wondered when $2^k + 1$ is prime, and proved that a necessary condition is for $k$ to be a power of 2. Thus we are led to:

**Definition 20.14.** The $n$th *Fermat number* is $F_n = 2^{2^n} + 1$.

The question becomes: when is $F_n$ prime?

Fermat noticed that $F_0 = 3, F_1 = 5, F_2 = 17, F_3 = 257$, and $F_4 = 65537$ are all prime. He conjectured that $F_n$ is prime for all $n$, but this was disproved by Euler in 1732, who proved that $F_5$ is divisible by 641 (Exercise 20.5). Knowledge of factors of Fermat numbers is changing almost daily, thanks to the prevalence of fast computers and special algorithms for primality testing of Fermat numbers: see References under 'Internet'. At the time of writing, the largest known composite Fermat number was $F_{18233954}$, with a factor $7.2^{18233956} + 1$, found in October 2020. At that time, 312 Fermat numbers were known to be composite.

No new Fermat primes have been found, so the only known Fermat primes are still those found by Fermat himself: 3, 5, 17, 257, and 65537. We sum up the current state of knowledge as:

**Proposition 20.15.** *If $p$ is a prime, then the regular $p$-gon is constructible for $p = 3, 5, 17, 257, 65537$.*

## 20.5 How to Construct a Regular 17-gon

Many constructions for the regular 17-gon have been devised, the earliest published being that of Huguenin (see Klein 1913) in 1803. For several of these constructions there are proofs of their correctness which use only synthetic geometry (ordinary Euclidean geometry without coordinates). A series of papers giving a construction for the regular 257-gon was published by F.J. Richelot (1832), under one of the longest titles I have ever seen. Bell (1965) tells of an over-zealous research student being sent away to find a construction for the 65537-gon, and reappearing with one twenty years later. This story, though apocryphal, is not far from the truth; Professor Hermes of Lingen spent ten years on the problem, and his manuscripts are still preserved at Göttingen.

One way to construct a regular 17-gon is to follow faithfully the above theory, which in fact provides a perfectly definite construction after a little extra calculation. With ingenuity it is possible to shorten the work. The construction that we now describe is taken from Hardy and Wright (1962).

Our immediate object is to find radical expressions for the zeros of the polynomial

$$\frac{t^{17}-1}{t-1} = t^{16} + \cdots + t + 1 \tag{20.1}$$

over $\mathbb{C}$. We know the zeros are $\zeta^k$, where $\zeta = e^{2\pi i/17}$ and $1 \leq k \leq 16$. To simplify notation, let

$$\theta = 2\pi/17$$

so that $\zeta^k = \cos k\theta + i \sin k\theta$.

Theorem 20.12 for $n = 17$ implies that the Galois group $\Gamma(\mathbb{Q}(\zeta)/\mathbb{Q})$ consists of the $\mathbb{Q}$-automorphisms $\rho_k$ defined by

$$\rho_k(\zeta) = \zeta^k \quad 1 \leq k \leq 16$$

and this is isomorphic to the multiplicative group $\mathbb{Z}_{17}^*$. By Theorem 19.8 $\mathbb{Z}_{17}^*$ is cyclic of order 16.

Galois theory now implies that $\zeta$ is constructible. In fact, there must exist a generator $\alpha$ for $\mathbb{Z}_{17}^*$. Then $\alpha^{16} = 1$ and no smaller power of $\alpha$ is 1. Consider the series of subgroups

$$1 = \langle \alpha^{16} \rangle \triangleleft \langle \alpha^8 \rangle \triangleleft \langle \alpha^4 \rangle \triangleleft \langle \alpha^2 \rangle \triangleleft \langle \alpha \rangle = \mathbb{Z}_{17}^* \tag{20.2}$$

The Galois correspondence leads to a tower of subfields from $\mathbb{Q}$ to $\mathbb{Q}(\zeta)$ in which each step is an extension of degree 2. By Theorem 20.1, $\zeta$ is constructible, so the regular 17-gon is constructible.

To convert this to an explicit construction we must find a generator for $\mathbb{Z}_{17}^*$. Experimenting with small values, $\alpha = 2$ is not a generator (it has order 8), but $\alpha = 3$ is a generator. In fact, the powers of 3 modulo 17 are:

| $m$ | 0 | 1 | 2 | 3 | 4 | 5 | 6 | 7 | 8 | 9 | 10 | 11 | 12 | 13 | 14 | 15 |
|---|---|---|---|---|---|---|---|---|---|---|---|---|---|---|---|---|
| $3^m$ | 1 | 3 | 9 | 10 | 13 | 5 | 15 | 11 | 16 | 14 | 8 | 7 | 4 | 12 | 2 | 6 |

Motivated by (20.2), define

$$\begin{aligned}
x_1 &= \zeta + \zeta^9 + \zeta^{13} + \zeta^{15} + \zeta^{16} + \zeta^8 + \zeta^4 + \zeta^2 \\
x_2 &= \zeta^3 + \zeta^{10} + \zeta^5 + \zeta^{11} + \zeta^{14} + \zeta^7 + \zeta^{12} + \zeta^6 \\
y_1 &= \zeta + \zeta^{13} + \zeta^{16} + \zeta^4 \\
y_2 &= \zeta^9 + \zeta^{15} + \zeta^8 + \zeta^2 \\
y_3 &= \zeta^3 + \zeta^5 + \zeta^{14} + \zeta^{12} \\
y_4 &= \zeta^{10} + \zeta^{11} + \zeta^7 + \zeta^6
\end{aligned}$$

By definition, these lie in various fixed fields in the aforementioned tower. Now

$$\zeta^k + \zeta^{17-k} = 2\cos k\theta \tag{20.3}$$

for $k = 1, \ldots, 16$, so

$$\begin{aligned}
x_1 &= 2(\cos\theta + \cos 8\theta + \cos 4\theta + \cos 2\theta) \\
x_2 &= 2(\cos 3\theta + \cos 7\theta + \cos 5\theta + \cos 6\theta) \\
y_1 &= 2(\cos\theta + \cos 4\theta) \\
y_2 &= 2(\cos 8\theta + \cos 2\theta) \\
y_3 &= 2(\cos 3\theta + \cos 5\theta) \\
y_4 &= 2(\cos 7\theta + \cos 6\theta)
\end{aligned} \tag{20.4}$$

Equation (20.1) implies that

$$x_1 + x_2 = -1$$

Now (20.4) and the identity

$$2\cos m\theta \cos n\theta = \cos(m+n)\theta + \cos(m-n)\theta$$

imply that

$$x_1 x_2 = 4(x_1 + x_2) = -4$$

using (20.3). Hence $x_1$ and $x_2$ are zeros of the quadratic polynomial

$$t^2 + t - 4 \tag{20.5}$$

Further, $x_1 > 0$ so that $x_1 > x_2$. By further trigonometric expansions,

$$y_1 + y_2 = x_1 \qquad y_1 y_2 = -1$$

and $y_1, y_2$ are the zeros of

$$t^2 - x_1 t - 1 \tag{20.6}$$

Further, $y_1 > y_2$. Similarly, $y_3$ and $y_4$ are the zeros of

$$t^2 - x_2 t - 1 \tag{20.7}$$

and $y_3 > y_4$. Now

$$\begin{aligned}
2\cos\theta + 2\cos 4\theta &= y_1 \\
4\cos\theta \cos 4\theta = 2\cos 5\theta + 2\cos 3\theta &= y_3
\end{aligned}$$

so

$$z_1 = 2\cos\theta \qquad z_2 = 2\cos 4\theta$$

are the zeros of

$$t^2 - y_1 t + y_3 \tag{20.8}$$

and $z_1 > z_2$.

Solving the series of quadratics (20.5–20.8) and using the inequalities to decide which zero is which, we obtain

$$\cos\theta = \frac{1}{16}\Bigg(-1+\sqrt{17}+\sqrt{34-2\sqrt{17}} \tag{20.9}$$
$$+\sqrt{68+12\sqrt{17}-16\sqrt{34+2\sqrt{17}}-2(1-\sqrt{17})\sqrt{34-2\sqrt{17}}}\Bigg)$$

where the square roots are the positive ones.

From this we can deduce a geometric construction for the 17-gon by constructing the relevant square roots. By using greater ingenuity it is possible to obtain an aesthetically more satisfying construction. The following method (Figure 30) is due to Richmond (1893).

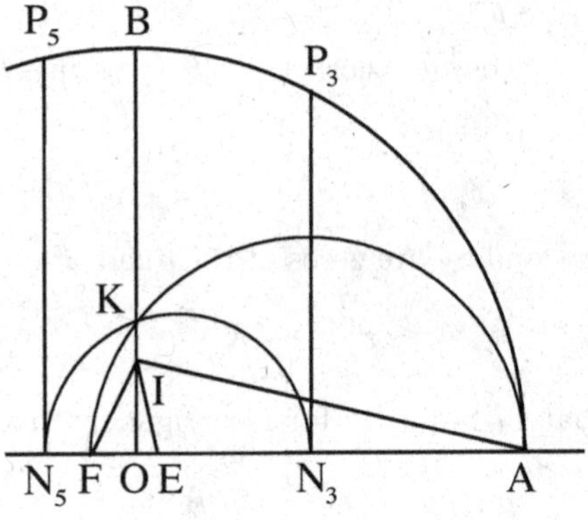


FIGURE 30: Richmond's construction for a regular 17-gon.

Let $\phi$ be the smallest positive acute angle such that $\tan 4\phi = 4$. Then $\phi, 2\phi$, and $4\phi$ are all acute. Expression (20.5) can be written

$$t^2 + 4t\cot 4\phi - 4$$

whose zeros are

$$2\tan 2\phi \qquad -2\cot 2\phi$$

Hence

$$x_1 = 2\tan 2\phi \qquad x_2 = -2\cot 2\phi$$

This implies that

$$y_1 = \tan\left(\phi + \frac{\pi}{4}\right) \qquad y_2 = \tan\left(\phi - \frac{\pi}{4}\right) \qquad y_3 = \tan\phi \qquad y_4 = -\cot\phi$$

so that

$$\begin{aligned} 2(\cos 3\theta + \cos 5\theta) &= \tan\phi \\ 4\cos 3\theta\cos 5\theta &= \tan\left(\phi - \tfrac{\pi}{4}\right) \end{aligned} \tag{20.10}$$

In Figure 30, let OA, OB be two perpendicular radii of a circle. Make OI $= \frac{1}{4}$OB and $\angle$OIE $= \frac{1}{4}\angle$OIA. Find F on AO produced to make $\angle$EIF $= \frac{\pi}{4}$. Let the circle on AF as diameter cut OB in K, and let the circle centre E through K cut OA in $N_3$ and $N_5$ as shown. Draw $N_3P_3$ and $N_5P_5$ perpendicular to OA. Then $\angle$OIA $= 4\phi$ and $\angle$OIE $= \phi$. Also

$$\begin{aligned} 2(\cos\angle AOP_3 + \cos\angle AOP_5) &= 2\frac{ON_3 - ON_5}{OA} \\ &= 4\frac{OE}{OA} + \frac{OE}{OI} = \tan\phi \end{aligned}$$

and

$$\begin{aligned} 4\cos\angle AOP_3 \cos\angle AOP_5 &= -4\frac{ON_3 \times ON_5}{OA \times OA} \\ &= -4\frac{OK^2}{OA^2} \\ &= -4\frac{OF}{OA} \\ &= -\frac{OF}{OI} = \tan\left(\phi - \frac{\pi}{4}\right) \end{aligned}$$

Comparing these with equation (20.10) we see that

$$\angle AOP_3 = 3\theta \qquad \angle AOP_5 = 5\theta$$

Hence A, $P_3$, $P_5$ are the zeroth, third, and fifth vertices of a regular 17-gon inscribed in the given circle. The other vertices are now easily found.

In Chapter 21 we return to topics associated with regular polygons, especially so-called cyclotomic polynomials. We end that chapter by investigating the construction of regular polygons when an angle-trisector is permitted, as well as the traditional ruler and compass.

---

## EXERCISES

20.1 Using only the operations 'ruler' and 'compass', show how to draw a parallel to a given line through a given point.

20.2 Verify the following approximate constructions for regular $n$-gons found by Oldroyd (1955):

(a) 7-*gon*. Construct $\cos^{-1}\frac{4+\sqrt{5}}{10}$ giving an angle of approximately $2\pi/7$.

(b) 9-*gon*. Construct $\cos^{-1}\frac{5\sqrt{3}-1}{10}$.

(c) 11-*gon.* Construct $\cos^{-1}\frac{8}{9}$ and $\cos^{-1}\frac{1}{2}$ and take their difference.

(d) 13-*gon.* Construct $\tan^{-1}1$ and $\tan^{-1}\frac{4+\sqrt{5}}{20}$ and take their difference.

20.3 Show that for $n$ odd the only known constructible $n$-gons are precisely those for which $n$ is a divisor of $2^{32}-1=4294967295$.

20.4 Work out the approximate size of $F_{18233954}$, which is known to be composite. Explain why it is no easy task to find factors of Fermat numbers.

20.5 Use the equations

$$641 = 5^4 + 2^4 = 5.2^7 + 1$$

to show that 641 divides $F_5$.

20.6 Show that

$$F_{n+1} = 2 + F_n F_{n-1} \dots F_0$$

and deduce that if $m \neq n$ then $F_m$ and $F_n$ are coprime. Hence show that there are infinitely many prime numbers.

20.7 List the values of $n \leq 100$ for which the regular $n$-gon can be constructed by ruler and compass.

20.8 Verify the following construction for the regular pentagon.

Draw a circle centre O with two perpendicular radii $\text{OP}_0$, OB. Let D be the midpoint of OB, join $\text{P}_0\text{D}$. Bisect $\angle\text{ODP}_0$ cutting $\text{OP}_0$ at N. Draw $\text{NP}_1$ perpendicular to $\text{OP}_0$ cutting the circle at $\text{P}_1$. Then $\text{P}_0$ and $\text{P}_1$ are the zeroth and first vertices of a regular pentagon inscribed in the circle.

20.9 Euclid's construction for an isosceles triangle with angles $4\pi/5, 4\pi/5, 2\pi/5$ depends on constructing the so-called golden section: that is, *To construct a given straight line so that the rectangle contained by the whole and one of the segments is equal to the square on the other segment.* The Greek term was 'extreme and mean ratio'. In Book 2 Proposition 11 of the *Elements* Euclid solves this problem as in Figure 31.

Let AB be the given line. Make ABDC a square. Bisect AC at E, and make EF = BE. Now find H such that AH = AF. Then the square on AH has the same area as the rectangle with sides AB and BH, as required.

Prove that Euclid was right.

20.10 Mark the following true or false.

(a) $2^n+1$ cannot be prime unless $n$ is a power of 2.

(b) If $n$ is a power of 2 then $2^n+1$ is always prime.

(c) The regular 771-gon is constructible using ruler and compass.

(d) The regular 768-gon is constructible using ruler and compass.

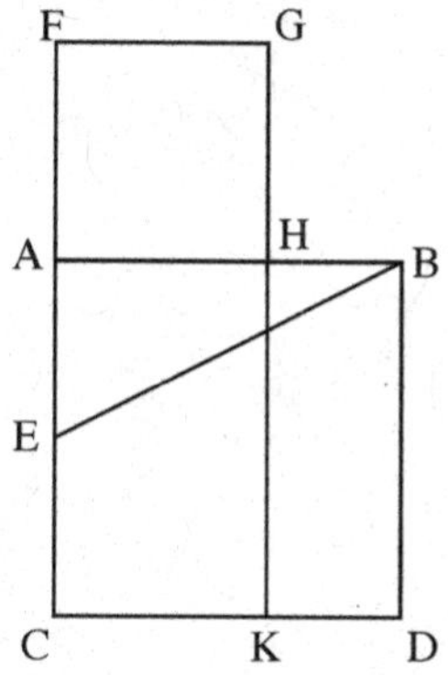


FIGURE 31: Cutting a line in extreme and mean ratio.

(e) The regular 51-gon is constructible using ruler and compass.

(f) The regular 25-gon is constructible using ruler and compass.

(g) For an odd prime $p$, the regular $p^2$-gon is never constructible using ruler and compass.

(h) If $n$ is an integer $> 0$ then a line of length $\sqrt{n}$ can always be constructed from $\mathbb{Q}$ using ruler and compass.

(i) If $n$ is an integer $> 0$ then a line of length $\sqrt[4]{n}$ can always be constructed from $\mathbb{Q}$ using ruler and compass.

(j) A point whose coordinates lie in a normal extension of $\mathbb{Q}$ whose degree is a power of 2 is constructible using ruler and compass.

(k) If $p$ is a prime, then $t^{p^2} - 1$ is irreducible over $\mathbb{Q}$.

# Chapter 21

## Circle Division

To halt the story of regular polygons at the stage of ruler-and-compass constructions would leave a small but significant gap in our understanding of the solution of polynomial equations by radicals. Our definition of 'radical extension' involves a slight cheat, which becomes evident if we ask what the expression of a root of unity looks like. Specifically, what does the radical expression of the primitive 11th root of unity

$$\zeta_{11} = \cos\frac{2\pi}{11} + \mathrm{i}\sin\frac{2\pi}{11}$$

look like?

As the theory stands, the best we can offer is

$$\sqrt[11]{1} \tag{21.1}$$

which is not terribly satisfactory, because the obvious interpretation of $\sqrt[11]{1}$ is 1, not $\zeta_{11}$. Gauss's theory of the 17-gon hints that there might be a more impressive answer. In place of $\sqrt[17]{1}$ Gauss has a marvellously complicated system of nested square roots, which we saw in equation (20.9), with a similar expression for $\sin\frac{2\pi}{17}$, leading to an even more impressive formula for $\zeta_{17} = \cos\frac{2\pi}{17} + \mathrm{i}\sin\frac{2\pi}{17}$.

Can something similar be done for the 11th root of unity? For *all* roots of unity? The answer to both questions is 'yes', and we are getting the history back to front, because Gauss gave that answer as part of his work on the 17-gon. Indeed, Vandermonde came very close to the same answer 25 years earlier, in 1771, and in particular he managed to find an expression by radicals for $\zeta_{11}$ that is less disappointing than (21.1). He, in turn, built on the epic investigations of Lagrange.

The technical term for this area is 'cyclotomy', from the Greek for 'circle cutting'. In particular, pursuing Gauss's and Vandermonde's line of enquiry will lead us to some fascinating properties of the 'cyclotomic polynomial' $\Phi_d(t)$, which is the minimal polynomial over $\mathbb{Q}$ of a primitive $d$th root of unity in $\mathbb{C}$.

From the viewpoint of Galois theory, the results of Vandermonde and Gauss are not surprising. A key step proves that if $\zeta$ is a primitive $n$th root of unity then the Galois group of $\mathbb{Q}(\zeta)/\mathbb{Q}$ is abelian. Since every abelian group is soluble, Theorem 18.22 now implies that the minimal polynomial of $\zeta$ over $\mathbb{Q}$ is soluble by radicals, so $\zeta$ can be expressed by radicals. Our main efforts are

DOI: 10.1201/9781003213949-21

therefore directed at finding the Galois group of $\mathbb{Q}(\zeta)/\mathbb{Q}$, which we accomplish in Section 21.6.

## 21.1 Genuine Radicals

Of course, we can 'solve' the entire problem at a stroke if we *define* $\sqrt[n]{1}$ to be the primitive $n$th root of unity

$$\cos\frac{2\pi}{n} + \mathrm{i}\sin\frac{2\pi}{n}$$

instead of defining it to be 1. In a sense, this is what Definition 15.1 does. However, there is a better solution, as we shall see. What makes the above interpretation of $\sqrt[n]{1}$ unsatisfactory? Consider the typical case of $\zeta_{17} = \sqrt[17]{1}$. The minimal polynomial of $\zeta_{17}$ is not $t^{17} - 1$, as the notation $\sqrt[17]{1}$ suggests; instead, it has degree 16, being equal to

$$t^{16} + t^{15} + \cdots t + 1$$

It would be reasonable to seek to determine the zeros of this 16th degree equation using radicals of degree 16 or less, but a 17th root seems rather out of place. Especially since we know from Gauss that in this case (nested) square roots are enough.

However, that is a rather special example. What about other $n$th roots of unity? Can they also be expressed by what we might informally call 'genuine' radicals, those not employing the $\sqrt[n]{1}$ trick? (We pin down this concept formally in Definition 21.2.) Classically, the answer was found to be 'yes' for $2 \leq n \leq 10$, as we now indicate.

When $n = 2$, the primitive square root of unity is $-1$. This lies in $\mathbb{Q}$, so no radicals are needed.

When $n = 3$, the primitive cube roots of unity are solutions of the *quadratic* equation

$$t^2 + t + 1 = 0$$

and so are of the form $\omega, \omega^2$ where

$$\omega = -\frac{1}{2} + \mathrm{i}\frac{\sqrt{3}}{2}$$

involving only a square root.

When $n = 4$, a primitive 4th root of unity is i, which again can be represented using only a square root, since $\mathrm{i} = \sqrt{-1}$.

When $n = 5$, we have to solve

$$t^4 + t^3 + t^2 + t + 1 = 0 \tag{21.2}$$

We know from Chapter 18 that any quartic can be solved by radicals; indeed only square and cube roots are required (in part because $\sqrt[4]{x} = \sqrt{\sqrt{x}}$). But we can do better. There is a standard trick that applies to equations of even degree that are *palindromic*—the list of coefficients is symmetric about the central term. We encountered this trick in Exercises 15.4 and 15.5: express the equations in terms of a new variable

$$u = t + \frac{1}{t} \tag{21.3}$$

Then

$$\begin{aligned} u^2 &= t^2 + 2 + \frac{1}{t^2} \\ u^3 &= t^3 + 3t + \frac{3}{t} + \frac{1}{t^3} \end{aligned}$$

and so on. Rewrite (21.2) by dividing by $t^2$:

$$t^2 + t + 1 + \frac{1}{t} + \frac{1}{t^2} = 0$$

which in terms of $u$ becomes

$$u^2 + u - 1 = 0$$

which is quadratic in $u$. Solving for $u$:

$$u = \frac{-1 \pm \sqrt{5}}{2}$$

Now we recover $t$ from $u$ by solving a second quadratic equation. From (21.3)

$$t^2 - ut + 1 = 0$$

so

$$t = \frac{u \pm \sqrt{u^2 - 4}}{2}$$

Explicitly, we get four zeros:

$$t = \frac{-1 \pm \sqrt{5} \pm \sqrt{-10 \pm 2\sqrt{5}}}{4} \tag{21.4}$$

with independent choices of the first two $\pm$ signs, and the third equalling the first. So we can express primitive 5th roots of unity using nothing worse than square roots.

Continuing in this way, we can find a radical expression for a primitive 6th root of unity (it is $-\omega$); a primitive 7th root of unity (use the $t + 1/t$ trick to reduce to a cubic); a primitive 8th root of unity ($\sqrt{\mathrm{i}}$ is one possibility, $\frac{1+\mathrm{i}}{\sqrt{2}}$ is perhaps better); a primitive 9th root of unity ($\sqrt[3]{\omega}$); and a primitive 10th root of unity ($-\zeta_5$). The first case that baffled mathematicians prior to 1771 was therefore the primitive 11th root of unity, which leads to a *quintic* if we try the $t+1/t$ trick. But in that year, Vandermonde obtained the explicit radical expression

$$\begin{aligned}\zeta_{11} = \frac{1}{5}\Bigg[&\sqrt[5]{\frac{11}{4}\left(89+25\sqrt{5}-5\sqrt{-5+2\sqrt{5}}+45\sqrt{-5-2\sqrt{5}}\right)}\\ &+\sqrt[5]{\frac{11}{4}\left(89+25\sqrt{5}+5\sqrt{-5+2\sqrt{5}}-45\sqrt{-5-2\sqrt{5}}\right)}\\ &+\sqrt[5]{\frac{11}{4}\left(89-25\sqrt{5}-5\sqrt{-5+2\sqrt{5}}-45\sqrt{-5-2\sqrt{5}}\right)}\\ &+\sqrt[5]{\frac{11}{4}\left(89-25\sqrt{5}+5\sqrt{-5+2\sqrt{5}}+45\sqrt{-5-2\sqrt{5}}\right)}\Bigg]\end{aligned}$$

He stated that his method would work for any primitive $n$th root of unity, but he did not give a proof. That was supplied by Gauss in 1796, (but with a gap in the proof, see below) and it was published in 1801 in his *Disquisitiones Arithmeticae*. It is not known whether Gauss was aware of Vandermonde's pioneering work.

---

## 21.2 Fifth Roots Revisited

Before proving a version of Gauss's theorem on the representability of roots of unity by genuine radicals, it helps to have an example. We can explain Vandermonde's approach in the simpler case $n = 5$, where explicit calculations are not too lengthy.

We need some notation:

**Definition 21.1.** The *group of units* $\mathbb{Z}_n^*$ of $\mathbb{Z}_n$ consists of the elements $a \in \mathbb{Z}_n$ such that $1 \leq a \leq n$ and $a$ is prime to $n$, under the operation of multiplication.

As before, we want to solve

$$t^4 + t^3 + t^2 + t + 1 = 0$$

by radicals. We know that the zeros are

$$\zeta \qquad \zeta^2 \qquad \zeta^3 \qquad \zeta^4$$

where $\zeta = \cos\frac{2\pi}{5} + \mathrm{i}\sin\frac{2\pi}{5}$. The exponents $1, 2, 3, 4$ can be considered as elements of the multiplicative group of the field $\mathbb{Z}_5$. Since 5 is prime, Theorem 20.12 shows that the Galois group of $\mathbb{Q}(\zeta)/\mathbb{Q}$ consists of the $\mathbb{Q}$-automorphisms

$$\rho_k : \zeta \mapsto \zeta^k \quad 1 \leq j \leq 4$$

and is therefore isomorphic to the group of units $\mathbb{Z}_5^*$, which is cyclic of order 4 by Theorem 19.8. Experiment quickly shows that it is generated by the element 2 (mod 5). Indeed, modulo 5 the powers of 2 are

$$2^0 = 1 \qquad 2^1 = 2 \qquad 2^2 = 4 \qquad 2^3 = 3 \tag{21.5}$$

Hilbert's Theorem 90, Theorem 18.19, leads us to consider the expression

$$\alpha_1 = \zeta + \mathrm{i}\zeta^2 - \zeta^4 - \mathrm{i}\zeta^3$$

and compute its fourth power. We find (suppressing some details) that

$$\alpha_1^2 = -(1+2\mathrm{i})(\zeta - \zeta^2 + \zeta^4 - \zeta^3)$$

so, squaring again,

$$\alpha_1^4 = -15 + 20\mathrm{i}$$

Therefore we can express $\alpha_1$ by radicals:

$$\alpha_1 = \sqrt[4]{-15+20\mathrm{i}}$$

We can play a similar game with

$$\alpha_3 = \zeta - \mathrm{i}\zeta^2 - \zeta^4 + \mathrm{i}\zeta^3$$

to get

$$\alpha_3 = \sqrt[4]{-15-20\mathrm{i}}$$

The calculation of $\alpha_1^4$ also draws attention to

$$\alpha_2 = \zeta - \zeta^2 + \zeta^4 - \zeta^3$$

and shows that $\alpha_2^2 = 5$, so

$$\alpha_2 = \sqrt{5}$$

Summarising:

$$\begin{aligned}
\alpha_0 &= \zeta + \zeta^2 + \zeta^4 + \zeta^3 &&= -1 \\
\alpha_1 &= \zeta + \mathrm{i}\zeta^2 - \zeta^4 - \mathrm{i}\zeta^3 &&= \sqrt[4]{-15+20\mathrm{i}} \\
\alpha_2 &= \zeta - \zeta^2 + \zeta^4 - \zeta^3 &&= \sqrt{5} \\
\alpha_3 &= \zeta - \mathrm{i}\zeta^2 - \zeta^4 + \mathrm{i}\zeta^3 &&= \sqrt[4]{-15-20\mathrm{i}}
\end{aligned}$$

Thus we find four *linear* equations in $\zeta, \zeta^2, \zeta^3, \zeta^4$. These equations are independent, and we can solve them. In particular,

$$\alpha_0 + \alpha_1 + \alpha_2 + \alpha_3$$

is equal to

$$\zeta(1+1+1+1)+\zeta^2(1+\mathrm{i}-1-\mathrm{i})+\zeta^4(1-1+1-1)+\zeta^3(1-\mathrm{i}-1+\mathrm{i}) = 4\zeta$$

Therefore

$$\zeta = \frac{1}{4}\left(-1-\sqrt{5}+\sqrt{\sqrt{-15+20\mathrm{i}}}+\sqrt{\sqrt{-15-20\mathrm{i}}}\right)$$

This expression is superficially different from (21.4), but in fact the two are equivalent. Both use nothing worse than square roots.

This calculation is too remarkable to be mere coincidence. It must work out nicely because of some hidden structure. What lies behind it?

The general idea behind Vandermonde's calculation, as isolated by Gauss, is the following. Recall Definition 21.1, which introduces the group of units $\mathbb{Z}_n^*$ of the ring $\mathbb{Z}_n$. This consists of all elements that have a multiplicative inverse (mod $n$), and it is a group under multiplication. When $n$ is prime, this consists of all nonzero elements. In general, it consists of those elements that are prime to $n$.

Consider the multiplicative group $\mathbb{Z}_5^*$. This group is cyclic of order 4, and the number 2 (modulo 5) is a generator. It has order 4 in $\mathbb{Z}_5^*$. The complex number i is a primitive 4th root of unity, so i has order 4 in the multiplicative group of 4th roots of unity, namely $1, \mathrm{i}, -1, -\mathrm{i}$. These two facts conspire to make the algebra work.

To see how, we apply a little Galois theory—a classic case of being wise after the event. In the previous section we observed that the Galois group $\Gamma$ of $\mathbb{Q}(\zeta)/\mathbb{Q}$ has order 4 and comprises the $\mathbb{Q}$-automorphisms generated by the maps

$$\rho_k : \zeta \mapsto \zeta^k$$

for $k = 1, 2, 3, 4$. The group $\Gamma$ is isomorphic to $\mathbb{Z}_5^*$ by the map $\rho_k \mapsto k \,(\text{mod } 5)$. Therefore $\rho_2$ has order 4 in $\Gamma$, hence generates $\Gamma$, and $\Gamma$ is cyclic of order 4.

The extension is normal, since it is a splitting field for an irreducible polynomial, and we are working over $\mathbb{C}$ so the extension is separable. By the Galois correspondence, any rational function of $\zeta$ that is fixed by $\rho_2$ is in fact a rational number.

Consider as a typical case the expression $\alpha_1$ above. Write this as

$$\alpha_1 = \zeta + \mathrm{i}\rho_2(\zeta) + \mathrm{i}^2\rho_2^2(\zeta) + \mathrm{i}^3\rho_2^3(\zeta)$$

Then

$$\rho_2(\alpha_1) = \rho_2(\zeta) + \mathrm{i}\rho_2^2(\zeta) + \mathrm{i}^2\rho_2^3(\zeta) + \mathrm{i}^3\zeta$$

since $\rho_4^4(\zeta) = \zeta$. Therefore

$$\rho_2(\alpha_1) = \mathrm{i}^{-1}\alpha_1$$

so

$$\rho_2(\alpha_1^4) = (\mathrm{i}^{-1}\alpha_1)^4 = \alpha_1^4$$

Thus $\alpha_1^4$ lies in the fixed field of $\rho_2$, that is, the fixed field of $\Gamma$, which is $\mathbb{Q}$...

Hold it.

The *idea* is right, but the argument has a flaw. The explicit calculation shows that $\alpha_1^4 = -15 + 20\mathrm{i}$, which lies in $\mathbb{Q}(\mathrm{i})$, not $\mathbb{Q}$. What was the mistake? The problem is that $\alpha_1$ is not an element of $\mathbb{Q}(\zeta)$. It belongs to the larger field $\mathbb{Q}(\zeta)(\mathrm{i})$, which equals $\mathbb{Q}(\mathrm{i}, \zeta)$. So we have to do the Galois theory for $\mathbb{Q}(\mathrm{i}, \zeta)/\mathbb{Q}$, not $\mathbb{Q}(\zeta)/\mathbb{Q}$.

This is fairly straightforward. Since 4 and 5 are coprime, the product $\xi = \mathrm{i}\zeta$ is a primitive 20th root of unity. Moreover, $\xi^5 = \mathrm{i}$ and $\xi^{16} = \zeta$. Therefore $\mathbb{Q}(\mathrm{i}, \zeta) = \mathbb{Q}(\xi)$. Since 20 is not prime, we do not know that this group is cyclic, so we have to work out its structure. In fact, it is the group of units $\mathbb{Z}_{20}^*$ of the ring $\mathbb{Z}_{20}$, which is isomorphic to $\mathbb{Z}_2 \times \mathbb{Z}_4$, not $\mathbb{Z}_8$. By considering the tower of fields

$$\mathbb{Q} \subseteq \mathbb{Q}(\mathrm{i}) \subseteq \mathbb{Q}(\xi)$$

and using the structure of $\mathbb{Z}_{20}^*$, it can be shown that the Galois group of $\mathbb{Q}(\xi)/\mathbb{Q}(\mathrm{i})$ is the subgroup of $\mathbb{Z}_{20}^*$ isomorphic to $\mathbb{Z}_4$, generated by the $\mathbb{Q}(\mathrm{i})$-automorphism $\tilde{\rho}_2$ that sends $\zeta$ to $\zeta^2$ and fixes $\mathbb{Q}(\mathrm{i})$. We prove a more general result in Lemma 21.4 below.

Having made the switch to $\mathbb{Q}(\xi)$, the above calculation shows that $\alpha_1^4$ lies in the fixed field of the Galois group $\Gamma(\mathbb{Q}(\xi)/\mathbb{Q}(\mathrm{i}))$. This fixed field is $\mathbb{Q}(\mathrm{i})$, because the extension is normal and separable. So without doing the explicit calculations, we can see in advance that $\alpha_1^4$ must lie in $\mathbb{Q}(\mathrm{i})$. The same goes for $\alpha_2^4, \alpha_3^4$, and (trivially) $\alpha_0^4$.

---

## 21.3 Vandermonde Revisited

Vandermonde did not follow up his idea in full generality, and thereby missed a major discovery. He could well have anticipated Gauss, possibly even Galois, if he had found a proof that his method was a completely general way to express roots of unity by genuine radicals, instead of just asserting that it was.

As preparation, we now establish Vandermonde's main point about the primitive 11th roots of unity. Any unproved assertions about Galois groups will be dealt with in the general case, see Section 21.4. Let $\zeta = \zeta_{11}$. Vandermonde started with the usual identity

$$\zeta^{10} + \zeta^9 + \cdots + \zeta + 1 = 0$$

and played the $u = \zeta + 1/\zeta$ trick to reduce the problem to a quintic, but with hindsight this step is not necessary and makes the underlying idea more obscure. Introduce a primitive 10th root of unity $\theta$, so that $\theta\zeta$ is a primitive 110th root of unity. Consider the field extension $\mathbb{Q}(\theta\zeta)/\mathbb{Q}(\theta)$, which turns out to be of degree 10, with a cyclic Galois group of order 10 that is isomorphic to $\mathbb{Z}_{11}^*$. A generator for $\mathbb{Z}_{11}^*$ is readily found; one possibility is the number 2, whose successive powers are

$$1, 2, 4, 8, 5, 10, 9, 7, 3, 6$$

Therefore $\Gamma = \Gamma(\mathbb{Q}(\theta\zeta)/\mathbb{Q}(\theta))$ consists of the $\mathbb{Q}(\theta)$-automorphisms $\sigma_k$, for $k = 1, \ldots, 10$, that map

$$\zeta \mapsto \zeta^k \qquad \theta \mapsto \theta$$

Let $l$ be any integer, $0 \leq l \leq 9$, and define

$$\begin{aligned} \alpha_l &= \zeta + \theta^l\zeta^2 + \theta^{2l}\zeta^4 + \cdots + \theta^{9l}\zeta^6 \\ &= \textstyle\sum_{j=0}^{9} \theta^{jl}\zeta^{2^j} \end{aligned} \tag{21.6}$$

Consider the effect of $\sigma_2$, which sends $\zeta \mapsto \zeta^2$ and fixes $\theta$. We have

$$\sigma_2(\alpha_l) = \sum_{j=0}^{9} \theta^{jl}\zeta^{2^{j+1}} = \theta^{-l}\alpha_l$$

so

$$\sigma_2(\alpha_l^{10}) = \theta^{-10l}\alpha_l^{10} = \alpha_l^{10}$$

and $\alpha_l^{10}$ lies in the fixed field of $\Gamma$, which is $\mathbb{Q}(\theta)$). Thus there is some polynomial $f_l(\theta)$, of degree $\leq 9$ over $\mathbb{Q}$, with

$$\alpha_l^{10} = f_l(\theta)$$

With effort, we can compute $f_l(\theta)$ explicitly. Short cuts help. At any rate,

$$\alpha_l = \sqrt[10]{f_l(\theta)} \tag{21.7}$$

We already know how to express $\theta$ by genuine radicals since it is a primitive 10th root of unity, so we have expressed $\alpha_l$ by radicals—in fact, only square roots and fifth roots are needed, since $\sqrt[10]{\ } = \sqrt[5]{\sqrt{\ }}$ and fifth roots of unity require only square roots.

Finally, equations (21.6) for the $\alpha_l$ can be interpreted as a system of 10 linear equations for the powers $\zeta, \zeta^2, \ldots, \zeta^{10}$ over $\mathbb{C}$. These equations are independent, so the system can be solved. Indeed, using elementary properties of 10th roots of unity, it can be shown that

$$\zeta^{2^j} = \frac{1}{10}\left(\sum_{l=0}^{9} \theta^{-jl}\alpha_l\right)$$

In particular,

$$\zeta = \frac{1}{10}\left(\sum_{l=0}^{9} \alpha_l\right) = \frac{1}{10}\left(\sum_{l=0}^{9} \sqrt[10]{f_l(\theta)}\right)$$

Thus we have expressed $\zeta_{11}$ in terms of radicals, using only square roots and fifth roots.

Vandermonde's answer also uses only square roots and fifth roots, and can be deduced from the above formula. Because he used a variant of the above strategy, his answer does not immediately look the same as ours, but it is equivalent. To go beyond Vandermonde, we must prove that his method works for *all* primitive $n$th roots of unity. This we now establish.

---

## 21.4 The General Case

The time has come to define what we mean by a 'genuine' radical expression. Recall from Definition 8.12 that the *radical degree* of the radical $\sqrt[n]{}$ is $n$, and define the radical degree of a radical expression to be the maximum radical degree of the radicals that appear in it.

**Definition 21.2.** A number $\alpha \in \mathbb{C}$ has a *genuine radical expression* if $\alpha$ belongs to a radical extension of $\mathbb{Q}$ formed by successive adjunction of $k$th roots of elements $\beta$, where at every step the polynomial $t^k - \beta$ is irreducible over the field to which the root is adjoined.

This definition rules out $\sqrt[11]{1}$ as a genuine radical expression for $\zeta_{11}$, but it permits $\sqrt{-1}$ as a genuine radical expression for i, and $\sqrt[3]{2}$ as a genuine radical expression for—well, $\sqrt[3]{2}$.

Our aim is to prove a theorem that was effectively stated by Vandermonde, and proved (modulo a gap that he probably could have filled) by Gauss. See the discussion after Lemma 21.4. The name 'Vandermonde–Gauss Theorem' is not standard, but it ought to be, so we shall use it.

**Theorem 21.3 (Vandermonde–Gauss Theorem).** *For any $n \geq 1$, any $n$th root of unity has a genuine radical expression.*

The aim of this section is to prove the Vandermonde–Gauss Theorem. In fact we prove something distinctly stronger: see Exercise 21.3. We prove the theorem by induction on $n$. It is easy to see that the induction step reduces to the case where $n$ is prime and the $n$th root of unity concerned is therefore primitive, because if $n$ is composite we can write it as $n = pq$ where $p$ is prime, and $\sqrt[n]{} = \sqrt[q]{\sqrt[p]{}}$.

Let $n = p$ be prime and focus attention on a primitive $p$th root of unity $\zeta_p$, which for simplicity we denote by $\zeta$. Explicitly,

$$\zeta = \cos\frac{2\pi}{p} + \mathrm{i}\sin\frac{2\pi}{p}$$

but we do not actually use this formula.

We already know the minimal polynomial of $\zeta$ over $\mathbb{Q}$, from Lemma 3.24. It is

$$m(t) = t^{p-1} + t^{p-2} + \cdots + t + 1 = \frac{t^p - 1}{t - 1}$$

Let

$$\theta = \cos\frac{2\pi}{p-1} + \mathrm{i}\sin\frac{2\pi}{p-1}$$

be a primitive $(p-1)$th root of unity. Since $p-1$ is composite (except when $p = 2, 3$) the minimal polynomial of $\theta$ over $\mathbb{Q}$ is *not* equal to

$$c(t) = t^{p-2} + t^{p-3} + \cdots + t + 1 = \frac{t^{p-1} - 1}{t - 1}$$

but instead it is some irreducible divisor of $c(t)$.

We work not with $\mathbb{Q}(\zeta)/\mathbb{Q}$, but with $\mathbb{Q}(\theta, \zeta)/\mathbb{Q}$. Since $p$ and $p-1$ are coprime, this extension is the same as $\mathbb{Q}(\theta\zeta)/\mathbb{Q}$ where $\theta\zeta$ is a primitive $p(p-1)$th root of unity. A general element of $\mathbb{Q}(\theta\zeta)$ can be written as a linear combination over $\mathbb{Q}(\theta)$ of the powers $1, \zeta, \zeta^2, \ldots, \zeta^{p-2}$. It is convenient to throw in $\zeta^{p-1}$ as well, but now we must always bear in mind the relation $1 + \zeta + \zeta^2 + \cdots + \zeta^{p-1} = 0$.

We base the deduction on the following result, which we prove below in Section 21.6 to avoid technical distractions.

**Lemma 21.4.** *The Galois group of $\mathbb{Q}(\theta\zeta)/\mathbb{Q}(\theta)$ is cyclic of order $p-1$. It comprises the $\mathbb{Q}(\theta)$-automorphisms of the form $\sigma_j$, $(j = 1, 2, \ldots p-1)$, where*

$$\sigma_j : \zeta \mapsto \zeta^j$$
$$\theta \mapsto \theta$$

The main technical issue in proving this theorem is that although we know that $\zeta, \zeta^2, \ldots, \zeta^{p-2}$ are linearly independent over $\mathbb{Q}$, we do not (yet) know that they are linearly independent over $\mathbb{Q}(\theta)$. Even Gauss omitted the proof of this fact from his discussion in the *Disquisitiones Arithmeticae*, but that may have been because to him it was obvious. He never published a proof of this particular fact, though he must have known one. So in a sense the first complete proof should probably be credited to Galois.

Assuming Lemma 21.4, we can follow Vandermonde's method in complete generality, using a few simple facts about roots of unity.

*Proof of the Vandermonde–Gauss Theorem.* We prove the theorem by induction on $n$. The cases $n = 1, 2$ are trivial since the roots of unity concerned are $1, -1$. As explained above, the induction step reduces to the case where $n$ is prime and the $n$th root of unity concerned is therefore primitive. Throughout the proof it helps to bear in mind the above examples when $n = 5, 11$.

We write $n = p$ to remind us that $n$ is prime. Let $\zeta$ be a primitive $p$th root of unity and let $\theta$ be a primitive $(p-1)$th root of unity as above. Then $\theta\zeta$ is a primitive $p(p-1)$th root of unity.

By Lemma 21.4, the Galois group of $\mathbb{Q}(\theta\zeta)/\mathbb{Q}$ is isomorphic to $\mathbb{Z}_p^*$, and is thus cyclic of order $p-1$ by Corollary 19.9. It comprises the automorphisms $\sigma_j$ for $j = 1, \ldots, p-1$. Since $\mathbb{Z}_p^*$ is cyclic, there exists a generator $a$. That is, every $j \in \mathbb{Z}_p^*$ can be expressed as a power $j = a^l$ of $a$. Then $\sigma_j = \sigma_a^l$; so $\sigma_a$ generates $\Gamma = \Gamma(\mathbb{Q}(\theta\zeta)/\mathbb{Q}(\theta))$.

By Lemma 21.4 and Proposition 17.18, $\mathbb{Q}(\theta\zeta)/\mathbb{Q}(\theta)$ is normal and separable, so in particular the fixed field of $\Gamma$ is $\mathbb{Q}(\theta)$ by Theorem 12.2(2). Since $\sigma_a$ generates $\Gamma$, any element of $\mathbb{Q}(\theta\zeta)$ that is fixed by $\sigma_a$ must lie in $\mathbb{Q}(\theta)$.

We construct elements fixed by $\sigma_a$ as follows. Define

$$\begin{aligned} \alpha_l &= \zeta + \theta^l\zeta^a + \theta^{2l}\zeta^{a^2} + \cdots + \theta^{(p-2)l}\zeta^{a^{p-2}} \\ &= \sum_{j=0}^{p-2} \theta^{jl}\zeta^{a^j} \end{aligned} \tag{21.8}$$

for $0 \leq l \leq p-2$. Then

$$\sigma_a(\alpha_l) = \sum_{j=0}^{p-2} \theta^{jl}\zeta^{a^{j+1}} = \theta^{-l}\alpha_l$$

Therefore

$$\sigma_a(\alpha_l^{p-1}) = (\theta^{-l}\alpha_l)^{p-1} = (\theta^{p-1})^{-l}\alpha_l^{p-1} = 1 \cdot \alpha_l^{p-1} = \alpha_l^{p-1}$$

so $\alpha_l^{p-1}$ is fixed by $\sigma_a$, hence lies in $\mathbb{Q}(\theta)$. Say

$$\alpha_l^{p-1} = \beta_l \in \mathbb{Q}(\theta)$$

Therefore

$$\alpha_l = \sqrt[p-1]{\beta_l} \qquad (0 \leq l \leq p-2)$$

Recall (Exercise 21.5) the following property of roots of unity:

$$1 + \theta^j + \theta^{2j} + \cdots + \theta^{(p-2)j} = \begin{cases} p-1 & \text{if } j = 0 \\ 0 & \text{if } 1 \leq j \leq p-2 \end{cases}$$

Therefore, from (21.8),

$$\begin{aligned} \zeta &= \tfrac{1}{p-1}[\alpha_0 + \alpha_1 + \cdots + \alpha_{p-2}] \\ &= \tfrac{1}{p-1}[\sqrt[p-1]{\beta_0} + \sqrt[p-1]{\beta_1} + \cdots + \sqrt[p-1]{\beta_{p-2}}] \end{aligned} \tag{21.9}$$

which expresses $\zeta$ by radicals over $\mathbb{Q}(\theta)$.

Now, $\theta$ is a primitive $(p-1)$th root of unity, so by induction $\theta$ is a radical expression over $\mathbb{Q}$ of maximum radical degree $\leq p-2$. Each $\beta_l$ is also a radical expression over $\mathbb{Q}$ of maximum radical degree $\leq p-2$, since $\beta_l$ is a polynomial in $\theta$ with rational coefficients. (Actually we can say more: if $p > 2$ then $p-1$ is even, so the maximum radical degree is $\max(2, (p-1)/2)$. Note that when $p = 3$ we require a square root, but $(p-1)/2 = 1$. See Exercise 21.3.)

Substituting the rational expressions in (21.9) we see that $\zeta$ is a radical expression over $\mathbb{Q}$ of maximum radical degree $\leq p-1$. (Again, this can be improved to $\max(2,(p-1)/2)$ for $p>2$, see Exercise 21.3.)

Therefore, in particular, (21.9) yields a genuine radical expression for $\zeta$ according to the definition, and the Vandermonde–Gauss Theorem is proved. □

We repeat that this proof assumes Lemma 21.4, which is proved below in Section 21.6.

---

## 21.5 Cyclotomic Polynomials

Before giving the proof of Lemma 21.4, some motivation will be useful.

Consider the case $n=12$. Let $\zeta = e^{\pi i/6}$ be a primitive 12th root of unity. We can classify its powers $\zeta^j$ according to their minimal power $d$ such that $(\zeta^j)^d = 1$. That is, we consider when they are *primitive* $d$th roots of unity. It is easy to see that in this case the primitive $d$th roots of unity are:

$$\begin{array}{ll}
d=1 & 1 \\
d=2 & \zeta^6 \ (=-1) \\
d=3 & \zeta^4, \zeta^8 \ (=\omega, \omega^2) \\
d=4 & \zeta^3, \zeta^9 \ (=\mathrm{i}, -\mathrm{i}) \\
d=6 & \zeta^2, \zeta^{10} \ (=-\omega, -\omega^2) \\
d=12 & \zeta, \zeta^5, \zeta^7, \zeta^{11}
\end{array}$$

We can factorise $t^{12}-1$ by grouping corresponding zeros:

$$\begin{aligned}
t^{12}-1 = {} & (t-1)\times \\
& (t-\zeta^6)\times \\
& (t-\zeta^4)(t-\zeta^8)\times \\
& (t-\zeta^3)(t-\zeta^9)\times \\
& (t-\zeta^2)(t-\zeta^{10})\times \\
& (t-\zeta)(t-\zeta^5)(t-\zeta^7)(t-\zeta^{11})
\end{aligned}$$

which simplifies to

$$t^{12}-1 = (t-1)(t+1)(t^2+t+1)(t^2+1)(t^2-t+1)F(t)$$

where

$$F(t) = (t-\zeta)(t-\zeta^5)(t-\zeta^7)(t-\zeta^{11})$$

whose explicit form is not immediately obvious. One way to work out $F(t)$ is to use trigonometry (Exercise 21.4). The other is to divide $t^{12}-1$ by all the other factors, which leads rapidly to

$$F(t) = t^4 - t^2 + 1$$

If we let $\Phi_d(t)$ be the factor corresponding to primitive $d$th roots of unity, we have proved that

$$t^{12} - 1 = \Phi_1(t)\Phi_2(t)\Phi_3(t)\Phi_4(t)\Phi_6(t)\Phi_{12}(t)$$

Our computations show that every factor $\Phi_j$ lies in $\mathbb{Z}[t]$. In fact, it turns out that the factors are all *irreducible* over $\mathbb{Z}$. This is obvious for all factors here except $t^4 - t^2 + 1$, where it can be proved by considering the factorisation $(t - \zeta)(t - \zeta^5)(t - \zeta^7)(t - \zeta^{11})$ (Exercise 21.5).

This calculation motivates:

**Definition 21.5.** The polynomial $\Phi_n(t)$ defined by

$$\Phi_n(t) = \prod_{a \in \mathbb{Z}_n, (a,n)=1} (t - \zeta^a) \tag{21.10}$$

is the $n$th *cyclotomic polynomial* over $\mathbb{C}$.

The most basic property of cyclotomic polynomials is the identity

$$t^n - 1 = \prod_{d|n} \Phi_d(t) \tag{21.11}$$

which is a direct consequence of their definition.

Recall that $\mathbb{Z}_n^*$ denotes the group of units of $\mathbb{Z}_n$; see Definition 21.1. The order of $\mathbb{Z}_n^*$ is given by an important number-theoretic function:

**Definition 21.6.** The *Euler function* $\phi(n)$ is the number of integers $a$, with $1 \leq a \leq n - 1$, such that $a$ is prime to $n$.

Definition 21.6 implies immediately that the order of $\mathbb{Z}_n^*$ is equal to $\phi(n)$.

**Proposition 21.7.** *The cyclotomic polynomial $\Phi_n(t)$ has degree $\phi(n)$.*

*Proof.* The degree is the number of linear factors in (21.10), which is the number of $a \in \mathbb{Z}_n$ such that $(a, n) = 1$. This is the definition of $\phi(n)$. □

The Euler function $\phi(n)$ has numerous interesting properties. In particular

$$\phi(p^k) = (p - 1)p^{k-1}$$

if $p$ is prime, and

$$\phi(r)\phi(s) = \phi(rs)$$

when $r, s$ are coprime. See Exercise 12.4.

The next step is to calculate the order of the Galois group of $\mathbb{Q}(\zeta)/\mathbb{Q}$:

**Theorem 21.8.** *Let $\zeta$ be a primitive $n$th root of unity in $\mathbb{C}$, and let $\Gamma = \Gamma(\mathbb{Q}(\zeta)/\mathbb{Q})$ be the Galois group of $\mathbb{Q}(\zeta)$ over $\mathbb{Q}$. Then $|\Gamma| = \phi(n)$.*

*Proof.* We prove:

(1) $|\Gamma| = \phi(n)$ when $n = p^k$ is a prime power.

(2) Suppose that $a, b$ are coprime. If $|\Gamma| = \phi(n)$ when $n = a$ and $n = b$, then $|\Gamma| = \phi(n)$ for $n = ab$.

The result then follows by induction on the number of distinct prime-power factors of $n$.

(1) Suppose that $n = p^k$. If $k = 0$ then $\Phi_n(t) = \Phi_1(t) = t - 1$, which is irreducible.

If $k > 0$ we use the Eisenstein trick. We know that

$$\Phi_{p^k}(t) = \frac{t^{p^k} - 1}{t^{p^{k-1}} - 1} = 1 + t^{p^{k-1}} + t^{2p^{k-1}} + \cdots + t^{(p-1)p^{k-1}}$$

Set $t = u + 1$. Now

$$\Phi_{p^k}(t) = 1 + (u+1)^{p^{k-1}} + (u+1)^{2p^{k-1}} + \cdots + (u+1)^{(p-1)p^{k-1}} = ph(t) + u^{(p-1)p^{k-1}}$$

where $h(t)$ has constant term 1. By Eisenstein's Criterion, $\Phi_n(t)$ is irreducible over $\mathbb{Q}$.

(2) Let $n = ab$ where $(a, b) = 1$. Now $\zeta^a$ is a primitive $b$th root of unity, and $\zeta^b$ is a primitive $a$th root of unity.

In the field extension $\mathbb{Q}(\zeta)/\mathbb{Q}$, consider the intermediate fields $M_a = \mathbb{Q}(\zeta^a)$ and $M_b = \mathbb{Q}(\zeta^b)$. The corresponding Galois groups are

$$\begin{aligned}
\mathbb{Q}^* &= \Gamma(\mathbb{Q}(\zeta)/\mathbb{Q}) = \Gamma \\
M_a^* &= \Gamma(\mathbb{Q}(\zeta)/\mathbb{Q}(\zeta^a)) \\
M_b^* &= \Gamma(\mathbb{Q}(\zeta)/\mathbb{Q}(\zeta^b)) \\
\mathbb{Q}(\zeta)^* &= \mathbf{1}
\end{aligned}$$

Since $\Gamma$ is abelian, all extensions of subfields of $\mathbb{Q}(\zeta)$ are normal. Therefore

$$\begin{aligned}
\Gamma(\mathbb{Q}(\zeta^a)/\mathbb{Q}) &\cong \Gamma/M_a^* \\
\Gamma(\mathbb{Q}(\zeta^b)/\mathbb{Q}) &\cong \Gamma/M_b^*
\end{aligned}$$

We are assuming that the theorem is true for $n = a, n = b$, so we know that

$$|\Gamma/M_a^*| = \phi(b) \qquad |\Gamma/M_b^*| = \phi(a)$$

Also

$$M_a^* \cap M_b^* = 1$$

because any $\mathbb{Q}$-automorphism fixing $\zeta^a$ and $\zeta^b$ must fix $\zeta$. So the subgroup $M_a^* M_b^* \subseteq \Gamma$ is isomorphic to $M_a^* \times M_b^*$, which has order $\phi(a)\phi(b) = \phi(n)$. Therefore $|\Gamma| \geq \phi(n)$.

However, we already know that $|\Gamma| \leq \phi(n)$, so $|\Gamma| = \phi(n)$. That is, condition (1) holds for $n = ab$, which completes the proof. □

## 21.6 Galois Group of $\mathbb{Q}(\zeta)/\mathbb{Q}$

In Theorem 20.12 we described the Galois group of $\mathbb{Q}(\zeta)/\mathbb{Q}$ when $\zeta$ is a primitive $p$th root of unity, $p$ prime. We now generalise this result to the composite case.

Let $f(t) = t^n - 1 \in \mathbb{Q}[t]$. The zeros in $\mathbb{C}$ are $1, \zeta, \zeta^1, \ldots, \zeta^{n-1}$ where $\zeta = e^{2\pi i/n}$ is a primitive $n$th root of unity. The splitting field of $f$ is clearly $\mathbb{Q}(\zeta)$. Theorem 9.9 implies that the extension $\mathbb{Q}(\zeta)/\mathbb{Q}$ is normal. By Proposition 9.14 it is separable.

We can now prove Lemma 21.4, along with several related results. This in particular completes the proof of the Vandermonde–Gauss Theorem 21.3.

**Theorem 21.9.** (1) *The Galois group $\Gamma(\mathbb{Q}(\zeta)/\mathbb{Q})$ consists of the $\mathbb{Q}$-automorphisms $\rho_k$ defined by*

$$\rho_k(\zeta) = \zeta^k$$

*where $0 \leq k \leq n-1$ and $k$ is prime to $n$.*

(2) *$\Gamma(\mathbb{Q}(\zeta)/\mathbb{Q})$ is isomorphic to $\mathbb{Z}_n^*$ and in particular it is an abelian group.*

(3) *Its order is $\phi(n)$.*

(4) *If $n$ is prime, $\mathbb{Z}_n^*$ is cyclic.*

*Proof.* (1) Let $\gamma \in \Gamma(\mathbb{Q}(\zeta)/\mathbb{Q})$. Since $\gamma(\zeta)$ is a zero of $t^n - 1$, $\gamma = \rho_k$ for some $k$.

If $k$ and $n$ have a common factor $d > 1$ then $\rho_k$ is not onto and hence not a $\mathbb{Q}$-automorphism.

If $k$ and $n$ are coprime, there exist integers $a, b$ such that $ak + bn = 1$. Then

$$\zeta = \zeta^{ak+bn} = \zeta^{ak}\zeta^{bn} = (\zeta^k)^a$$

so $\zeta$ lies in the image of $\rho_k$. Therefore $\rho_k$ is a $\mathbb{Q}$-automorphism.

(2) Clearly $\rho_j\rho_k = \rho_{jk}$, so the map $\rho_k \mapsto k$ is an isomorphism from $\Gamma(\mathbb{Q}(\zeta)/\mathbb{Q})$ to $\mathbb{Z}_n^*$.

(3) $|\Gamma(\mathbb{Q}(\zeta)/\mathbb{Q})| = |\mathbb{Z}_n^*| = \phi(n)$.

(4) This follows from Corollary 19.9. □

**Corollary 21.10.** *The cyclotomic polynomial $\Phi_n(t)$ is irreducible over $\mathbb{Q}$ for all $n$.*

*Proof.* All zeros of $\Phi_n(t)$ are conjugate under the Galois group, so $\Phi_n(t)$ is irreducible over $\mathbb{Q}$ by Proposition 17.27. □

We can use the identity (21.11) recursively to compute $\Phi_n(t)$. Thus

$$\Phi_1(t) = t - 1$$

so

$$t^2 - 1 = \Phi_2(t)\Phi_1(t)$$

which implies that

$$\Phi_2(t) = \frac{t^2 - 1}{\Phi_1(t)} = \frac{t^2 - 1}{t - 1} = t + 1$$

Similarly

$$\Phi_3(t) = \frac{t^3 - 1}{t - 1} = t^2 + t + 1$$

and

$$\Phi_4(t) = \frac{t^4 - 1}{(t - 1)(t + 1)} = t^2 + 1$$

and so on. Table 21.6 shows the first 15 cyclotomic polynomials, computed in this manner.

| $n$ | $\Phi_n(t)$ |
|---|---|
| 1 | $t - 1$ |
| 2 | $t + 1$ |
| 3 | $t^2 + t + 1$ |
| 4 | $t^2 + 1$ |
| 5 | $t^4 + t^3 + t^2 + t + 1$ |
| 6 | $t^2 - t + 1$ |
| 7 | $t^6 + t^5 + t^4 + t^3 + t^2 + t + 1$ |
| 8 | $t^4 + 1$ |
| 9 | $t^6 + t^3 + 1$ |
| 10 | $t^4 - t^3 + t^2 - t + 1$ |
| 11 | $t^{10} + t^9 + t^8 + t^7 + t^6 + t^5 + t^4 + t^3 + t^2 + t + 1$ |
| 12 | $t^4 - t^2 + 1$ |
| 13 | $t^{12} + t^{11} + t^{10} + t^9 + t^8 + t^7 + t^6 + t^5 + t^4 + t^3 + t^2 + t + 1$ |
| 14 | $t^6 - t^5 + t^4 - t^3 + t^2 - t + 1$ |
| 15 | $t^8 - t^7 + t^5 - t^4 + t^3 - t + 1$ |

At the moment all we have proved is that $\Phi_n(t)$ has rational coefficients. The table suggests that the coefficients are actually integers. (It also suggests that these integers are either 0, 1, or –1, but this is false: Exercise 21.12 shows that when $n = 105$, $\Phi_{105}$ is a counterexample. This is the smallest $n$ for which $\Phi_n$ has a coefficient that is not 0, 1, or –1.) We now prove that $\Phi_n(t) \in \mathbb{Z}[t]$ using the recursive procedure above. First, we prove:

**Lemma 21.11.** *Suppose that $p(t) \in \mathbb{Z}[t], q(t) \in \mathbb{Q}[t]$, and both $p$ and $q$ are monic. If $p(t)q(t) \in \mathbb{Z}[t]$ then $q(t) \in \mathbb{Z}[t]$.*

*Proof.* Let

$$\begin{aligned} p(t) &= t^r + a_{r-1}t^{r-1} + \cdots + a_0 \\ q(t) &= t^s + b_{s-1}t^{s-1} + \cdots + b_0 \end{aligned}$$

where $a_0, \ldots, a_r \in \mathbb{Z}$ and $b_0, \ldots, b_s \in \mathbb{Q}$.

Expanding the product $p(t)q(t)$ we find that

$$\begin{aligned} p(t)q(t) = \; & t^{r+s} \\ & +t^{r+s-1}(1.b_{s-1} + a_{r-1}) \\ & +t^{r+s-2}(1.b_{s-2} + a_{r-1}b_{s-1} + a_{r-2}.1) \\ & \vdots \end{aligned}$$

Since $p(t)q(t) \in \mathbb{Z}[t]$, all coefficients on the right-hand side belong to $\mathbb{Z}$. Therefore

$$\begin{aligned} & b_{s-1} + a_{r-1} \in \mathbb{Z} \\ & b_{s-2} + a_{r-1}b_{s-1} + a_{r-2} \in \mathbb{Z} \\ & \vdots \end{aligned}$$

The first equation implies that $b_{s-1} \in \mathbb{Z}$, then the second implies that $b_{s-2} \in \mathbb{Z}$. Inductively, $b_j \in \mathbb{Z}$ for all $j$. Therefore $q(t) \in \mathbb{Z}[t]$. □

**Theorem 21.12.** *For all $n$, the cyclotomic polynomial $\Phi_n(t)$ has integer coefficients.*

*Proof.* Use induction on $n$. When $n = 1$ we have $\Phi_1(t) = t - 1 \in \mathbb{Z}[t]$. Now assume that the result is valid for all $m < n$; we prove it for $n$. Let

$$g(t) = (t^n - 1)/\Phi_n(t)$$

so that

$$t^n - 1 = g(t)\Phi_n(t)$$

By (21.10), both $g(t)$ and $\Phi_n(t)$ are monic. By (21.11),

$$g(t) = \prod_{d|n, d \neq n} \Phi_d(t)$$

By induction each factor $\Phi_d(t) \in \mathbb{Z}[t]$, so $g(t) \in \mathbb{Z}[t]$. By Lemma 21.11, $\Phi_n(t) \in \mathbb{Z}[t]$. This completes the induction. □

Using results from algebraic number theory, we can prove a stronger result: if any $\mathbb{Z}$-linear combination of powers of $\zeta$ is rational then it is an integer. (The coefficients of $\Phi_n(t)$ are linear combinations of this kind by expanding (21.10).) See for example Stewart and Tall (2015b) Theorem 3.5.

## 21.7 Constructions Using a Trisector

The main reason or the limitation to rule and compass in geometric constructions is tradition: it is what Euclid assumes. However, the ancient Greeks themselves considered other geometric instruments. We remarked in Chapter 7 that Archimedes knew how to trisect angles using a marked ruler, and Dudley (1987) surveys many other methods. For a final flourish, we apply our results to the construction of regular polygons when an angle-trisector is permitted, as well as the traditional ruler and compass. The results are instructive, amusing, and slightly surprising. For example, the regular 7-gon can now be constructed. It is not immediately clear why the angle $\frac{2\pi}{7}$ arises from trisections. Other regular polygons, such as the 13-gon and 19-gon, also become constructible. On the other hand, the regular 11-gon still cannot be constructed.

The main point is the link between trisection and irreducible cubic equations. The trigonometric solution of cubics, Exercise 1.8, shows that an angle-trisector can be used to solve some cubic equations: those with three distinct real roots. Specifically, we use the trigonometric identity $\cos 3\theta = 4\cos^3\theta - 3\cos\theta$ to solve the cubic equation $t^3+pt+q=0$ when $27q^2+4p^3<0$. This is the condition for three distinct real roots. The method is as follows.

The inequality $27pq^2+4p^3<0$ implies that $p<0$, so we can find $a, b$ such that $p=-3a^2, q=-a^2b$. The cubic becomes

$$t^3 - 3a^2t = a^2b$$

and the inequality becomes $a > |b|/2$.

Substitute $t = 2a\cos\theta$, and observe that

$$t^3 - 3a^2t = 8a^3\cos^3\theta - 6a^3\cos\theta = 2a^3\cos 3\theta$$

The cubic thus reduces to

$$\cos 3\theta = \frac{b}{2a}$$

which we can solve using $\cos^{-1}$ because $|\frac{b}{2a}| \leq 1$, getting

$$\theta = \frac{1}{3}\cos^{-1}\frac{b}{2a}$$

There are three possible values of $\theta$, the other two being obtained by adding $\frac{2\pi}{3}$ or $\frac{4\pi}{3}$. Finally, eliminate $\theta$ to get

$$t = 2a\cos\left(\frac{1}{3}\cos^{-1}\frac{b}{2a}\right)$$

where $a = \sqrt{\frac{-p}{3}}, b = \frac{3q}{p}$.

Conversely, solving cubics with real coefficients and three distinct real roots lets us trisect angles. So when a trisector is made available, the constructible numbers now lie in a series of extensions, starting with $\mathbb{Q}$, such that each successive extension has degree 2 or 3.

The use of a trisector motivates a generalisation of Fermat primes, named after the mathematician James Pierpont.

**Definition 21.13.** A *Pierpont prime* is a prime $p$ of the form

$$p = 2^a 3^b + 1$$

where $a \geq 1, b \geq 0$.

(Here we exclude $a = 0$ because in this case $2^a 3^b + 1 = 3^b + 1$ is even.)

The Pierpont primes up to 100 are 3, 5, 7, 13, 17, 19, 37, 73, and 97. So they appear to be more common than Fermat primes, a point to which we return later.

Andrew Gleason (1988) proved the following theorem characterising those regular $n$-gons that can be constructed when the traditional instruments of Euclid are supplemented by an angle-trisector. He also gave explicit constructions of that kind for the regular 7-gon and 13-gon.

**Theorem 21.14.** *The regular $n$-gon can be constructed using ruler, compass, and trisector, if and only if $n$ is of the form $2^r 3^s p_1 \cdots p_k$ where $r, s \geq 0$ and the $p_j$ are distinct Pierpont primes $> 3$.*

*Proof.* First, suppose that the regular $n$-gon can be constructed using ruler, compass, and trisector. As remarked above, this implies that the primitive $n$th root of unity $\zeta = e^{2\pi i/n}$ lies in the largest field in some series of extensions, which starts with $\mathbb{Q}$, such that each successive extension has degree 2 or 3. Therefore

$$[\mathbb{Q}(\zeta) : \mathbb{Q}] = 2^c 3^d$$

for $c, d \in \mathbb{N}$.

The degree $[\mathbb{Q}(\zeta) : \mathbb{Q}]$ equals $\phi(n)$, where $\phi$ is the Euler function. This is the degree of the cyclotomic polynomial $\Phi_n(t)$, which is irreducible over $\mathbb{Q}$. Therefore a necessary condition for constructibility with ruler, compass, and trisector is $\phi(n) = 2^a 3^b$ for $a, b \in \mathbb{N}$. What does this imply about $n$?

Write $n$ as a product of distinct prime powers $p_j^{m_j}$. Then $\phi(p_j^{m_j})$ must be of the form $2^{a_j} 3^{b_j}$. Since $\phi(p^m) = (p-1)p^{m-1}$ when $p$ is prime, we require $(p_j - 1 p_j^{m_j - 1}$ to be of the form $2^{a_j} 3^{b_j}$.

Either $m_j = 1$ or $p_j = 2, 3$. If $p_j = 2$ then $\phi(p_j^{m_j}) = 2^{m_j - 1}$ and any $m_j$ can occur. If $p_j = 3$ then $\phi(p_j^{m_j}) = 2 \cdot 3^{m_j - 1}$ and again any power of 3 can occur. Otherwise $m_j = 1$ so $\phi(p_j^{m_j}) = \phi(p_j) = p_j - 1$, and $p_j = 2^{a_j} 3^{b_j} + 1$. Thus $p_j$ is a Pierpont prime.

We have now proved the theorem in one direction: in order for the regular $n$-gon to be constructible by ruler, compass, and trisector, $n$ must be a product of powers of 2, powers of 3, and distinct Pierpont primes $> 3$. We claim that the converse is also true.

The proof is a simple application of Galois theory. Let $p = 2^a 3^b + 1$ be an odd prime. Let $\zeta = e^{2\pi i/p}$. Then $[\mathbb{Q}(\zeta) : \mathbb{Q}] = p - 1 = 2^a 3^b$. The extension $\mathbb{Q}(\zeta)/\mathbb{Q}$ is normal and separable, so the Galois correspondence is a bijection, and the Galois group $\Gamma = \Gamma(\mathbb{Q}(\zeta)/\mathbb{Q})$ has order $m = 2^a 3^b$. By Theorem 21.9 it is abelian, isomorphic to $\mathbb{Z}_m^*$. Therefore it has a series of normal subgroups

$$1 = \Gamma_0 \lhd \Gamma_1 \lhd \cdots \lhd \Gamma_r = \Gamma$$

where each factor $\Gamma_{j+1}/\Gamma_j$ is isomorphic either to $\mathbb{Z}_2$ or $\mathbb{Z}_3$. In fact, $r = a + b$.

Let

$$\theta = \zeta + \zeta^1 = \zeta + \zeta^{p-1} = \zeta + \overline{\zeta} = 2\cos 2\pi/p$$

where the bar indicates complex conjugate. Then $\theta \in \mathbb{R}$. Consider the tower of subfields

$$\mathbb{Q} \subseteq \mathbb{Q}(\theta) \subseteq \mathbb{Q}(\zeta)$$

Clearly $\mathbb{Q}(\theta) \subseteq \mathbb{R}$. We have $\zeta + \zeta^{-1} = \theta, \zeta \cdot \zeta^{-1} = 1$, so $\zeta$ and $\zeta^{-1}$ are the zeros of $t^2 - \theta t + 1$ over $\mathbb{Q}(\theta)$. Therefore $[\mathbb{Q}(\zeta) : \mathbb{Q}(\theta)] \leq 2$, but $\zeta \notin \mathbb{R} \supseteq \mathbb{Q}(\theta)$ so $[\mathbb{Q}(\zeta) : \mathbb{Q}(\theta)] = 2$.

The group $\Lambda$ of order 2 generated by complex conjugation is a subgroup of $\Gamma$, and it is a normal subgroup since $\Gamma$ is abelian. We claim that the fixed field $\Lambda^\dagger = \mathbb{Q}(\theta) = \mathbb{Q}(\zeta) \cap \mathbb{R}$. We have $\mathbb{Q}(\zeta) \subseteq \mathbb{R}$ so $\mathbb{Q}(\zeta) \subseteq \Lambda^\dagger$. Since $[\mathbb{Q}(\zeta) : \mathbb{Q}(\theta)] = 2$ the only subfield properly containing $\mathbb{Q}(\zeta)$ is $\mathbb{Q}(\theta)$, and this is not fixed by $\Lambda$. Therefore $\mathbb{Q}(\zeta) = \Lambda^\dagger$. (It is easy to see that in fact, $\mathbb{Q}(\theta) = \mathbb{Q}(\zeta) \cap \mathbb{R}$.)

Therefore the Galois group of $\mathbb{Q}(\theta)/\mathbb{Q}$ is isomorphic to the quotient group $\Delta = \Gamma/\Lambda$, so it is cyclic of order $m/2 = 2^{a-1}3^b$. It has a series of normal subgroups

$$1 = \Delta_0 \lhd \Delta_1 \lhd \cdots \lhd \Delta_{r-1} = \Delta$$

where each factor $\Delta_{j+1}/\Delta_j$ is isomorphic either to $\mathbb{Z}_2$ or $\mathbb{Z}_3$.

The corresponding fixed subfields $K_j = \Delta_j^\dagger$ form a tower

$$\mathbb{Q}(\theta) = K_0 \supseteq K_1 \supseteq \cdots \supseteq K_{r-1} = \mathbb{Q}$$

and each degree $[K_j : K_{j+1}]$ is either 2 or 3. So $K_j$ can be obtained from $K_{j+1}$ by adjoining either:

a root of a quadratic over $K_{j+1}$, or

a root of an irreducible cubic over $K_{j+1}$ with all three roots real

(the latter because $\mathbb{Q}(\theta) \subseteq \mathbb{R}$).

In the quadratic case, any $z \in K_j$ can be constructed from $K_{j+1}$ by ruler and compass. In the cubic case, any $z \in K_j$ can be constructed from $K_{j+1}$ by trisector (plus ruler and compass for field operations). By backwards induction from $K_{r-1} = \mathbb{Q}$, we see that any element of $K_0$ can be constructed from $\mathbb{Q}$ by ruler, compass, and trisector. Finally, any element of $\mathbb{Q}(\zeta)$ can be constructed from $\mathbb{Q}$ by ruler, compass, and trisector. In particular, $\zeta$ can be so constructed, which gives a construction for a regular $p$-gon. $\square$

This is a remarkable result, since at first sight there is no obvious link between regular polygons with (say) 7, 13, or 19 sides and angle-trisection. They appear to need division of an angle by 7, 13, or 19. So we give further details for the first two cases, the 7-gon and the 13-gon.

$p = 7$: Let $\zeta = e^{2\pi i/7}$. Recall the basic relation

$$1 + \zeta + \zeta^2 + \zeta^3 + \zeta^4 + \zeta^5 + \zeta^6 = 0 \tag{21.12}$$

Define

$$\begin{aligned} r_1 &= \zeta + \zeta^6 = 2\cos\frac{2\pi}{7} \in \mathbb{R} \\ r_2 &= \zeta^2 + \zeta^5 = 2\cos\frac{4\pi}{7} \in \mathbb{R} \\ r_3 &= \zeta^3 + \zeta^4 = 2\cos\frac{6\pi}{7} \in \mathbb{R} \end{aligned}$$

Compute the elementary symmetric functions of the $r_j$. By (21.13)

$$r_1 + r_2 + r_3 = -1$$

Next,

$$\begin{aligned} r_1 r_2 r_3 &= (\zeta + \zeta^6)(\zeta^2 + \zeta^5)(\zeta^3 + \zeta^4) \\ &= \zeta^6 + \zeta^0 + \zeta^2 + \zeta^3 + \zeta^4 + \zeta^5 + \zeta^0 + \zeta \\ &= 1 + 1 - 1 = 1 \end{aligned}$$

Finally,

$$\begin{aligned} r_1 r_2 + r_1 r_3 + r_2 r_3 &= (\zeta + \zeta^6)(\zeta^2 + \zeta^5) + (\zeta + \zeta^6)(\zeta^3 + \zeta^4) + (\zeta^2 + \zeta^5)(\zeta^3 + \zeta^4) \\ &= \zeta^3 + \zeta^6 + \zeta + \zeta^4 + \zeta^4 + \zeta^5 + \zeta^2 + \zeta^3 + \zeta^5 + \zeta^6 + \zeta + \zeta^2 \\ &= -2 \end{aligned}$$

Therefore the $r_j$ are roots of the cubic $t^3 + 2t^2 + t - 1 = 0$. This is irreducible and the roots $r_j$ are real, so they can be constructed using a trisector (plus ruler and compass for field operations). An explicit construction can be found in Gleason (1988) and Conway and Guy (1996) page 200.

$p = 13$: Let $\zeta = e^{2\pi i/13}$. Recall the basic relation

$$1 + \zeta + \zeta^2 + \cdots + \zeta^{12} = 0 \tag{21.13}$$

Define $r_j = \zeta^j + \zeta^{-j} = 2\cos\frac{2\pi j}{13}$ for $1 \le j \le 6$.

It turns out that 2 is primitive root modulo 13. That is, the powers of 2 (mod 13) are, in order,

$$1 \quad 2 \quad 4 \quad 8 \quad 3 \quad 6 \quad 12 \quad 11 \quad 9 \quad 5 \quad 10 \quad 7$$

and then repeat: these are all the nonzero elements of $\mathbb{Z}_{13}$.

Add powers of $\zeta$ corresponding to every third number in this sequence:

$$\begin{aligned} s_1 &= \zeta + \zeta^8 + \zeta^{12} + \zeta^5 = r_1 + r_5 \\ s_2 &= \zeta^2 + \zeta^3 + \zeta^{11} + \zeta^{10} = r_2 + r_3 \\ s_3 &= \zeta^4 + \zeta^6 + \zeta^9 + \zeta^7 = r_4 + r_6 \end{aligned}$$

Tedious but routine calculations, or computer algebra, show that the $s_j$ are the three roots of the cubic

$$t^3 + t^2 - 4t + 1 = 0$$

which is irreducible and has all roots real (exercise). Therefore the $s_j$ can be constructed using trisector, ruler, and compass.

Then, for example,

$$\begin{aligned} r_1 + r_5 &= s_1 \\ r_1 r_5 &= (\zeta + \zeta^{12})(\zeta^5 + \zeta^8) \\ &= \zeta^6 + \zeta^9 + \zeta^4 + \zeta^7 = s_3 \end{aligned}$$

so $r_1, r_5$ are roots of a quadratic over $\mathbb{Q}(s_1, s_2, s_3)$. The same goes for the other pairs of $r_j$. Therefore we can construct the $r_j$ by ruler and compass from the $s_j$. Finally, we can construct $\zeta$ from the $r_j$ by solving a quadratic, hence by ruler and compass.

An explicit construction can again be found in Gleason (1988) and Conway and Guy (1996) page 200.

Earlier it was stated that the Pierpont primes $p = 2^a 3^b + 1$ form a much richer set than the Fermat primes. It is worth expanding on that statement. It is generally believed that the only Fermat primes are the known ones, 2, 3, 5, 17, 257, and 65537, though this has not been proved. In contrast, Gleason (1988) conjectured that Pierpont primes are so common that there should be *infinitely many*; he suggested that there should be about $9k$ of them less than $10^k$. More formally, the number of Pierpont primes less than $N$ should be asymptotic to a constant times $\log N$. This conjecture remains open, but with modern computer algebra it is easy to explore larger values. For example, a quick, unsystematic search turned up the Pierpont prime

$$\begin{aligned} 2^{148}3^{95} + 1 = \ &756\ 760\ 676\ 272\ 923\ 020\ 551\ 154\ 471\ 073 \\ &240\ 459\ 834\ 492\ 063\ 891\ 235\ 892\ 290\ 277 \\ &703\ 256\ 956\ 240\ 171\ 581\ 788\ 957\ 704\ 193 \end{aligned}$$

with 90 digits. There are 789 Pierpont primes up to $10^{100}$. Currently, the largest known Pierpont prime is $3 \times 2^{16408818} + 1$ with 4,939,547 decimal digits, which was proved prime in 2020.

## EXERCISES

21.1 Prove that, in the notation of Section 21.4,

$$\zeta^j = \frac{1}{p-1}\left(\sum_{l=0}^{p-2} \theta^{-jl}\alpha_l\right)$$

21.2 Prove that $\Phi_{24}(t) = t^8 - t^4 + 1$.

21.3 Show that the zeros of the $d$th cyclotomic polynomial can be expressed by radicals of degree at most $\max(2, (d-1)/2)$. (The 2 occurs because of the case $d = 3$.)

21.4 Use trigonometric identities to prove directly from the definition that $\Phi_{12}(t) = t^4 - t^2 + 1$.

21.5 Prove that $\Phi_{12}(t)$ is irreducible over $\mathbb{Q}$.

21.6 Prove that if $\theta$ is a primitive $(p-1)$th root of unity, then

$$1 + \theta^j + \theta^{2j} + \cdots + \theta^{(p-2)j} = \begin{cases} p-1 & \text{if } j = 0 \\ 0 & \text{if } j \leq l \leq p-2 \end{cases}$$

21.7 Prove that the coefficients of $\Phi_p(t)$ are all contained in $\{-1, 0, 1\}$ when $p$ is prime.

21.8 Prove that the coefficients of $\Phi_{p^k}(t)$ are all contained in $\{-1, 0, 1\}$ when $p$ is prime and $k > 1$.

21.9 If $m$ is odd, prove that $\Phi_{2m}(t) = \Phi_m(-t)$, and deduce that the coefficients of $\Phi_{2p^k}(t)$ are contained in $\{-1, 0, 1\}$ when $p$ is an odd prime and $k > 1$.

21.10 If $p, q$ are distinct odd primes, find a formula for $\Phi_{pq}(t)$ and deduce that the coefficients of $\Phi_{pq}(t)$ are all contained in $\{-1, 0, 1\}$.

21.11 Relate $\Phi_{pa}(t)$ and $\Phi_{p^k a}(t)$ when $a, p$ are odd, $p$ is prime, $p$ and $a$ are coprime, and $k > 1$. Deduce that if the coefficients of $\Phi_{pa}(t)$ are all contained in $\{-1, 0, 1\}$, so are those of $\Phi_{p^k a}(t)$.

21.12 Show that the smallest $n$ such that the coefficients of $\Phi_m(t)$ might *not* all be contained in $\{-1, 0, 1\}$ is $n = 105$. If you have access to symbolic algebra software, or have an evening to spare, lots of paper and are willing to be very careful checking your arithmetic, compute $\Phi_{105}(t)$ and see if some coefficient is not contained in $\{-1, 0, 1\}$.

21.13 Let $\phi(n)$ be the Euler function. Prove that

$$\phi(p^k) = (p-1)p^{k-1}$$

if $p$ is prime, and

$$\phi(r)\phi(s) = \phi(rs)$$

when $r, s$ are coprime. Deduce a formula for $\phi(n)$ in terms of the prime factorisation of $n$.

21.14 Prove that

$$\phi(n) = n \prod_{p \text{ prime},\ p|n} \left(1 - \frac{1}{p}\right)$$

21.15 If $a$ is prime to $n$, where both are integers, prove that $a^{\phi(n)} \equiv 0 \pmod{n}$.

21.16 Prove that for any $m \in \mathbb{N}$ the equation $\phi(n) = m$ has only finitely many solutions $n$. Find examples to show that there may be more than one solution.

21.17 Experiment, make an educated guess, and prove a formula for $\sum_{d|n} \phi(d)$.

21.18 If $n$ is odd, prove that $\phi(4n) = 2\phi(n)$.

21.19 Check that

$$\begin{aligned} 1+2 &= \frac{3}{2}\phi(3) \\ 1+3 &= \frac{4}{2}\phi(4) \\ 1+2+3+4 &= \frac{5}{2}\phi(5) \\ 1+5 &= \frac{6}{2}\phi(6) \\ 1+2+3+4+5+6 &= \frac{7}{2}\phi(7) \end{aligned}$$

What is the theorem? Prove it.

21.20* Prove that if $g \in \mathbb{Z}_{24}^*$ then $g^2 = 1$, so $g$ has order 2 or is the identity. Show that 24 is the largest value of $n$ for which every non-identity element of $\mathbb{Z}_n^*$ has order 2. Which are the others?

21.21 Outline how to construct a regular 19-gon using ruler, compass, and trisector, along the lines discussed for the 7-gon and 13-gon.

21.22 Extend the list of Pierpont primes up to 1000.

21.23 If you have access to a computer algebra package, use it to extend the list of Pierpont primes up to 1,000,000. Experiment to find the largest Pierpont prime that you can.

21.24 (1) Prove that $2^a3^b + 1$ is composite if $a$ and $b$ have an odd common factor greater than 1.

(2) Prove that $2^a3^b+1$ is divisible by 5 if and only if $a-b \equiv 2 \pmod 4$.

(3) Prove that $2^a3^b+1$ is divisible by 7 if and only if $a+2b \equiv 0 \pmod 3$.

(4) Find similar necessary and sufficient conditions for $2^a3^b + 1$ to be divisible by 11, 13, 17, 19.

(5) Prove that $2^a3^b + 1$ is never divisible by 23.

(*Hint*: For (2, 3, 4, 5) prove that if $p$ is prime then $2^a3^b+1 \equiv 0 \pmod p$ if and only if $2^a + 3^{-b} \equiv 0 \pmod p$, and look at powers of 2 and 3 modulo $p$.)

21.25 Mark the following true or false.

(a) Every root of unity in $\mathbb{C}$ has a expression by genuine radicals.

(b) A primitive 11th root of unity in $\mathbb{C}$ can be expressed in terms of rational numbers using only square roots and fifth roots.

(c) Any two primitive roots of unity in $\mathbb{C}$ have the same minimal polynomial over $\mathbb{Q}$.

(d) The Galois group of $\Phi_n(t)$ over $\mathbb{Q}$ is cyclic for all $n$.

(e) The Galois group of $\Phi_n(t)$ over $\mathbb{Q}$ is abelian for all $n$.

(f) The coefficients of any cyclotomic polynomial are all equal to $0, \pm 1$.

(g) The regular 483729409-gon can be constructed using ruler, compass, and trisector. (*Hint*: This number is prime, and you may assume this without further calculation.) Use computer algebra if you wish.

(h) The regular $(3 \times 2^{16408818} + 1)$-gon can be constructed using ruler, compass, and trisector. (*Hint*: This number is prime, and you may assume this without further calculation.)

(i) The regular $(3 \times 2^{16408818})$-gon can be constructed using ruler, compass, and trisector.

# Chapter 22

# Calculating Galois Groups

In order to apply Galois theory to specific polynomials, it is necessary to compute the corresponding Galois group. This was the weak point in the memoir that Galois submitted to the French Academy of Sciences, as Poisson and Lacroix pointed out in their referees' report.

However, the computation is possible—at least in principle. It becomes practical only with modern computers. It is neither simple nor straightforward, and until now we have emulated Galois and strenuously avoided it. Instead we have either studied special equations whose Galois group is relatively easy to find, resorted to special tricks, or obtained results that require only partial knowledge of the Galois group. The time has now come to face up squarely to the problem. This chapter contains fairly complete discussions for cubic and quartic polynomials. It also provides a general algorithm for equations of any degree, which is of theoretical importance but is too cumbersome to use in practice. More practical methods exist, but they go beyond the scope of this book: see Soicher and McKay (1985), Fieker and Klüners (2013), and the two references for Hulpke (Internet). The packages Maple and GAP can compute Galois groups for relatively small degrees. A wealth of information can be found online.

## 22.1 Transitive Subgroups

We know that the Galois group $\Gamma(f)$ of a polynomial $f$ with no multiple zeros of degree $n$ is (isomorphic to) a subgroup of the symmetric group $\mathbb{S}_n$. In classical terminology, $\Gamma(f)$ permutes the roots of the equation $f(t) = 0$. Renumbering the roots changes $\Gamma(f)$ to some conjugate subgroup of $\mathbb{S}_n$, so we need consider only the conjugacy classes of subgroups. However, $\mathbb{S}_n$ has rather a lot of conjugacy classes of subgroups, even for moderate $n$ (say $n \geq 6$). So the list of cases rapidly becomes unmanageable.

However, if $f$ is irreducible (which we may always assume when solving $f(t) = 0$ by resolving $f$ into irreducible factors) we can place a fairly stringent restriction on the subgroups that can occur. To state it we need:

DOI: 10.1201/9781003213949-22

**Definition 22.1.** Let $G$ be a permutation group; that is, a subgroup of the group of all permutations on a set $S$. We say that $G$ is *transitive* (or *transitive on $S$*) if for all $s, t \in S$ there exists $\gamma \in G$ such that $\gamma(s) = t$.

To prove $G$ transitive it is enough to show that for some fixed $s_0 \in S$, and any $s \in S$, there exists $\gamma \in G$ such that $\gamma(s_0) = s$. For if this holds, then given $t \in S$ there also exists $\delta \in G$ such that $\delta(s_0) = t$, so $(\delta\gamma^{-1})(s) = t$.

**Examples 22.2.** (1) The Klein four-group $\mathbb{V}$ is transitive on $\{1, 2, 3, 4\}$. The element 1 is mapped to:

1 by the identity

2 by $(1\,2)(3\,4)$

3 by $(1\,3)(2\,4)$

4 by $(1\,4)(2\,3)$

(2) The cyclic group generated by $\alpha = (1\,2\,3\,4)$ is transitive on $\{1, 2, 3, 4\}$. In fact, $\alpha^i$ maps 1 to $i$ for $i = 1, 2, 3, 4$.
(3) The cyclic group generated by $\beta = (1\,2\,3)$ is not transitive on $\{1, 2, 3, 4\}$. There is no power of $\beta$ that maps 1 to 4.

**Proposition 22.3.** *The Galois group of an irreducible polynomial $f$ is transitive on the set of zeros of $f$.*

*Proof.* If $\alpha$ and $\beta$ are two zeros of $f$ then they have the same minimal polynomial, namely $f$. By Theorem 17.4 and Proposition 11.4 there exists $\gamma$ in the Galois group such that $\gamma(\alpha) = \beta$. □

Listing the (conjugacy classes of) transitive subgroups of $\mathbb{S}_n$ is not as formidable as listing all (conjugacy classes of) subgroups. The transitive subgroups, up to conjugacy, have been classified for low values of $n$ by Conway, Hulpke, and MacKay (1998). The GAP data library

gap-system.org/Datalib/trans.html

contains all transitive subgroups of $\mathbb{S}_n$ for $n \leq 30$. The methods used can be found in Hulpke (1996). There is only one such subgroup when $n = 2$, two when $n = 3$, and five when $n = 4, 5$. The magnitude of the task becomes apparent when $n = 6$: in this case there are 16 transitive subgroups up to conjugacy. The number drops to seven when $n = 7$; in general prime $n$ lead to fewer conjugacy classes of transitive subgroups than composite $n$ of similar size.

## 22.2 Bare Hands on the Cubic

As motivation, we begin with a cubic equation over $\mathbb{Q}$, where the answer can be obtained by direct 'bare hands' methods. Consider a cubic polynomial

$$f(t) = t^3 - s_1t^2 + s_2t - s_3 \in \mathbb{Q}[t]$$

The coefficient $s_j$ are the elementary symmetric polynomials in the zeros $\alpha_1, \alpha_2, \alpha_3$, as in Section 18.2. If $f$ is reducible then the calculation of its Galois group is easy: it is the trivial group **1** if all zeros are rational, and $\mathbb{S}_2$ otherwise. Thus we may assume that $f$ is irreducible over $\mathbb{Q}$.

Let $\Sigma$ be the splitting field of $f$,

$$\Sigma = \mathbb{Q}(\alpha_1, \alpha_2, \alpha_3)$$

By Proposition 22.3 the Galois group of $f$ is a transitive subgroup of $\mathbb{S}_3$, hence is either $\mathbb{S}_3$ or $\mathbb{A}_3$. Suppose for argument's sake that it is $\mathbb{A}_3$. What does this imply about the zeros $\alpha_1, \alpha_2, \alpha_3$? By the Galois correspondence, the fixed field $\mathbb{A}_3^\dagger$ of $\mathbb{A}_3$ is $\mathbb{Q}$. Now $\mathbb{A}_3$ consists of the identity, and the two cyclic permutations (123) and (132). Any expression in $\alpha_1, \alpha_2, \alpha_3$ that is invariant under cyclic permutations must therefore lie in $\mathbb{Q}$. Two obvious expressions of this type are

$$\phi = \alpha_1^2\alpha_2 + \alpha_2^2\alpha_3 + \alpha_3^2\alpha_1$$

and

$$\psi = \alpha_1^2\alpha_3 + \alpha_2^2\alpha_1 + \alpha_3^2\alpha_2$$

Indeed it can, with a little effort, be shown that

$$\mathbb{A}_3^\dagger = \mathbb{Q}(\phi, \psi)$$

(see Exercise 22.3). In other words, the Galois group of $f$ is $\mathbb{A}_3$ if and only if $\phi$ and $\psi$ are rational.

This is useful only if we can calculate $\phi$ and $\psi$, which we now do. Because $\mathbb{S}_3$ is generated by $\mathbb{A}_3$ together with the transposition (12), which interchanges $\phi$ and $\psi$, it follows that both $\phi + \psi$ and $\phi\psi$ are symmetric polynomials in $\alpha_1, \alpha_2, \alpha_3$. By Theorem 18.11 they are therefore polynomials in $s_1, s_2$, and $s_3$. We can compute these polynomials explicitly, as follows. We have

$$\phi + \psi = \sum_{i \neq j} \alpha_i^2\alpha_j$$

Compare this with

$$s_1s_2 = (\alpha_1 + \alpha_2 + \alpha_3)(\alpha_1\alpha_2 + \alpha_2\alpha_3 + \alpha_3\alpha_1) = \sum_{i \neq j} \alpha_i^2\alpha_j + 3\alpha_1\alpha_2\alpha_3$$

Since $\alpha_1\alpha_2\alpha_3 = s_3$ we deduce that

$$\phi + \psi = s_1 s_2 - 3s_3$$

Similarly

$$\begin{aligned}\phi\psi &= \alpha_1^4\alpha_2\alpha_3 + \alpha_2^4\alpha_3\alpha_1 + \alpha_3^4\alpha_1\alpha_2 + \alpha_1^3\alpha_2^3 + \alpha_2^3\alpha_3^3 + \alpha_3^3\alpha_1^3 + 3\alpha_1^2\alpha_2^2\alpha_3^2 \\ &= s_3(\alpha_1^3 + \alpha_2^3 + \alpha_3^3) + 3s_3^2 + \sum_{i<j} \alpha_i^3\alpha_j^3\end{aligned}$$

Now

$$\begin{aligned}s_1^3 &= (\alpha_1 + \alpha_2 + \alpha_3)^3 \\ &= (\alpha_1^3 + \alpha_2^3 + \alpha_3^3) + 3\sum_{i\neq j} \alpha_i^2\alpha_j + 6\alpha_1\alpha_2\alpha_3\end{aligned}$$

so that

$$\alpha_1^3 + \alpha_2^3 + \alpha_3^3 = s_1^3 - 6s_3 - 3(s_1s_2 - 3s_3)$$

Moreover,

$$\begin{aligned}s_2^3 &= (\alpha_1\alpha_2 + \alpha_2\alpha_3 + \alpha_3\alpha_1)^3 \\ &= \sum_{i<j} \alpha_i^3\alpha_j^3 + 3\sum_{i,j,k} \alpha_i^3\alpha_j^2\alpha_k + 6\alpha_1^2\alpha_2^2\alpha_3^2 \\ &= \sum_{i<j} \alpha_i^3\alpha_j^3 + 3s_3\left(\sum_{i\neq j} \alpha_i^2\alpha_j\right) + 6s_3^2\end{aligned}$$

Therefore

$$\begin{aligned}\sum_{i<j} \alpha_i^3\alpha_j^3 &= s_2^3 - 3s_3(s_1s_2 - 3s_3) - 6s_3^2 \\ &= s_2^3 - 3s_1s_2s_3 + 3s_3^2\end{aligned}$$

Putting all these together,

$$\begin{aligned}\phi\psi &= s_3(s_1^3 - 3s_1s_2 + 3s_3) + s_2^3 + 3s_3^2 - 3s_1s_2s_3 + 3s_3^2 \\ &= s_1^3s_3 + 9s_3^2 - 6s_1s_2s_3 + s_2^3\end{aligned}$$

Hence $\phi$ and $\psi$ are the roots of the quadratic equation

$$t^2 - at + b = 0$$

where

$$\begin{aligned}a &= s_1s_2 - 3s_3 \\ b &= s_1^3s_3 + 9s_3^2 - 6s_1s_2s_3 + s_2^3\end{aligned}$$

By the formula for quadratics, this equation has rational zeros if and only if $\sqrt{a^2 - 4b} \in \mathbb{Q}$. Direct calculation shows that

$$a^2 - 4b = s_1^2 s_2^2 + 18 s_1 s_2 s_3 - 27 s_3^2 - 4 s_1^3 s_3 - 4 s_2^3$$

We denote this expression by $\Delta$, because it turns out to be the discriminant of $f$. Thus we have proved:

**Proposition 22.4.** *Let $f(t) = t^3 - s_1 t^2 + s_2 t - s_3 \in \mathbb{Q}[t]$ be irreducible over $\mathbb{Q}$. Then its Galois group is $\mathbb{A}_3$ if*

$$\Delta = s_1^2 s_2^2 + 18 s_1 s_2 s_3 - 27 s_3^2 - 4 s_1^3 s_3 - 4 s_2^3 \tag{22.1}$$

*is a perfect square in $\mathbb{Q}$ and is $\mathbb{S}_3$ otherwise.*

**Examples 22.5.** (1) Let $f(t) = t^3 + 3t + 1$. This is irreducible, and

$$s_1 = 0 \qquad s_2 = 3 \qquad s_3 = -1$$

We find that $\Delta = -27 - 4.27 = -135$, which is not a square. Hence the Galois group is $\mathbb{S}_3$.
(2) Let $f(t) = t^3 - 3t - 1$. This is irreducible, and

$$s_1 = 0 \qquad s_2 = -3 \qquad s_3 = 1$$

Now $\Delta = 81$, which is a square. Hence the Galois group is $\mathbb{A}_3$.

There is a valuable lesson here: the Galois group depends on number-theoretic properties of the coefficients, not just algebraic ones.

## 22.3 The Discriminant

More elaborate versions of the above method can be used to treat quartics or quintics, but in this form the calculations are very unstructured. See Exercise 22.6 for quartics. In this section we provide an interpretation of the expression $\Delta$ in (22.1): it is the discriminant, (8.7), which generalises to any field. We show that the discriminant distinguishes between polynomials of degree $n$ whose Galois groups are, or are not, contained in $\mathbb{A}_n$.

**Theorem 22.6.** *Let $f \in K[t]$, where the characteristic of $K$ is not 2. Then*

(1) $\Delta(f) \in K$.

(2) $\Delta(f) = 0$ *if and only if $f$ has a multiple zero.*

(3) If $\Delta(f) \neq 0$ then $\Delta(f)$ *is a perfect square in* $K$ *if and only if the Galois group of* $f$*, interpreted as a group of permutations of the zeros of* $f$*, is contained in the alternating group* $\mathbb{A}_n$.

*Proof.* Let $\sigma \in \mathbb{S}_n$, acting by permutations of the $\alpha_j$. It is easy to check that if $\sigma$ is applied to $\delta$ then it changes it to $\pm\delta$, the sign being + if $\sigma$ is an even permutation and - if $\sigma$ is odd. (Indeed in many algebra texts the sign of a permutation is defined in this manner.) Therefore $\delta \in \mathbb{A}_n^\dagger$. Further, $\Delta(f) = \delta^2$ is unchanged by any permutation in $\mathbb{S}_n$, hence lies in $K$. This proves (1).

Part (2) follows from the definition of $\Delta(f)$.

Let $G$ be the Galois group of $f$, considered as a subgroup of $\mathbb{S}_n$. If $\Delta(f)$ is a perfect square in $K$ then $\delta \in K$, so $\delta$ is fixed by $G$. Now odd permutations change $\delta$ to $-\delta$, and since $\text{char}(K) \neq 2$ we have $\delta \neq -\delta$. Therefore all permutations in $G$ are even, that is, $G \subseteq \mathbb{A}_n$. Conversely, if $G \subseteq \mathbb{A}_n$ then $\delta \in G^\dagger = K$. Therefore $\Delta(f)$ is a perfect square in $K$. □

In order to apply Theorem 22.6, we must calculate $\Delta(f)$ explicitly. Because it is a symmetric polynomial in the zeros $\alpha_j$, it must be given by some polynomial in the elementary symmetric polynomials $s_k$. Brute force calculations show that if $f$ is a cubic polynomial then

$$\Delta(f) = s_1^2 s_2^2 + 18 s_1 s_2 s_3 - 27 s_3^2 - 4 s_1^3 s_3 - 4 s_2^3$$

which is precisely the expression $\Delta$ obtained in Proposition 22.4. Proposition 22.4 is thus a corollary of Theorem 22.6.

## 22.4 General Algorithm for the Galois Group

We now describe a method which, in principle, will compute the Galois group of any polynomial. The practical obstacles involved in carrying it out are considerable for equations of even modestly high degree, but it does have the virtue of showing that the problem possesses an algorithmic solution. More efficient algorithms have been invented, but to describe them would take us too far afield: see previous references in this chapter.

Suppose that

$$f(t) = t^n - s_1 t^{n-1} + \cdots + (-1)^n s_n$$

is a monic irreducible polynomial over a field $K$, having distinct zeros $\alpha_1, \ldots, \alpha_n$ in a splitting field $\Sigma$. That is, we assume $f$ is separable. The $s_k$ are the elementary symmetric polynomials in the $\alpha_j$. The idea is to consider not just how an element $\gamma$ of the Galois group $G$ of $f$ acts on $\alpha_1, \ldots, \alpha_n$, but how $\gamma$ acts on arbitrary 'linear combinations'

$$\beta = x_1 \alpha_1 + \cdots + x_n \alpha_n$$

To make this action computable we form polynomials having zeros $\gamma(\beta)$ as $\gamma$ runs through $G$. To do so, let $x_1, \dots, x_n$ be independent indeterminates, let $\beta$ be defined as above, and for every $\sigma \in \mathbb{S}_n$ define

$$\begin{aligned}\sigma_x(\beta) &= x_{\sigma(1)}\alpha_1 + \cdots + x_{\sigma(n)}\alpha_n \\ \sigma_\alpha(\beta) &= x_1\alpha_{\sigma(1)} + \cdots + x_n\alpha_{\sigma(n)}\end{aligned}$$

By rearranging terms, we see that $\sigma_\alpha(\beta) = \sigma_x^{-1}(\beta)$.

(The notation here reminds us that $\sigma_x$ acts on the $x_j$, whereas $\sigma_\alpha$ acts on the $\alpha_j$.)

Since $f$ has distinct zeros, $\sigma_x(\beta) \neq \tau_x(\beta)$ if $s \neq \tau$. Define the polynomial

$$Q = \prod_{\sigma \in \mathbb{S}_n} (t - \sigma_x(\beta)) = \prod_{\sigma \in \mathbb{S}_n} (t - \sigma_\alpha(\beta))$$

If we use the second expression for $Q$, expand in powers of $t$, collect like terms, and write all symmetric polynomials in the $\alpha_j$ as polynomials in the $s_k$, we find that

$$Q = \sum_{j=0}^{n!} \left( \sum_i g_i(s_1, \dots, s_n) x_1^{i_1} \dots x_n^{i_n} \right) t^j$$

where the $g_i$ are explicitly computable functions of $s_1, \dots, s_n$. In particular $Q \in K[t, x_1, \dots, x_n]$. (In the second sum above, $i$ ranges over all $n$-tuples of nonnegative integers $(i_1, \dots, i_n)$ with $i_1 + \cdots + i_n + j = n$)

Next we split $Q$ into a product of irreducibles,

$$Q = Q_1 \dots Q_k$$

in $K[t, x_1, \dots, x_n]$. In the ring $\Sigma[t, x_1, \dots, x_n]$ we can write

$$Q_j = \prod_{\sigma \in S_j} (t - \sigma_x(\beta))$$

where $\mathbb{S}_n$ is the disjoint union of the subsets $S_j$. We choose the labels so that the identity of $\mathbb{S}_n$ is contained in $S_1$, and then $t - \beta$ divides $Q_1$ in $\Sigma[t, x_1, \dots, x_n]$.

If $\sigma \in \mathbb{S}_n$ then

$$Q = \sigma_x Q = (\sigma_x Q_1) \cdots (\sigma_x Q_k)$$

Hence $\sigma_x$ permutes the irreducible factors $Q_j$ of $Q$. Define

$$\mathbf{G} = \{\sigma \in \mathbb{S}_n : \sigma_x Q_1 = Q_1\}$$

a subgroup of $\mathbb{S}_n$. Then we have the following characterisation of the Galois group of $f$:

**Theorem 22.7.** *The Galois group $G$ of $f$ is isomorphic to the group* $\mathbf{G}$.

*Proof.* The subset $S_1$ of $\mathbb{S}_n$ is in fact equal to $\mathbf{G}$, because

$$\begin{aligned} S_1 &= \{\sigma : t - \sigma_x\beta \text{ divides } Q_1 \text{ in } \Sigma[t, x_1, \ldots, x_n]\} \\ &= \{\sigma : t - \beta \text{ divides } \sigma_x^{-1}Q_1 \text{ in } \Sigma[t, x_1, \ldots, x_n]\} \\ &= \{\sigma : \sigma_x^{-1}Q_1 = Q_1\} = \mathbf{G} \end{aligned}$$

Let $P$ be the group of all permutations of the zeros of $f$, and define

$$H = \prod_{\sigma\in P}(t - \sigma_\alpha(\beta)) = \prod_{\sigma\in P}(t - \sigma_x(\beta))$$

Clearly $H \in K[t, x_1, \ldots, x_n]$. Now $H$ divides $Q$ in $\Sigma[t, x_1, \ldots, x_n]$ so $H$ divides $Q$ in $\Sigma(x_1, \ldots, x_n)[t]$. Therefore $H$ divides $Q$ in $K(x_1, \ldots, x_n)[t]$ so that $H$ divides $Q$ in $K[t, x_1, \ldots, x_n]$ by the analogue of Gauss's Lemma for $K(x_1, \ldots, x_n)[t]$, which can be proved in a similar manner to Lemma 3.19.

Thus $H$ is a product of some of the irreducible factors $Q_j$ of $Q$. Because $y - \beta$ divides $H$ we know that $Q_1$ is one of these factors. Therefore $Q_1$ divides $H$ in $K[t, x_1, \ldots, x_n]$ so $\mathbf{G} \subseteq G$.

Conversely, let $\gamma \in G$ and apply the automorphism $\gamma$ to the relation $(t - \beta)|Q_1$. Since $Q_1$ has coefficients in $K$, we get $(t - \gamma_\alpha(\beta))|Q_1$. Now $t - \gamma_\alpha(\beta) = t - \gamma_x^{-1}(\beta) = \gamma_x^{-1}(t - \beta)$, so $\gamma_x^{-1}(t - \beta)|Q_1$. Equivalently, $(t - \beta)|\gamma_x(Q_1)$. But $Q_1$ is the unique irreducible factor of $Q$ that is divisible by $t - \beta$, so $\gamma_x(Q_1) = Q_1$, so $\gamma \in \mathbf{G}$.

□

**Example 22.8.** Suppose that $\alpha$, $\beta$ are the zeros of a quadratic polynomial $t^2 - At + B = 0$, where $A = \alpha + \beta$ and $B = \alpha\beta$. The polynomial $Q$ takes the form

$$\begin{aligned} Q &= (t - \alpha x - \beta y)(t - \alpha y - \beta x) \\ &= t^2 - t(\alpha x + \beta y + \alpha y + \beta x) + [(\alpha^2 + \beta^2)xy + \alpha\beta(x^2 + y^2)] \\ &= t^2 - t(Ax + Ay) + [(A^2 - 2B)xy + B(x^2 + y^2)] \end{aligned}$$

This is either irreducible or has two linear factors. The condition for irreducibility is that

$$A^2(x + y)^2 - 4[(A^2 - 2B)xy + B(x^2 + y^2)]$$

is not a perfect square. But this is equal to

$$(A^2 - 4B)(x - y)^2$$

which is a perfect square if and only if $A^2 - 4B$ is a perfect square. Thus the Galois group $G$ is trivial if $A^2 - 4B$ is a perfect square and is cyclic of order 2 if $A^2 - 4B$ is not a perfect square.

It is of course much simpler to prove this directly, but the calculation illustrates how the theorem works.

## EXERCISES

22.1 Let $f \in K[t]$ where char $(K) \neq 2$. If $\Delta(f)$ is not a perfect square in $K$ and $G$ is the Galois group of $f$, show that $G \cap \mathbb{A}_n$ has fixed field $K(\delta)$.

22.2* Find an expression for the discriminant of a quartic polynomial. (*Hint:* You may assume without proof that this is the same as the discriminant of its resolvent cubic.)

22.3 In thè notation of Proposition 22.4, show that $\mathbb{A}_3^\dagger = \mathbb{Q}(\phi, \psi)$.

22.4 Show that the discriminant is given by the Vandermonde determinant (see Exercise 2.5)

$$\begin{vmatrix} 1 & 1 & \dots & 1 \\ \alpha_1 & \alpha_2 & \dots & \alpha_n \\ \alpha_1^2 & \alpha_2^2 & \dots & \alpha_n^2 \\ \vdots & \vdots & \ddots & \vdots \\ \alpha_1^{n-1} & \alpha_2^{n-1} & \dots & \alpha_n^{n-1} \end{vmatrix}$$

Multiply this matrix by its transpose and take the determinant to show that $\Delta(f)$ is equal to

$$\begin{vmatrix} \lambda_0 & \lambda_1 & \dots & \lambda_{n-1} \\ \lambda_1 & \lambda_2 & \dots & \lambda_n \\ \vdots & \vdots & \ddots & \vdots \\ \lambda_{n-1} & \lambda_n & \dots & \lambda_{2n-2} \end{vmatrix}$$

where $\lambda_k = \alpha_1^k + \cdots + \alpha_n^k$. Hence, using Exercise 18.17, compute $\Delta(f)$ when $f$ is of degree 2, 3, or 4. Check your result is the same as that obtained previously.

22.5* If $f(t) = t^n + at + b$, show that

$$\Delta(f) = \mu_{n+1} n^n b^{n-1} - \mu_n (n-1)^{n-1} a^n$$

where $\mu_n$ is 1 if $n$ is a multiple of 4 and is $-1$ otherwise.

22.6* Show that any transitive subgroup of $\mathbb{S}_4$ is conjugate to one of $\mathbb{S}_4, \mathbb{A}_4, \mathbb{D}_4, \mathbb{V}$, or $\mathbb{Z}_4$, defined as follows:

$$\begin{aligned} \mathbb{A}_4 &= \text{alternating group of degree 4} \\ \mathbb{V} &= \{1, (1\,2)(3\,4), (1\,3)(2\,4), (1\,4)(2\,3)\} \\ \mathbb{D}_4 &= \text{group generated by } \mathbb{V} \text{ and } (1\,2) \\ \mathbb{Z}_4 &= \text{group generated by } (1\,2\,3\,4) \end{aligned}$$

22.7* Let $f$ be a monic irreducible quartic polynomial over a field $K$ of characteristic $\neq 2, 3$ with discriminant $\Delta$. Let $g$ be its resolvent cubic, defined (after a Tschirnhaus transformation) by (18.3), and let $M$ be a splitting field for $g$ over $K$. Show that:

(a) $\Gamma(f) \cong \mathbb{S}_4$ if and only if $\Delta$ is not a square in $K$ and $g$ is irreducible over $K$.

(b) $\Gamma(f) \cong \mathbb{A}_4$ if and only if $\Delta$ is a square in $K$ and $g$ is irreducible over $K$.

(c) $\Gamma(f) \cong \mathbb{D}_4$ if and only if $\Delta$ is not a square in $K, g$ is reducible over $K$, and $f$ is irreducible over $M$.

(d) $\Gamma(f) \cong \mathbb{V}$ if and only if $\Delta$ is a square in $K$ and $g$ is reducible over $K$.

(e) $\Gamma(f) \cong \mathbb{Z}_4$ if and only if $\Delta$ is not a square in $K, g$ is reducible over $K$, and $f$ is reducible over $M$.

22.8 Prove that $\{(1\,2\,3), (4\,5\,6), (1\,4)\}$ generates a transitive subgroup of $\mathbb{S}_6$. Generalise this result to $\mathbb{S}_{2n}$.

22.9 Mark the following true or false.

(a) Every nontrivial normal subgroup of $\mathbb{S}_n$ is transitive.

(b) Every nontrivial subgroup of $\mathbb{S}_n$ is transitive.

(c) Every transitive subgroup of $\mathbb{S}_n$ is normal.

(d) Every transitive subgroup of $\mathbb{S}_n$ has order divisible by $n$.

(e) The Galois group of any irreducible cubic polynomial over a field of characteristic zero is isomorphic either to $\mathbb{S}_3$ or to $\mathbb{A}_3$.

(f) If $K$ is a field of characteristic zero in which every element is a perfect square, then the Galois group of any irreducible cubic polynomial over $K$ is isomorphic to $\mathbb{A}_3$.

# Chapter 23

# Algebraically Closed Fields

Back to square one.

In Chapter 2 we proved the Fundamental Theorem of Algebra, Theorem 2.4, using some basic point-set topology and simple estimates. It is also possible to give an 'almost' algebraic proof, in which the only extraneous information required is that every polynomial of odd degree over $\mathbb{R}$ has a real zero. This follows immediately from the continuity of polynomials over $\mathbb{R}$ and the fact that an odd degree polynomial changes sign somewhere between $-\infty$ and $+\infty$.

We now present this almost-algebraic proof, which applies to a slight generalisation. The main property of $\mathbb{R}$ that we require is that $\mathbb{R}$ is an ordered field, with a relation $\leq$ that satsfies the usual properties. So we start by defining an ordered field. Then we develop some group theory, a far-reaching generalisation of Cauchy's Theorem due to the Norwegian mathematician Ludwig Sylow, about the existence of certain subgroups of prime power order in any finite group. Finally, we combine Sylow's Theorem with the Galois correspondence to prove the main theorem, which we set in the general context of an 'algebraically closed' field.

## 23.1 Ordered Fields and Their Extensions

As remarked in Chapter 2, the first proof of the Fundamental Theorem of Algebra was given by Gauss in his doctoral dissertation of 1799. His title (in Latin) was *A New Proof that Every Rational Integral Function of One Variable can be Resolved into Real Factors of the First or Second Degree.* Gauss was being polite in using the word 'new', because his was the first genuine proof. Even his proof, from the modern viewpoint, has gaps; but these are topological in nature and not hard to fill. In Gauss's day they were not considered to be gaps at all. Gauss came up with several different proofs of the Fundamental Theorem of Algebra; among them is a topological proof that can be found in Hardy (1960) page 492.

As discussed in Chapter 2, many other proofs are now known. Several of them use complex analysis. The one in Titchmarsh (1960) page 118 is probably the proof most commonly encountered in an undergraduate course. Less well

DOI: 10.1201/9781003213949-23

known is a proof by Clifford (1968) page 20, which is almost entirely algebraic. His idea is to show that any irreducible polynomial over $\mathbb{R}$ is of degree 1 or 2. The proof we give here follows this route: it is essentially due to Legendre, but his original proof had gaps, which we fill using Galois theory. It is unreasonable to ask for a *purely* algebraic proof of the theorem, since the real numbers (and hence the complex numbers) are defined in terms of analytic concepts such as Cauchy sequences, Dedekind cuts, or completeness in an ordering.

We begin by abstracting some properties of the reals.

**Definition 23.1.** An *ordered field* is a field $K$ with a relation $\leq$ such that:

(1) $k \leq k$ for all $k \in K$.

(2) $k \leq l$ and $l \leq m$ implies $k \leq m$ for all $k, l, m \in K$.

(3) $k \leq l$ and $l \leq k$ implies $k = l$ for all $k, l \in K$.

(4) If $k, l \in K$ then either $k \leq l$ or $l \leq k$.

(5) If $k, l, m \in K$ and $k \leq l$ then $k + m \leq l + m$.

(6) If $k, l, m \in K$ and $k \leq l$ and $0 \leq m$ then $km \leq lm$.

The relation $\leq$ is an ordering on $K$. The associated relations $<$, $\geq$, $>$ are defined in terms of $\leq$ in the obvious way, as are the concepts 'positive' and 'negative'.

Examples of ordered fields are $\mathbb{Q}$ and $\mathbb{R}$. We need two simple consequences of the definition of an ordered field.

**Lemma 23.2.** *Let $K$ be an ordered field. Then $k^2 \geq 0$ for any $k \in K$, and the characteristic of $K$ is zero.*

*Proof.* If $k \geq 0$ then $k^2 \geq 0$ by (6). So by (3) and (4) we may assume $k < 0$. If now we had $-k < 0$ it would follow that

$$0 = k + (-k) < k + 0 = k$$

a contradiction. So $-k \geq 0$, whence $k^2 = (-k)^2 \geq 0$. This proves the first statement.

We now know that $1 = 1^2 > 0$, so for any finite $n$ the number

$$n \cdot 1 = 1 + \cdots + 1 > 0$$

implying that $n \cdot 1 \neq 0$ and $K$ must have characteristic 0. □

We quote the following properties of $\mathbb{R}$.

**Lemma 23.3.** $\mathbb{R}$, *with the usual ordering, is an ordered field. Every positive element of $\mathbb{R}$ has a square root in $\mathbb{R}$. Every odd degree polynomial over $\mathbb{R}$ has a zero in $\mathbb{R}$.*

These are all proved in any course in analysis, and depend on the fact that a polynomial function on $\mathbb{R}$ is continuous.

## 23.2 Sylow's Theorem

Next, we set up the necessary group theory. Sylow's Theorem is based on the concept of a $p$-group:

**Definition 23.4.** Let $p$ be a prime. A finite group $G$ is a *$p$-group* if its order is a power of $p$.

For example, the dihedral group $\mathbb{D}_4$ is a 2-group. If $n \geq 3$, then the symmetric group $\mathbb{S}_n$ is never a $p$-group for any prime $p$.

The $p$-groups have many pleasant properties (and many unpleasant ones, but we shall not dwell on their Dark Side). Recall Definition 14.13 of the centre of a group. A very useful property of $p$-groups is:

**Theorem 23.5.** *If $G \neq \mathbf{1}$ is a finite $p$-group, then $G$ has non-trivial centre.*

*Proof.* The class equation (14.2) of $G$ reads

$$p^n = |G| = 1 + |C_2| + \cdots + |C_r|$$

and Corollary14.12 implies that $|C_j| = p^{n_j}$ for some $n_j \geq 0$. Now $p$ divides the right-hand side of the class equation, so that at least $p-1$ values of $|C_j|$ must be equal to 1. But if $x$ lies in a conjugacy class with only one element, then $g^{-1}xg = x$ for all $g \in G$, that is, $gx = xg$. Hence $x \in Z(G)$. Therefore $Z(G) \neq 1$. □

From this we easily deduce:

**Lemma 23.6.** *If $G$ is a finite $p$-group of order $p^n$, then $G$ has a series of normal subgroups*

$$1 = G_0 \subseteq G_1 \subseteq \ldots \subseteq G_n = G$$

*such that $|G_j| = p^j$ for all $j = 0, \ldots, n$.*

*Proof.* Use induction on $n$. If $n = 0$ all is clear. If not, let $Z = Z(G) \neq 1$ by Theorem 23.5. Since $Z$ is an abelian group of order $p^m$ it has an element of order $p$. The cyclic subgroup $K$ generated by such an element has order $p$ and is normal in $G$ since $K \subseteq Z$. Now $G/K$ is a $p$-group of order $p^{n-1}$, and by induction there is a series of normal subgroups

$$K/K = G_1/K \subseteq \ldots \subseteq G_n/K$$

where $|G_j/K| = p^{j-1}$. But then $|G_j| = p^j$ and $G_j \lhd G$. If we let $G_0 = 1$, the result follows. □

**Corollary 23.7.** *Every finite $p$-group is soluble.*

*Proof.* The quotients $G_{j+1}/G_j$ of the series afforded by Lemma 23.6 are of order $p$, hence cyclic and in particular abelian. □

In 1872 Sylow discovered some fundamental theorems about the existence of $p$-groups inside given finite groups. We shall need one of his results in this chapter. We state all of his results, though we shall prove only the one that we require, statement (1).

**Theorem 23.8 (Sylow's Theorem).** *Let $G$ be a finite group of order $p^a r$ where $p$ is prime and does not divide $r$. Then*

(1) *$G$ possesses at least one subgroup of order $p^a$.*

(2) *All such subgroups are conjugate in $G$.*

(3) *Any $p$-subgroup of $G$ is contained in one of order $p^a$.*

(4) *The number of subgroups of $G$ of order $p^a$ leaves remainder 1 on division by $p$.*

This result motivates:

**Definition 23.9.** If $G$ is a finite group of order $p^a r$ where $p$ is prime and does not divide $r$, then a *Sylow $p$-subgroup* of $G$ is a subgroup of $G$ of order $p^a$.

In this terminology Theorem 23.8 says that for finite groups Sylow $p$-subgroups exist for all primes $p$, are all conjugate, are the maximal $p$-subgroups of $G$, and occur in numbers restricted by condition (4).

*Proof of Theorem* 23.8(1). Use induction on $|G|$. The theorem is obviously true for $|G| = 1$ or 2. Let $C_1, \ldots, C_s$ be the conjugacy classes of $G$, and let $c_j = |C_j|$. The class equation of $G$ is

$$p^a r = c_1 + \cdots + c_s \tag{23.1}$$

Let $Z_j$ denote the centraliser in $G$ of some element $x_j \in C_j$, and let $n_j = |Z_j|$. By Lemma 14.11

$$n_j = p^a r / c_j \tag{23.2}$$

Suppose first that some $c_j$ is greater than 1 and not divisible by $p$. Then by (23.2) $n_j < p^a r$ and is divisible by $p^a$. Hence by induction $Z_j$ contains a subgroup of order $p^a$. Therefore we may assume that for all $j = 1, \ldots, s$ either $c_j = 1$ or $p|c_j$. Let $z = |Z(G)|$. As in Theorem 23.5, $z$ is the number of values of $i$ such that $c_j = 1$. So $p^a r = z + kp$ for some integer $k$. Hence $p$ divides $z$, and $G$ has a non-trivial centre $Z$ such that $p$ divides $|Z|$. By Lemma 14.14, the group $Z$ has an element of order $p$, which generates a subgroup $P$ of $G$ of order $p$. Since $P \subseteq Z$ it follows that $P \lhd G$. By induction $G/P$ contains a subgroup $S/P$ of order $p^{a-1}$, whence $S$ is a subgroup of $G$ of order $p^a$ and the theorem is proved. □

**Example 23.10.** Let $G = \mathbb{S}_4$, so that $|G| = 24$. According to Sylow's theorem $G$ must have subgroups of orders 3 and 8. Subgroups of order 3 are easy to find: any 3-cycle, such as (123) or (134) or (234), generates such a group. We shall find a subgroup of order 8. Let $\mathbb{V}$ be the Klein four-group, which is normal in $G$. Let $\tau$ be any 2-cycle, generating a subgroup $T$ of order 2. Then $\mathbb{V} \cap T = 1$, and $\mathbb{V}T$ is a subgroup of order 8. (It is isomorphic to $\mathbb{D}_4$.)

Analogues of Sylow's theorem do not work as soon as we go beyond prime powers. Exercise 23.1 illustrates this point.

---

## 23.3 The Algebraic Proof

With Sylow's Theorem under our belt, all that remains is to set up a little more Galois-theoretic machinery.

**Lemma 23.11.** *Let $K$ be a field of characteristic zero, such that for some prime $p$ every finite extension $M$ of $K$ with $M \neq K$ has $[M : K]$ divisible by $p$. Then every finite extension of $K$ has degree a power of $p$.*

*Proof.* Let $N$ be a finite extension of $K$. The characteristic is zero so $N/K$ is separable. Passing to a normal closure we may assume $N/K$ is also normal, so the Galois correspondence is bijective. Let $G$ be the Galois group of $N/K$ and let $P$ be a Sylow $p$-subgroup of $G$. By Theorem 12.2(3) the fixed field $P^\dagger$ has degree $[P^\dagger : K]$ equal to the index of $P$ in $G$ , but this is prime to $p$. By hypothesis, $P^\dagger = K$, so $P = G$. Then $[N : K] = |G| = p^n$ for some $n$. □

**Theorem 23.12.** *Let $K$ be an ordered field in which every positive element has a square root and every odd-degree polynomial has a zero. Then $K(\mathrm{i})$ is algebraically closed, where $\mathrm{i}^2 = -1$.*

*Proof.* $K$ cannot have any extensions of finite odd degree greater than 1. For suppose $[M : K] = r > 1$ where $r$ is odd. Let $\alpha \in M \backslash K$ have minimal polynomial $m$. Then $\partial m$ divides $r$, so is odd. By hypothesis $m$ has a zero in $K$, so is reducible, contradicting Lemma 5.5. Hence every finite extension of $K$ has even degree over $K$. The characteristic of $K$ is 0 by Lemma 23.2, so by Lemma 23.11 every finite extension of $K$ has 2-power degree.

Let $M \neq K(\mathrm{i})$ be any finite extension of $K(\mathrm{i})$ where $\mathrm{i}^2 = -1$. By taking a normal closure we may assume $M/K$ is normal, so the Galois group of $M/K$ is a 2-group. Using Lemma 23.6 and the Galois correspondence, we can find an extension $N$ of $K(\mathrm{i})$ of degree $[N : K(\mathrm{i})] = 2$. By the formula for solving quadratic equations, $N = K(\mathrm{i}, \alpha)$ where $\alpha^2 \in K(\mathrm{i})$. But if $a, b \in K$ then recall equation (2.3):

$$\sqrt{a + b\mathrm{i}} = \sqrt{\frac{a + \sqrt{a^2 + b^2}}{2}} + \mathrm{i}\sqrt{\frac{-a + \sqrt{a^2 + b^2}}{2}}$$

where the square root of $a^2 + b^2$ is the positive one, and the signs of the other two square roots are chosen to make their product equal to $b$. The square roots exist in $K$ since the elements inside them are positive, as is easily checked.

Therefore $\alpha \in K(\mathrm{i})$, so that $N = K(\mathrm{i})$, which contradicts the assumption on $N$. Therefore $M = K(\mathrm{i})$, and $K(\mathrm{i})$ has no finite extensions of degree $> 1$. Hence any irreducible polynomial over $K(\mathrm{i})$ has degree 1, otherwise a splitting field would have finite degree $> 1$ over $K(\mathrm{i})$. Therefore $K(\mathrm{i})$ is algebraically closed. □

**Corollary 23.13 (Fundamental Theorem of Algebra).** *The field $\mathbb{C}$ of complex numbers is algebraically closed.*

*Proof.* Put $\mathbb{R} = K$ in Theorem 23.12 and use Lemma 23.3. □

---

## EXERCISES

23.1 Show that $\mathbb{A}_5$ has no subgroup of order 15.

23.2 Show that a subgroup or a quotient of a $p$-group is again a $p$-group. Show that an extension of a $p$-group by a $p$-group is a $p$-group.

23.3 Show that $\mathbb{S}_n$ has trivial centre if $n \geq 3$.

23.4 Prove that every group of order $p^2$ (with $p$ prime) is abelian. Hence show that there are exactly two non-isomorphic groups of order $p^2$ for any prime number $p$.

23.5 Show that a field $K$ is algebraically closed if and only if $L/K$ algebraic implies $L = K$.

23.6 Show that every algebraic extension of $\mathbb{R}$ is isomorphic to $\mathbb{R}/\mathbb{R}$ or $\mathbb{C}/\mathbb{R}$.

23.7 Show that $\mathbb{C}$, with the traditional field operations, cannot be given the structure of an ordered field. If we allow different field operations, can the set $\mathbb{C}$ be given the structure of an ordered field?

23.8 Prove the theorem whose statement is the title of Gauss's doctoral dissertation mentioned at the beginning of the chapter. ('Rational integral function' was his term for 'polynomial'.)

23.9 Suppose that $K/\mathbb{Q}$ is a finitely generated extension. Prove that there exists a $\mathbb{Q}$-monomorphism $K \to \mathbb{C}$. (*Hint:* Use cardinality considerations to adjoin transcendental elements, and algebraic closure of $\mathbb{C}$ to adjoin algebraic elements.) Is the theorem true for $\mathbb{R}$ rather than $\mathbb{C}$?

23.10 Mark the following true or false.

(a) Every soluble group is a $p$-group.

(b) Every Sylow subgroup of a finite group is soluble.

(c) Every simple $p$-group is abelian.

(d) The field $\mathbb{A}$ of algebraic numbers defined in Exercise 17.6 is algebraically closed.

(e) There is no ordering on $\mathbb{C}$ making it into an ordered field.

(f) Every ordered field has characteristic zero.

(g) Every field of characteristic zero can be ordered.

(h) In an ordered field, every square is positive.

(i) In an ordered field, every positive element is a square.

# Chapter 24

## Transcendental Numbers

Our discussion of the three geometric problems of antiquity—trisecting the angle, duplicating the cube, and squaring the circle—left one key fact unproved. To complete the proof of the impossibility of squaring the circle by a ruler-and-compass construction, crowning three thousand years of mathematical effort, we must prove that $\pi$ is transcendental. (In this chapter the word 'transcendental' means 'transcendental over $\mathbb{Q}$'.) The proof we give is analytic, which should not really be surprising since $\pi$ is best defined analytically. The techniques involve symmetric polynomials, integration, differentiation, and some manipulation of inequalities, together with a healthy lack of respect for apparently complicated expressions.

It is not at all obvious that transcendental real (or complex) numbers exist. That they do was first proved by Liouville in 1844, by considering the approximation of reals by rationals. It transpires that algebraic numbers cannot be approximated by rationals with more than a certain 'speed' (see Exercises 24.5–24.7). To find a transcendental number reduces to finding a number that can be approximated more rapidly than the known bound for algebraic numbers. Liouville showed that this is the case for the real number

$$\xi = \sum_{n=1}^{\infty} 10^{-n!}$$

but no 'naturally occurring' number was proved transcendental until Charles Hermite, in 1873, proved that e, the base of natural logarithms, is. Using similar methods, Ferdinand Lindemann demonstrated the transcendence of $\pi$ in 1882.

Meanwhile Georg Cantor, in 1874, had produced a revolutionary proof of the existence of transcendental numbers, without actually constructing any. His proof (see Exercises 24.1–24.4) used set-theoretic methods, and was one of the earliest triumphs of Cantor's theory of infinite cardinals. When it first appeared, the mathematical world viewed it with great suspicion, but nowadays it scarcely raises an eyebrow.

We shall prove four theorems in this chapter. In each case the proof proceeds by contradiction, and the final blow is dealt by the following simple result:

**Lemma 24.1.** *Let $f : \mathbb{Z} \to \mathbb{Z}$ be a function such that $f(n) \to 0$ as $n \to +\infty$. Then there exists $N \in \mathbb{Z}$ such that $f(n) = 0$ for all $n \geq N$.*

DOI: 10.1201/9781003213949-24

*Proof.* Since $f(n) \to 0$ as $n \to +\infty$, there exists $N \in \mathbb{Z}$ such that $|f(n)-0| < \frac{1}{2}$ whenever $n \geq N$, for some integer $N$. Since $f(n)$ is an integer, this implies that $f(n) = 0$ for $n \geq N$. □

---

## 24.1 Irrationality

Lindemann's proof is ingenious and intricate. To prepare the way we first prove some simpler theorems of the same general type. These results are not needed for Lindemann's proof, but familiarity with the ideas is. The first theorem was initially proved by Johann Heinrich Lambert in 1770 using continued fractions, although it is often credited to Legendre.

**Theorem 24.2.** *The real number $\pi$ is irrational.*

*Proof.* Consider the integral

$$I_n = \int_{-1}^{1} (1-x^2)^n \cos(\alpha x)\mathrm{d}x$$

Integrating by parts, twice, and performing some fairly routine calculations, lead to a recurrence relation

$$\alpha^2 I_n = 2n(2n-1)I_{n-1} - 4n(n-1)I_{n-2} \tag{24.1}$$

if $n \geq 2$. After evaluating the cases $n = 0, 1$, induction on $n$ yields

$$\alpha^{2n+1} I_n = n!(P_n \sin(\alpha) + Q_n \cos(\alpha)) \tag{24.2}$$

where $P_n$ and $Q_n$ are polynomials in $\alpha$ of degree $< 2n+1$ with integer coefficients. The term $n!$ comes from the factor $2n(2n-1)$ of (24.1).

Assume, for a contradiction, that $\pi$ is rational, so that $\pi = a/b$ where $a, b \in \mathbb{Z}$ and $b \neq 0$. Let $\alpha = \pi/2$ in (24.2). Then

$$J_n = a^{2n+1} I_n / n!$$

is an integer. By the definition of $I_n$,

$$J_n = \frac{a^{2n+1}}{n!} \int_{-1}^{1} (1-x^2)^n \cos\frac{\pi}{2}x\,\mathrm{d}x$$

The integrand is $> 0$ for $-1 < x < 1$, so $J_n > 0$. Hence $J_n \neq 0$ for all $n$. But

$$\begin{aligned} |J_n| &\leq \frac{|a|^{2n+1}}{n!} \int_{-1}^{1} \cos\frac{\pi}{2}x\,\mathrm{d}x \\ &\leq 2|a|^{2n+1}/n! \end{aligned}$$

Hence $J_n \to 0$ as $n \to +\infty$. This contradicts Lemma 24.1, so the assumption that $\pi$ is rational is false. □

The next, slightly stronger, result was proved by Legendre in his *Éléments de Géométrie* of 1794, which, as we remarked in the Historical Introduction, greatly influenced the young Galois.

**Theorem 24.3.** *The real number $\pi^2$ is irrational.*

*Proof.* Assume if possible that $\pi^2 = a/b$ where $a, b \in \mathbb{Z}$ and $b \neq 0$. Define

$$f(x) = x^n(1-x)^n/n!$$

and

$$G(x) = b^n\left(\pi^{2n}f(x) - \pi^{2n-2}f''(x) + \cdots + (-1)^n\pi^0 f^{(2n)}(x)\right)$$

where the superscripts on $f$ indicate derivatives. We claim that any derivative of $f$ takes integer values at 0 and 1. Recall Leibniz's rule for differentiating a product:

$$\frac{\mathrm{d}^m}{\mathrm{d}x^m}(uv) = \sum \binom{m}{r}\frac{\mathrm{d}^r u}{\mathrm{d}x^r}\frac{\mathrm{d}^{m-r}v}{\mathrm{d}x^{m-r}}$$

If both factors $x^n$ or $(1-x)^n$ are differentiated fewer than $n$ times, then the value of the corresponding term is 0 whenever $x = 0$ or 1. If one factor is differentiated $n$ or more times, then the denominator $n!$ is cancelled out. Hence $G(0)$ and $G(1)$ are integers. Now

$$\begin{aligned}\frac{\mathrm{d}}{\mathrm{d}x}\left[G'(x)\sin(\pi x) - \pi G(x)\cos(\pi x)\right] &= \left[G''(x) + \pi^2 G(x)\right]\sin(\pi x)\\ &= b^n\pi^{2n+2}f(x)\sin(\pi x)\end{aligned}$$

since $f(x)$ is a polynomial in $x$ of degree $2n$, so that $f^{(2n+2)}(x) = 0$. And this expression is equal to

$$\pi^2 a^n \sin(\pi x)f(x)$$

Therefore

$$\begin{aligned}\pi\int_0^1 a^n\sin(\pi x)f(x)\mathrm{d}x &= \left[\frac{G'(x)\sin(\pi x)}{\pi} - G(x)\cos(\pi x)\right]_0^1\\ &= G(0) + G(1)\end{aligned}$$

which is an integer. As before the integral is not zero. But

$$\begin{aligned}\left|\int_0^1 a^n\sin(\pi x)f(x)\mathrm{d}x\right| &\leq |a|^n\int_0^1 |\sin(\pi x)||f(x)|\mathrm{d}x\\ &\leq |a|^n\int_0^1 \frac{|x^n(1-x)^n|}{n!}\,\mathrm{d}x\\ &\leq \frac{1}{n!}\int_0^1 |(ax)^n(1-x)^n|\mathrm{d}x\end{aligned}$$

which tends to 0 as $n$ tends to $+\infty$. The usual contradiction completes the proof. □

## 24.2 Transcendence of e

We move from irrationality to the far more elusive transcendence. Hermite's original proof was simplified by Karl Weierstrass, Hilbert, Adolf Hurwitz, and Paul Gordan, and it is the simplified proof that we give here. The same holds for the proof of Lindemann's theorem in the next section.

**Theorem 24.4 (Hermite).** *The real number* e *is transcendental.*

*Proof.* Assume that e is not transcendental. Then

$$a_m \mathrm{e}^m + \cdots + a_1 \mathrm{e} + a_0 = 0$$

where without loss of generality we may suppose that $a_j \in \mathbb{Z}$ for all $j$ and $a_0 \neq 0$. Define

$$f(x) = \frac{x^{p-1}(x-1)^p(x-2)^p \ldots (x-m)^p}{(p-1)!}$$

where $p$ is an arbitrary prime number. Then $f$ is a polynomial in $x$ of degree $mp + p - 1$. Put

$$F(x) = f(x) + f'(x) + \cdots + f^{(mp+p-1)}(x)$$

and note that $f^{(mp+p)}(x) = 0$. Calculate:

$$\frac{\mathrm{d}}{\mathrm{d}x}(\mathrm{e}^{-x}F(x)) = \mathrm{e}^{-x}(F'(x) - F(x)) = -\mathrm{e}^{-x}f(x)$$

Hence for any $j$

$$\begin{aligned} a_j \int_0^j \mathrm{e}^{-x} f(x)\mathrm{d}x &= a_j \left[-\mathrm{e}^{-x}F(x)\right]_0^j \\ &= a_j F(0) - a_j \mathrm{e}^{-j} F(j)x \end{aligned}$$

Multiply by $\mathrm{e}^j$ and sum over $j$ to get

$$\begin{aligned} \sum_{j=0}^{m} \left( a_j \mathrm{e}^j \int_0^j \mathrm{e}^{-x} f(x)\mathrm{d}x \right) &= F(0) \sum_{j=0}^{m} a_j \mathrm{e}^j - \sum_{j=0}^{m} a_j F(j) \\ &= -\sum_{j=0}^{m} \sum_{i=0}^{mp+p-1} a_j f^{(i)}(j) \end{aligned} \tag{24.3}$$

from the equation supposedly satisfied by e.

We claim that each $f^{(i)}(j)$ is an integer, and that this integer is divisible by $p$ unless $j = 0$ and $i = p - 1$. To establish the claim we use Leibniz's rule again; the only non-zero terms arising when $j \neq 0$ come from the factor $(x-j)^p$ being differentiated exactly $p$ times. Since $p!/(p-1)! = p$, all such terms are integers divisible by $p$. In the exceptional case $j = 0$, the first non-zero term occurs when $i = p-1$, and then

$$f^{(p-1)}(0) = (-1)^p \dots (-m)^p$$

Subsequent non-zero terms are all multiples of $p$. The value of equation (24.3) is therefore

$$K_p + a_0(-1)^p \dots (-m)^p$$

for some $K \in \mathbb{Z}$. If $p > \max(m, |a_0|)$, then the integer $a_0(-1)^p \dots (-m)^p$ is not divisible by $p$. So for sufficiently large primes $p$ the value of equation (6.3) is an integer not divisible by $p$, hence not zero.

Now we estimate the integral. If $0 \leq x \leq m$ then

$$|f(x)| \leq m^{mp+p-1}/(p-1)!$$

so

$$\left|\sum_{j=0}^{m} a_j e^j \int_0^j e^{-x} f(x) dx\right| \leq \sum_{j=0}^{m} |a_j e^j| \int_0^j \frac{m^{mp+p-1}}{(p-1)!} dx$$
$$\leq \sum_{j=0}^{m} |a_j e^j| j \frac{m^{mp+p-1}}{(p-1)!}$$

which tends to 0 as $p$ tends to $+\infty$.

This is the usual contradiction. Therefore e is transcendental. □

## 24.3 Transcendence of $\pi$

The proof that $\pi$ is transcendental involves the same sort of trickery as the previous results but is far more elaborate. At several points in the proof we use properties of symmetric polynomials from Chapter 18.

**Theorem 24.5 (Lindemann).** *The real number $\pi$ is transcendental.*

*Proof.* Suppose for a contradiction that $\pi$ is a zero of some non-zero polynomial over $\mathbb{Q}$. Then so is $i\pi$. Let $\theta_1(x) \in \mathbb{Q}[x]$ be a polynomial with zeros $\alpha_1 = i\pi, \alpha_2, \dots, \alpha_n$. By a famous theorem of Euler,

$$e^{i\pi} + 1 = 0$$

so

$$(e^{\alpha_1} + 1)(e^{\alpha_2} + 1) \dots (e^{\alpha_n} + 1) = 0 \qquad (24.4)$$

We now construct a polynomial with integer coefficients whose zeros are the exponents $\alpha_{i_1} + \cdots + \alpha_{j_r}$ of e that appear in the expansion of the product in (24.4). For example, terms of the form

$$e^{\alpha_s}.e^{\alpha_t}.1.1.1\ldots 1$$

give rise to exponents $\alpha_s + \alpha_t$. Taken over all pairs $s, t$ we get exponents of the form $\alpha_1 + \alpha_2, \ldots, \alpha_{n-1} + \alpha_n$. The elementary symmetric polynomials of these are symmetric in $\alpha_1, \ldots, \alpha_n$, so by Theorem 18.11 they can be expressed as polynomials in the elementary symmetric polynomials of $\alpha_1, \ldots, \alpha_n$. These in turn are expressible in terms of the coefficients of the polynomial $\theta_1$ whose zeros are $\alpha_1, \ldots, \alpha_n$. Hence the pairs $\alpha_s + \alpha_t$ satisfy a polynomial equation $\theta_2(x) = 0$ where $\theta_2$ has rational coefficients. Similarly the sums of $k$ of the $\alpha$'s are zeros of a polynomial $\theta_k(x)$ over $\mathbb{Q}$. Then

$$\theta_1(x)\theta_2(x)\ldots\theta_n(x)$$

is a polynomial over $\mathbb{Q}$ whose zeros are the exponents of e in the expansion of equation (24.4). Dividing by a suitable power of $x$ and multiplying by a suitable integer we obtain a polynomial $\theta(x)$ over $\mathbb{Z}$, whose zeros are the non-zero exponents $\beta_1, \ldots, \beta_r$ of e in the expansion of equation (24.4).

Now (24.4) takes the form

$$e^{\beta_1} + \cdots + e^{\beta_r} + e^0 + \cdots + e^0 = 0$$

that is,

$$e^{\beta_1} + \cdots + e^{\beta_r} + k = 0 \qquad (24.5)$$

where $k \in \mathbb{Z}$. The term $1 \cdot 1 \cdots 1$ occurs in the expansion, so $k > 0$.

Suppose that

$$\theta(x) = cx^r + c_1x^{r-1} + \cdots + c_r$$

We know that $c_r \neq 0$ since 0 is not a zero of $\theta$. Define

$$f(x) = \frac{c^s x^{p-1}[\theta(x)]^p}{(p-1)!}$$

where $s = rp - 1$ and $p$ is any prime number. Define also

$$F(x) = f(x) + f'(x) + \cdots + f^{(s+p+r-1)}(x)$$

and note that $f^{(s+p+r)}(x) = 0$. As before

$$\frac{d}{dx}[e^{-x}F(x)] = -e^{-x}f(x)$$

Hence

$$e^{-x}F(x) - F(0) = -\int_0^x e^{-y}f(y)dy$$

Putting $y = \lambda x$ we get

$$F(x) - e^x F(0) = -x \int_0^1 \exp[(1-\lambda)x] f(\lambda x) d\lambda$$

Let $x$ range over $\beta_1, \ldots, \beta_r$ and sum: by (24.5)

$$\sum_{j=1}^{r} F(\beta_j) + kF(0) = -\sum_{j=1}^{r} \beta_j \int_0^1 \exp[(1-\lambda)\beta_j] f(\lambda\beta_j) d\lambda \tag{24.6}$$

We claim that for all sufficiently large $p$ the left-hand side of (24.6) is a non-zero integer. To prove the claim, observe that

$$\sum_{j=1}^{r} f^{(t)}(\beta_j) = 0$$

if $0 < t < p$. Each derivative $f^{(t)}(\beta_j)$ with $t \geq p$ has a factor $p$, since we must differentiate $[\theta(x)]^p$ at least $p$ times to obtain a non-zero term. For any such $t$,

$$\sum_{j=1}^{r} f^{(t)}(\beta_j)$$

is a symmetric polynomial in the $\beta_j$ of degree $\leq s$. Thus by Theorem 18.11 it is a polynomial of degree $\leq s$ in the coefficients $c_i/c$. The factor $c^s$ in the definition of $f(x)$ makes this into an integer. So for $t \geq p$

$$\sum_{j=1}^{r} f^{(t)}(\beta_j) = pk_t$$

for suitable $k_t \in \mathbb{Z}$.

Now we look at $F(0)$. Computations show that

$$f^{(t)}(0) = \begin{cases} 0 & (t \leq p-2) \\ c^s c_r^p & (t = p-1) \\ l_t p & (t \geq p) \end{cases}$$

for suitable $l_t \in \mathbb{Z}$. Consequently the left-hand side of (24.6) is

$$mp + kc^s c_r^p$$

for some $m \in \mathbb{Z}$. Now $k \neq 0$, $c \neq 0$, and $c_r \neq 0$. If we take

$$p > \max(k, |c|, |c_r|)$$

then the left-hand side of (24.6) is an integer not divisible by $p$, so is non-zero.

The last part of the proof is routine: we estimate the size of the right-hand side of (24.6). Now

$$|f(\lambda\beta_j)| \leq \frac{|c|^s|\beta_j|^{p-1}(m(j))^p}{(p-1)!}$$

where

$$m(j) = \sup_{0\leq\lambda\leq 1} |\theta(\lambda\beta_j)|$$

Therefore

$$\left| -\sum_{j=1}^{r} \beta_j \int_0^1 \exp[(1-\lambda)\beta_j] f(\lambda\beta_j)\mathrm{d}\lambda \right| \leq \sum_{j=1}^{r} \frac{|\beta_j|^p|c^s||m(j)|^p B}{(p-1)!}$$

where

$$B = \left| \max_j \int_0^1 \exp[(1-\lambda)\beta_j]\mathrm{d}\lambda \right|$$

Thus the expression tends to 0 as $p$ tends to $+\infty$. By the standard contradiction, $\pi$ is transcendental. □

---

## EXERCISES

The first four exercises outline Cantor's proof of the existence of transcendental numbers, using what are now standard results on infinite cardinals.

24.1 Prove that $\mathbb{R}$ is uncountable, that is, there is no bijection $\mathbb{Z} \to \mathbb{R}$.

24.2 Define the *height* of a polynomial

$$f(t) = a_0 + \cdots + a_n t^n \in \mathbb{Z}[t]$$

to be

$$h(f) = n + |a_0| + \cdots + |a_n|$$

Prove that there is only a finite number of polynomials over $\mathbb{Z}$ of given height $h$.

24.3 Show that any algebraic number satisfies a polynomial equation over $\mathbb{Z}$. Using Exercise 24.2 show that the algebraic numbers form a countable set.

24.4 Combine Exercises 24.1 and 24.3 to show that transcendental numbers exist.

The next three exercises give Liouville's proof of the existence of transcendental numbers.

24.5* Suppose that $x$ is irrational and that

$$f(x) = a_n x^n + \cdots + a_0 = 0$$

where $a_0, \ldots, a_n \in \mathbb{Z}$. Show that if $p, q \in \mathbb{Z}$ and $q \neq 0$, and $f(p/q) \neq 0$, then

$$|f(p/q)| \geq 1/q^n$$

24.6* Now suppose that $x - 1 < p/q < x + 1$ and $p/q$ is nearer to $x$ than any other zero of $f$. There exists $M$ such that $|f'(y)| < M$ if $x - 1 < y < x + 1$. Use the mean value theorem to show that

$$|p/q - x| \geq M^{-1} q^{-n}$$

Hence show that for any $r > n$ and $K > 0$ there exist only finitely many $p$ and $q$ such that

$$|p/q - x| < K q^{-r}$$

24.7 Use this result to prove that $\sum_{n=1}^{\infty} 10^{-n!}$ is transcendental.

24.8 Prove that $z \in \mathbb{C}$ is transcendental if and only if its real part is transcendental or its imaginary part is transcendental.

24.9 Mark the following true or false.

(a) $\pi$ is irrational.

(b) All irrational numbers are transcendental.

(c) Any nonzero rational multiple of $\pi$ is transcendental.

(d) $\pi + \mathrm{i}\sqrt{5}$ is transcendental.

(e) e is irrational.

(f) If $\alpha$ and $\beta$ are real and transcendental then so is $\alpha + \beta$.

(g) If $\alpha$ and $\beta$ are real and transcendental then so is $\alpha + i\beta$.

(h) Transcendental numbers form a subring of $\mathbb{C}$.

(i) The field $\mathbb{Q}(\pi)$ is isomorphic to $\mathbb{Q}(t)$ for any indeterminate $t$.

(j) $\mathbb{Q}(\pi)$ and $\mathbb{Q}(\mathrm{e})$ are non-isomorphic fields.

(k) $\mathbb{Q}(\pi)$ is isomorphic to $\mathbb{Q}(\pi^2)$.

# Chapter 25

## What Did Galois Do or Know?

This is not a scholarly book on the history of mathematics, but it does contain a substantial amount of historical material, intended to locate the topic in its context and to motivate Galois theory as currently taught at undergraduate level. (At the research frontiers, the entire subject is even more general and more abstract.)

There is a danger in this approach: it can mix up history as it actually happened with how we reformulate the ideas now. This can easily be misinterpreted, distorting our view of the past and propagating historical myths. Peter Neumann makes this point very effectively in his admirable English translation of Galois's writings, Neumann (2011). The book covers both Galois's published papers and those of his unpublished manuscripts that have survived—very few, even when brief scraps are included.

To set the record straight, we now take a look at what this material tells us about what Galois actually did, what he knew, and what he might have been able to prove. Placing the material at the end of this book allows us to refer back to all of the historical and mathematical material.

The folklore story is: Galois proved that $\mathbb{A}_5$ is simple, indeed, the smallest simple group other than cyclic groups of prime order. From this he deduced that the quintic is not soluble by radicals. However, as Neumann states, Galois made the first statement without proof (and it is questionable whether he possessed one), while the link to the second does not appear explicitly in any of his extant manuscripts. The central issue, and our main focus here, is the relation between solving the quintic by radicals and the alternating group $\mathbb{A}_5$. It would be easy to imagine, and has often been asserted, that Galois viewed these topics in the same way as they have been presented in earlier chapters, and that in particular that the key issue, for him, was to prove that $\mathbb{A}_5$ is simple.

Not so.

However, history is seldom straightforward, especially when sources are fragmentary and limited. Closely related statements do appear, enough to justify Galois's stellar reputation among mathematicians and to credit him with the most penetrating insights of his period into the solution of equations by radicals and its relation to groups of permutations. As Neumann writes: 'The [First] memoir on the conditions for solubility of equations by radicals is undoubtedly Galois's most important work. It is here that he presented his original approach to the theory of equations that has now become known as Galois Theory.'

DOI: 10.1201/9781003213949-25 

## 25.1 List of the Relevant Material

Galois's published papers are five in number, and only one, 'Analysis of a memoir on the algebraic solution of equations', is relevant here. After Galois died, his manuscripts went to a literary executor, his friend Auguste Chevalier. Chevalier passed them on to Liouville, who eventually brought Galois's work to the attention of the mathematical community, probably encouraged by the brother, Alfred Galois. Liouville's daughter Mme de Blignières gave them to the French Academy of Sciences in 1905 or 1906, where they were organised into 25 'dossiers' and bound into a single volume. Parts were published or analysed by Chevalier, Liouville, Jules Tannery, and Émile Picard. Bourgne and Azra (1962) published a complete edition. The first and currently the only complete English translation is Neumann (2011). This also contains a printed version of the French originals, in parallel with the translation for ease of comparison. Scans of the manuscripts are available on the internet at

bibliotheque-institutdefrance.fr

The documents referred to below (the dossier numbers are those assigned by the Academy) are:

Analysis of a memoir on the algebraic solution of equations, *Bulletin des Sciences Mathématiques, Physiques et Chimiques* **13** (April 1830) 271–272.

Testamentary Letter, 29 May 1832, to Chevalier.

First Memoir, sent to the Academy.

Second Memoir, sent to the Academy.

Dossier 8: Torn fragment related to the First Memoir.

Dossier 10: Publication project and note on Abel.

Dossier 15: Fragments on permutations and equations.

Several other documents refer to groups and algebraic equations, and there are some on other topics altogether.

## 25.2 The First Memoir

The document called the First Memoir is the one that Galois sent to the Academy on 17 January 1831; it is actually his third submission, the other two having been lost. In the opening paragraph to the First Memoir, which functions as an abstract of the contents, Galois states that he will present

> ... a general *condition satisfied by every equation that is soluble by radicals*, and which conversely ensures their solubility. An application is made just to equations of which the degree is a prime number. Here is the theorem given by our analysis:
> In order that an equation of prime degree ... be soluble by radicals, it is *necessary* and it is *sufficient* that all the roots be rational functions of any two of them.

He adds that his theory has other applications, but 'we reserve them for another occasion.'

In this abstract, there is no mention of the quintic as such, although its degree 5 is prime, so his main theorem obviously applies to it. It is not mentioned in the rest of the paper either. There is also no mention of the concept of a group. It is hard not to have some sympathy for Poisson and Lacroix, the referees: it looks like they did a professional job, and spotted a key weakness in the theorem upon which Galois places so much emphasis. (Admittedly, this is not difficult.) Namely: although Galois's condition 'all the roots be rational functions of any two of them' is indeed necessary and sufficient for solubility by radicals, it is hard to think of any practical way to verify it for any specific equation.

The Historical Introduction mentioned the referees' statement that 'one could not derive from [Galois's condition] any good way of deciding whether a given equation of prime degree is soluble or not by radicals,' and the remark by Tignol (1988) that Galois's memoir 'did not yield any workable criterion to determine whether an equation is solvable by radicals.' I also wrote: 'What the referees wanted was some kind of condition on the *coefficients* that determined solubility; what Galois gave them was a condition on the *roots*.' But I think that a stronger criticism is in order: apparently, there is no algorithmic procedure to check whether the condition on the roots is valid. Or to prove that it is not. How, for example, would we use it to prove the quintic insoluble?

It turns out that this judgement is not entirely correct, but further work is needed to see why. It is implicit in a table that Galois includes titled 'Example of Theorem VII', and I'll come back to that shortly. But he does not make the connection explicit.

---

## 25.3 What Galois Proved

Before discussing possible reasons for the (to our eyes) curious omission of the application to quintics, we review the results that Galois does include in the First Memoir. These alone would establish his reputation.

The work is short, succinct, and clearly written. A modern reader will have no difficulty in following the reasoning, once they get used to the terminology. Galois develops several key ideas needed to prove his necessary and sufficient condition for solubility by radicals, which we *now* recognise as the core concepts of Galois Theory. It is clear that he recognised the importance of these ideas, but, once again, he does not say so in the paper.

After a few preliminaries, which would have been familiar to anyone working in the area, Galois presents his first key theorem:

**Proposition I** *Let an equation be given of which the m roots are* $a, b, c, \ldots$. *There will always be a group of permutations of the letters* $a, b, c, \ldots$ *which will enjoy the following property:*

> *That every function of the roots invariant* [a footnote explains this term] *under the substitutions of this group will be rationally known;*
>
> *Conversely, that every function of the roots that is rationally determinable will be invariant under the substitutions.*

This is his definition of what we now call the Galois group. It also makes the central point about the Galois correspondence, expressed in terms of the roots rather than the modern interpretation in terms of the subfield they generate.

Next, he studies how the group can be decomposed by adjoining the roots of auxiliary equations; that is, extending the field. He deduces that when a $p$th root is extracted, for (without loss of generality) prime $p$, the group must have what we would now express as a normal subgroup of index $p$. This leads to the next big result, initially posed as a question:

**Proposition V** *Under what circumstances is an equation soluble by radicals?*

Galois writes '... to solve an equation it is necessary to reduce its group successively to the point where it does not contain more than a single permutation.' He analyses what happens when the reduction is performed by adjoining 'radical quantities'. He concludes, slightly obscurely, that the group of the equation must have a normal subgroup of prime index, which in turn has a normal subgroup of prime index, and so on, until we reach the group with a single element. In short: the equations is soluble by radicals if and only if its group is soluble. But he fails to state this as an explicit proposition.

Galois goes on to illustrate the result for the general quartic equation, obtaining essentially what we found in Section 18.5 of Chapter 18. This of course was a known result, and Lagrange had already related it to permutation groups in his *Traité de la Résolution des Équations Numériques de Tous les Degrés*. But instead of continuing to the quintic, and proving that the group is not soluble, Galois does something that is in some ways more interesting, but answers another (closely related) question instead:

**Proposition VII** *What is the group of an equation of prime degree n that is soluble by radicals?*

His answer is that if the roots are suitably numbered, the group of the equation can contain only substitutions of the form

$$x_k \mapsto x_{ak+b} \tag{25.1}$$

where the roots are the $x_k$, the symbols $a, b$ denote constants, and $ak + b$ is to be computed modulo $n$.

To modern eyes, what he *should* have remarked at this point is that when $n = 5$ the group of all such substitutions has $4.5 = 20$ elements (we need $0 \neq a \in \mathbb{Z}_5$ and $b \in \mathbb{Z}_5$), so it cannot equal $\mathbb{S}_5$, the group of the general quintic. Moreover, Galois definitely *knew* that for any $m$ the group of the general equation of degree $m$ is the symmetric group $\mathbb{S}_m$. He states as much in the discussion of his Proposition I:

> In the case of algebraic equations, this group is nothing other than the collection of the $1.2.3 \dots m$ possible permutations on the $m$ letters, because in this case, only the symmetric functions are rationally determinable.

By 'algebraic equation' he meant what we now call the 'general polynomial equation'. Galois distinguished 'numerical' and 'literal' equations: those in which the coefficients are specific numbers, and those in which they are arbitrary symbols. He is clearly thinking of literal equations here. But to a casual reader this statement is somewhat confusing.

Anyway, Galois does no such thing. Instead, he in effect observes that once you have two numbers of the form $ak+b, a'k+b'$, you can generate all numbers of this form. Whence the criterion that given any two roots, the others are all rationally expressible.

## 25.4 What is Galois up to?

Taking inspiration and historical information from Neumann (2011), I now think there is a sensible explanation of what at first sight seems to be a strange series of omissions and obscurities, in which Galois wanders all round a key idea without ever putting his finger on it. Namely: Galois wasn't interested in discussing the quintic. He was after something quite different.

We know that he had taken on board the work of Ruffini and Abel, because Dossier 10 refers to Abel's proof that the quintic is insoluble, and Dossier 8 states:

> It is today a commonly known truth that general equations of degree greater than the $4^{\text{th}}$ cannot be solved by radicals.
>
> This truth has become commonly known to some extent by hearsay and even though most geometers do not know the proofs of it given by Ruffini, Abel, *etc.*, proofs founded upon the fact that such a solution is already impossible for the fifth degree.

This being so, why should Galois place any emphasis on the quintic? I think he had his sights set on something more ambitious: to say something *new* about solutions by radicals.

The first piece of evidence is the continuation of the above quotation: 'In the first instance it would seem that the [theory] of solution of equations by radicals would end there.' Unfortunately the text on that side of the paper ends at this point, and the other side merely lists titles of four memoirs.

Another is Dossier 9, which includes:

> The proposed goal is to determine the characteristics for the solubility of equations by radicals... that is the question to which we offer a complete solution.

He then acknowledges that in practice 'the calculations are impracticable,' but attempts to justify the importance of the result nonetheless:

> ... most of the time in algebraic analysis one is led to equations all of whose properties one knows beforehand: properties by means of which it will always be easy to answer the question by the rules we shall expound ... I will cite, for example, the equations which give the division of elliptic functions and which the celebrated Abel has solved ...

Galois refers to these 'modular equations' from the theory of elliptic functions elsewhere, and they presumably played a major role in his thinking.

Dossier 10 states:

> ... Abel did not know the particular circumstances of solution by radicals ... he has left nothing on the general discussion of the problem which has occupied us. Once and for all, what is remarkable in our theory [is to be able to answer yes or no in all cases, *crossed out*].

Over and over again Galois places emphasis not on proving equations such as the general quintic insoluble, but on finding equations that *are* soluble. The title of the First Memoir says it all: 'Memoir on the conditions for solubility of equations by radicals.' So does that of the Second Memoir: 'On primitive equations which are soluble by radicals.' Galois is not interested in impossibility proofs. To him, they are old hat; they do not lead anywhere new. This,

I suspect, is why he does not use the quintic as an example in the First Memoir; it is most definitely why his main general result is Proposition VII. In modern terms, he is telling us that an equation of prime degree $n$ is soluble by radicals if and only if its Galois group is conjugate to a subgroup of the affine general linear group $\mathbb{AGL}(1,n)$, which consists of the transformations (25.1). These are the equations that Galois considers interesting; this is the theorem of which he is justly proud, since it constitutes a major advance and characterises soluble equations.

It is also worth remarking that the form in which Galois states Proposition VII does not involve the notion of a group. It would be immediately comprehensible to any algebraist of the period, without having to explain to them the new—and rather unorthodox—concept of a group. This is reminiscent of the way that Isaac Newton used classical geometry rather than calculus to prove many statements in his *Principia Mathematica*, even though he probably used calculus to derive them in the first place. Ironically, by trying—for once—to make his ideas more accessible, Galois obscured their importance.

## 25.5 Alternating Groups, Especially $\mathbb{A}_5$

Neumann (2011) discusses several myths about Galois. Prominent among them is the claim that he proved the alternating group $\mathbb{A}_n$ is simple when $n \geq 5$. However, these groups are not mentioned in any of the works of Galois published by Liouville in 1846, which was the main source for professional mathematicians. There is no mention even of $\mathbb{A}_5$, and even the symmetric groups are mentioned only to illustrate Proposition I of the First Memoir (see the quotation in Section 25.3) and as an example for Proposition V when the degree is 4.

One reason why Galois did not mention the simplicity of $\mathbb{A}_n$ or even of $\mathbb{A}_5$ is that he didn't need it. His necessary and sufficient condition for solubility—having a group conjugate to a subgroup of $\mathbb{AGL}(1,n)$—was all he needed. We can prove that $\mathbb{A}_5$ cannot occur rather easily: its order is 60 while that of $\mathbb{AGL}(1,5)$ is only 20. Simplicity is not the issue. However, Galois doesn't even say that: insolubility is also not the issue, for him.

But ...

As Neumann recognises, Galois does give brief mention to alternating groups in a few manuscripts. One is Dossier 15, which consists of a series of short headings. It looks suspiciously like the outline of a lecture course. Could it be the one on advanced algebra that he offered on 13 January 1831? It might be a plan for a memoir, or even a book, for all we know. Crossed out, we find the words:

Example. Alternate groups (Two similar groups). Properties of the alternate groups.

By 'two similar groups' Galois is referring to two cosets with the same structure: this was his way to say 'normal subgroup of index 2', no doubt in $\mathbb{S}_n$. The same text appears slightly later, also crossed out. Later still we find 'New proof of the theorem relative to the alternate groups', not crossed out. This is followed shortly by 'One may suppose that the group contains only even substitutions', which can be interpreted as 'without loss of generality we may assume the group is contained in the alternating group'.

There is a simple way to set this up, which was known to every algebraist, and Galois would have learned it at his mother's knee. It uses the quantity $\delta$ defined in (1.13). This changes sign if any two roots are interchanged; that is, it is invariant under $\mathbb{A}_n$ but not $\mathbb{S}_n$. However, its square $\Delta = \delta^2$ is a symmetric function of the roots and therefore can be expressed as a function of the coefficients. It is the discriminant of the equation, so named because its traditional role is to provide a computable algebraic test for the existence of a multiple root. Indeed, $\Delta = 0$ if and only if the equation has a multiple root.

Since $\Delta$ is a rational function of the coefficients, we can adjoin $\delta$ by taking a square root. As far as solving equations by radicals goes, this is harmless, and it reduces the group to its intersection with $\mathbb{A}_n$. Probably Galois had something like this in mind.

The same document includes a reference to Cauchy's work on permutations, including

> Theorem. If a function on $m$ indeterminates is given by an equation of degree $m$ all of whose coefficients [are symmetric functions, permanent or alternating, of these indeterminates], this function will be symmetric, permanent or alternating, with respect to all letters or at least with respect to $m-1$ among them.
>
> Theorem. No algebraic equation of degree higher than 4 may be solved or reduced.

So there is no doubt that Galois was *aware* of the link between $\mathbb{S}_5, \mathbb{A}_5$, and the quintic.

## 25.6 Simple Groups Known to Galois

What about simple groups? Neumann points out that Galois definitely knew about simple groups (his term is 'indecomposable'). But the examples he cites are the projective special linear groups $\mathbb{PSL}(2,p)$ for prime $p$. His Second Memoir was clearly heading in that direction, and this fact is stated explicitly in the letter to Chevalier: '[this group] is not further decomposable unless $p = 2$ or $p = 3$.'

This bring us to another statement in the letter to Chevalier, which Neumann reasonably considers a 'mysterious assertion'. Namely:

> The smallest number of permutations which can have an indecomposable [that is, simple] group, when this number is not prime, is 5.4.3.

That is, the smallest order for a simple group is 60. Neumann argues persuasively that Galois was thinking of $\mathbb{PSL}(2,5)$, not $\mathbb{A}_5$. Agreed, these groups are isomorphic, but Galois writes extensively about what we now call $\mathbb{PSL}(2,p)$, and says virtually nothing about $\mathbb{A}_n$.

Neumann also provides a fascinating discussion of whether Galois actually possessed a proof that the smallest order for a simple group is 60.

> He was so insightful that, perhaps, yes, he could have known it. Nevertheless, I very much doubt it. How could he have excluded orders such as 30, 32, 36, 40, 48, 56? With Sylow's theorems and some calculation, such orders can be excluded... but... it seems unlikely that Galois had Sylow's theorems available to him. Besides, there is no hint in any of the extant manuscripts and scraps of the kind of case-by-case analysis that is needed...

It is of course *conceivable* that Galois knew the results we now call Sylow's Theorem. He was very clever, and his known insights into group theory are impressive. However, even granting that, the viewpoint needed to prove Sylow's Theorem seems too sophisticated for the period. The biggest problem is that it is difficult to imagine him failing to tell anyone about such discoveries, and some hint ought to have survived among his papers. In their absence, Neumann's last point is especially telling. On the other hand, and grasping at straws, Galois's affairs were somewhat chaotic. Like most mathematicians, he probably threw a lot of scraps away, especially 'rough work'. In the Historical Introduction we saw that when at school he did a lot of work in his head, instead of on paper—and was criticised for it. So the absence of evidence is not evidence of absence.

---

## 25.7 Speculations about Proofs

It is worth examining just what a mathematician of the period would have needed to prove Galois's statement about the smallest order for a simple group. The answer is 'less than we might expect.' What follows illustrates what might have been possible given a little ingenuity, and it shows in particular that Sylow's theorems are *not* required. The trickiest case for orders less than 60 is 56, but a proof is not particularly difficult if you assemble the right results. Instead, we use only a few basic theorems in group theory, all of which have easy proofs. We make no claim that Galois was aware of any of this material, since there is no documentary evidence, but the proofs are well within his capabilities. Today the usual approach is to apply Sylow's theorems, which are probably beyond what Galois could have known. This may be what has led some historians to question whether Galois had a proof of his assertion. However, we show that using techniques that Galois could easily have discovered, it is possible to prove, in just 4–5 pages with all details, not only that the smallest non-cyclic simple group has order 60, but that, up to isomorphism, there is a unique simple group of order 60. This must be isomorphic to the alternating group $\mathbb{A}_5$, and also to $\mathbb{PSL}(2,5)$, since both are simple with order 60.

Galois definitely knew about what we now call subgroups, cosets, conjugacy, and normal subgroups: equivalent ideas are clearly present in his papers. We know that he read Lagrange, so he must have known Lagrange's theorem: the order of a subgroup (or element) divides the order of the group.

He could have defined the normaliser $N_G(H)$ of a subgroup of $G$, which is the set of all $g \in G$ such that $g^{-1}Hg = H$. This is obviously a subgroup, and $H \lhd N_G(H)$. Moreover, it is evident that the number of distinct conjugates of $H$ is equal to the index $|G : N_G(H)|$. The index of a subgroup $K \subseteq G$, usually denoted $|G : H|$, is equal to $|G|/|H|$ for finite groups and is the number of distinct cosets (left or right) of $H$ in $G$. Galois knew about cosets (though he called them 'groups'.)

Galois would also have been aware of what we now call the centraliser $C_G(g)$ of an element $g \in G$: the set of all $h \in G$ such that $h^{-1}gh = g$. This too is a subgroup, and the number of distinct conjugates of $g$ is equal to the index $|G : C_G(H)|$. This line of thinking leads inevitably to the *class equation* discussed in Chapter 14 (14.2). We rewrite it in the form:

$$|G| = 1 + \sum_{g_i} |G : C_G(g_i)| \tag{25.2}$$

where $\{g_i\}$ is a set of representatives of the non-identity conjugacy classes of $G$. The extra 1 takes care of the identity. As we will see, the class equation is a surprisingly powerful tool when investigating simple groups of small order.

Indeed, using the class equation, Galois would easily have been able to prove Theorem 14.15, published in 1845 by Cauchy. This is a limited converse to Lagrange's theorem: if a prime number $p$ divides the order of a finite group, the group has an element of order $p$. The class equation is the key to the proof, as we saw in Chapter 14.

It turns out that for putative simple groups of small order, Cauchy's Theorem works fairly well as a substitute for Sylow's theorems. Some systematic counting of elements then goes a long way. However, it is a bit of a scramble. The main results we need are:

**Lemma 25.1.** *Let $G$ be a non-cyclic finite simple group. Then:*

(1) *The normaliser of any proper subgroup of $G$ is a proper subgroup of $G$.*

(2) *The centraliser of any element of $G$ is a proper subgroup of $G$.*

(3) *No prime $p$ can divide the indices of all proper subgroups of $G$.*

(4) *There cannot exist a unique proper subgroup of $G$ of given order $k > 1$.*

*Proof.* (1) If not, the subgroup is normal.

(2) If not, the element generates a cyclic normal subgroup.

(3) If such a $p$ exists, the class equation takes the form

$$1 + c_1 + \cdots + c_k = |G|$$

where the $c_j$ are the indices of centralisers of non-identity elements, which by (2) are proper subgroups. Therefore $p|c_j$ for all $j$. Also $p$ divides $|G|$ since $p$ divides $c_1$, which divides $|G|$. So the class equation taken (mod $p$) implies that $1 \equiv 0 \pmod{p}$, a contradiction.

(4) Suppose that $H$ is the unique subgroup of order $k$. The order of any conjugate $g^{-1}Hg$ is also $k$, so $g^{-1}Hg = H$ for all $g \in G$. Therefore $H \lhd G$, a contradiction. □

We need one further idea. Galois's definition of 'normal' immediately implies that a subgroup of index 2 is normal. More generally, a little thought about the conjugates of a subgroup leads to a useful generalisation:

**Lemma 25.2.** *Let $G$ be a finite simple group with a nontrivial proper subgroup $H$ of index $m$. Then $G$ is isomorphic to a subgroup of $\mathbb{S}_m$. In particular, $|G| \leq m!$.*

*Proof.* Let $N = N_G(H)$. The number of distinct conjugates of $H$ is the index $k$ of $N$ in $G$, and $k \leq m$. Let $\mathcal{H}$ be the set of conjugates of $H$, so that $|\mathcal{H}| = k$. We can identify the group of permutations of $\mathcal{H}$ with $\mathbb{S}_k$. Any $g \in G$ induces a permutation $\phi(g)$ of $|\mathcal{H}|$ where

$$\phi(g)(h^{-1}Hh) = (hg)^{-1}H(hg) = g^{-1}h^{-1}Hhg$$

and $\phi : G \to \mathbb{S}_k$ is clearly a homomorphism. Let $K = \ker(\phi)$. Then $K \lhd G$. Since $G$ is simple, either $K = \mathbf{1}$ or $K = G$. If $K = G$ then $g^{-1}Hg = H$ for all $g \in G$, so $H \lhd G$, which cannot happen since $H$ is nontrivial and proper. Therefore $K = \mathbf{1}$ and $G$ is isomorphic to a subgroup of $\mathbb{S}_k$. Since $\mathbb{S}_k$ is isomorphic to a subgroup of $\mathbb{S}_m$, the group $G$ is isomorphic to a subgroup of $\mathbb{S}_m$. □

Armed with these weapons, Galois would easily have been able to prove:

**Theorem 25.3.** *Let $p, q$ be distinct primes and $k \geq 2$. A finite non-cyclic simple group cannot have order $p^k$, $pq$, $2p^k$, $3p^k$, $4p^k$, or $4p$ for $p \geq 7$.*

*Proof.* (1) Order $p^k$ is ruled out by Lemma 25.1, since $p$ divides the index of any proper subgroup. This is how we proved Theorem 23.5, but there we obtained a further consequence: the group has non-trivial centre.
(2) Suppose $G$ is simple of order $pq$. By Cauchy's Theorem it has subgroups $H$ of order $p$ and $K$ of order $q$. All nontrivial proper subgroups have order $p$ or $q$. Each of $H, K$ must equal its normaliser, otherwise it would be a normal subgroup. Therefore $H$ has $q$ conjugates, which intersect pairwise in the identity, and $K$ has $p$ conjugates, which intersect pairwise in the identity. Therefore $G$ has 1 element of order 1, at least $(p-1)q$ elements of order $p$, and at least $p(q-1)$ elements of order $q$. These total $2pq-p-q+1 = pq+(p-1)(q-1)$ elements, a contradiction since $p, q > 1$.
(3) Suppose $G$ is simple of order $2p^k$. There is no subgroup of index 2, so every proper subgroup has index divisible by $p$, contrary to Lemma 25.1(3).
(4) Suppose $G$ is simple of order $3p^k$. Since $3p^k \geq 8$, Lemma 25.2 implies that there is no subgroup of index $\leq 3$. Therefore every proper subgroup has index divisible by $p$, contrary to Lemma 25.1(3).
(5) Suppose $G$ is simple of order $4p^k$. If $p = 2$ apply part (1). Otherwise $4p^k \geq 36$. By Lemma 25.2 there is no subgroup of index $\leq 4$, so every proper subgroup has index divisible by $p$, contrary to Lemma 25.1(3).
(6) Suppose $G$ is simple of order $4p$. Since $p \geq 7$ we have $|G| > 24$, so by Lemma 25.2 there is no proper subgroup of index $\leq 4$. In particular there is no subgroup of order $p$, contrary to Cauchy's Theorem. □

We now present a proof, using nothing that could not easily have been known to Galois, of his mysterious statement:

**Theorem 25.4.** *There is no non-cyclic simple group of order less than* 60.

*Proof.* Let $G$ be a non-cyclic simple group of order less than 60. This rules out groups of prime order, and Theorem 25.3 rules out many other orders. Only six orders survive:

$$20 \quad 30 \quad 40 \quad 42 \quad 45 \quad 56$$

and we dispose of these in turn. Throughout, we apply Lemma 25.1(1, 2) without further comment.

**Order 20** By Lemma 25.2 $G$ has no subgroups of index $\leq 3$. Therefore the possible orders of nontrivial proper subgroups are 2, 4, 5 only. By Cauchy's Theorem there exist elements of orders 2 and 5.

The class equation does not lead directly to a contradiction, so we argue as follows. Let $N$ be the normaliser of any order-5 subgroup $H$. This is a proper subgroup. Since all proper subgroups have order 1, 2, 4, or 5, we have $|N| = 5$. Therefore $H$ has $20/5 = 4$ distinct conjugates. Since 5 is prime, these conjugates intersect only in the identity. Each non-identity element of $\mathbb{Z}_5$ has order 5, so there are 4 elements of order 5 in each order-5 subgroup. Therefore together these conjugates contain $4.4 = 16$ elements of order 5.

There is also at least one element of order 2. Its normaliser has order 2 or 4, so cannot contain an element of order 5. It therefore has 5 distinct conjugates by any order-5 element. Therefore $G$ has at least $1+16+5 = 22$ elements, contradiction.

**Order 30** Since $30 > 4!$, Lemma 25.2 implies that $G$ has no subgroups of index $\leq 4$. Therefore the possible orders of nontrivial proper subgroups are 2, 3, 5, 6 only. By Cauchy's Theorem there exist elements of orders 2, 3, and 5.

The class equation can be used here, but there is a simpler argument. The normaliser of any $\mathbb{Z}_5$ subgroup has order 5, hence index 6. Thus there are at least $6.4 = 24$ elements of order 5. The normaliser of any $\mathbb{Z}_3$ subgroup has order 3 or 6, hence index 10 or 5. Thus there are at least $5.2 = 10$ elements of order 3. But $24 + 10 = 34 > 30$, a contradiction.

**Order 40** Lemma 25.2 implies that $G$ has no subgroups of index $\leq 4$. Therefore the possible orders of nontrivial proper subgroups are 2, 4, 5, 8 only. By Cauchy's Theorem there exist elements of orders 2 and 5.

The normaliser of any $\mathbb{Z}_5$ subgroup has order 5, hence index 8. Thus there are at least $8.4 = 32$ elements of order 5. Each has centraliser of order 5, so its conjugacy class has 8 elements. Any further order-5 element gives rise to 32 more elements for the same reason, not conjugate to the above, which is impossible. So we have found all order-5 elements and their conjugacy classes.

The centraliser of any element of order $2^k$ has order 2, 4, or 8, hence index 20, 10, or 5.

The class equation therefore becomes

$$40 = 1 + 32 + 5a + 10b + 20c$$

so

$$7 = 5a + 10b + 20c$$

which is impossible since $5 \nmid 7$.

**Order 42** Lemma 25.2 implies that $G$ has no subgroups of index $\leq 4$. Therefore the possible orders of nontrivial proper subgroups are 2, 3, 6, 7 only. Their indices are 21, 14, 7, and 6. The class equation takes the form

$$42 = 1 + 6a + 7b + 14c + 21d$$

where $a$ arises from elements of order 7. Consider this (mod 7) to deduce that $a \equiv 1 \pmod 7$. If $a = 1$ then there is a unique $\mathbb{Z}_7$ subgroup. But this contradicts Lemma 25.1(4). Otherwise $a \geq 8$, which yields at least $6.8 = 48$ elements of order 7, contradiction.

**Order 45** Lemma 25.2 implies that $G$ has no subgroups of index $\leq 4$. Therefore the possible orders of nontrivial proper subgroups are 3, 5, 9 only. Their indices are 15, 9, and 5.

The centraliser of any order-5 element has order 5, index 9. So there are at least $9.4 = 36$ elements of order 5.

The centraliser of any order-3 element has order 3 or 9, index 15 or 5. So there are at least $2.5 = 10$ elements of order 3, giving at least $36 + 10 = 46$ elements, contradiction.

**Order 56** Lemma 25.2 implies that $G$ has no subgroups of index $\leq 4$. Therefore the possible orders of nontrivial proper subgroups are 2, 4, 7, 8 only. Their indices are 28, 14, 8, and 7.

The normaliser of any $\mathbb{Z}_7$ subgroup has order 7, index 8, yielding at least $6.8 = 48$ elements of order 7.

The normaliser of any $\mathbb{Z}_2$ subgroup has order 2, 4, or 8, index 28, 14, or 7, yielding at least 7 elements of order 2.

Together with the identity, these give all 56 elements. Therefore there are exactly 48 order-7 elements and 7 order-2 elements.

The centraliser of any order-7 element must have order 7, index 8. So there are 6 conjugacy classes of order-7 elements.

The centraliser of any order-2 element must have order 2, 4, or 8, index 28, 14, or 7.

The class equation takes the form

$$56 = 1 + 48 + 7a + 14b + 28c$$

so $a = 1, b = c = 0$ and there are precisely 7 order-2 elements, all conjugate to each other. Their centralisers have order 8, so do not contain any order-7 element; therefore each has the same centraliser. This is the unique order-8 subgroup, contradicting Lemma 25.1(4). □

Galois would have had little difficulty with these orders. If he needed scrap paper calculations, they would have been short, and easily lost or thrown away. However, history relies on written evidence, and there is no documentary evidence that Galois ever proved Theorem 25.4. However, the above proof makes it plausible that Galois *could have* known how to prove that the smallest non-cyclic simple group has order 60.

## 25.8 $\mathbb{A}_5$ is Unique

Indeed, he could have gone further, proving that all simple groups of order 60 are isomorphic: in particular, isomorphic to to $\mathbb{A}_5$.

**Theorem 25.5.** *Let $G$ be a simple group of order* 60. *Then $G \cong \mathbb{A}_5$.*

*Proof.* First, we claim that $G$ has a subgroup of order 12. For a contradiction, assume it has no subgroup of order 12.

If $G$ has a subgroup of index $\leq 4$ then Lemma 25.2 implies that $|G| \leq 24$, a contradiction. Therefore

$$\text{proper subgroups have orders } 1, 2, 3, 4, 5, 6, \text{ or } 10 \qquad (25.3)$$

(Order 12 is ruled out by the assumption we hope to contradict.)

This applies in particular to all normalisers and centralisers of nontrivial proper subgroups. Indices of centralisers are therefore

$$1, 6, 10, 12, 15, 20, 30$$

of which only 1 and 15 are odd. By the class equation at least one conjugacy class contains 15 elements. So there exists an element with centraliser $H$ of order 4. $H$ is abelian.

By Cauchy's Theorem there exists an element of order 3, generating $K \cong \mathbb{Z}_3$. Its normaliser has order 3 or 6, so it has at least 10 distinct conjugates. These are disjoint, so contain at least 20 order-3 elements.

The normaliser of $H$ has order divisible by $|H| = 4$, but the only such subgroup has order 4 by (25.3). Therefore the normaliser of $H$ is $H$, of index 15, and $H$ has 15 distinct conjugates. If all conjugates of $H$ are disjoint, that gives $15.3 = 45$ elements of order 2 or 4, giving a total of at least $1+45+20 = 66$, contradiction. Therefore there exists $h \in H \cap g^{-1}Hg$ where $H \neq g^{-1}Hg$. Now the centraliser of $h$ contains both $H$ and $g^{-1}Hg$, so it has order $> 4$, a contradiction.

This proves that there is at least one subgroup $S$ of order 12, that is, with index $60/12 = 5$. Lemma 25.2 now implies that $G$ is isomorphic to a subgroup of $\mathbb{S}_5$. However, the only subgroup of $\mathbb{S}_5$ of order 60 is $\mathbb{A}_5$. □

---

## EXERCISES

25.1 In the proof of Lemma 25.2, prove that the kernel $K$ of $\phi$ is the intersection of the normalisers of the conjugates of $H$, and that this subgroup is equal to the intersection of the conjugates of the normaliser of $H$.

25.2 Using the methods of this chapter, prove that a simple group cannot have order $5p^k$ where $k \geq 2$ and $p \geq 5$ is prime. Deduce that a simple group cannot have order 80.

25.3* Using the methods of this chapter, prove that there is no simple group of order 84.

(*Hint*: $84 = 2^2.3.7$ so there are elements of orders $2, 3, 7$. Show that if $G$ is simple of order 84 then its nontrivial proper subgroups must have orders $2, 3, 4, 6, 7$, or 12. In particular, all nontrivial proper normalisers must have one of these orders. Deduce that if $H_7$ is a subgroup of order 7 then its has precisely 12 distinct conjugates, giving at least 72 distinct elements of order 7. Deduce that if $H_3$ is a subgroup of order 3 then its has at least 7 distinct conjugates, giving at least 14 distinct elements of order 3. Thus $|G| \geq 1 + 14 + 72 = 87 > 84$, a contradiction.)

25.4* Extend the list of impossible orders for non-cyclic simple groups from 61 upwards, as far as you can, using the methods of this chapter.

(More advanced methods such as Sylow's Theorem show that the next possible order is 168, so there are plenty of orders to try. Orders 72 and 90 seem to be the most difficult cases of order 100 or less, and may be beyond the methods of this chapter. See if Sylow's Theorem helps.)

25.5 The group $\mathbb{GL}_3(\mathbb{F}_2)$ consists of invertible $3 \times 3$ matrices over the field $\mathbb{F}_2$ with two elements. Prove that the order of this group is 168.

25.6 The group $\mathbb{PSL}_2(\mathbb{F}_7)$ consists of all $2 \times 2$ matrices over the field $\mathbb{F}_7$ with 7 elements that have determinant 1, quotiented by the normal subgroup $\{I, -I\}$ where $I$ is the identity matrix. Prove that the order of this group is 168.

25.7 It can be shown by direct (and tedious, unless done by computer) computation that the group $\mathbb{PSL}_2(\mathbb{F}_7)$ has six conjugacy classes, of sizes $1, 21, 24, 24, 42, 56$. Assuming this, prove that $\mathbb{PSL}_2(\mathbb{F}_7)$ is simple.

25.8* If you know how to use GAP, go to

groupprops.subwiki.org/wiki/Groups_of_order_168

and use GAP to investigate the 57 non-isomorphic groups of order 168.

# Chapter 26

# Further Directions

In this final chapter we give brief descriptions of other directions in which the ideas of Galois theory have been developed: the inverse Galois problem, differential Galois theory, and a brief introduction to $p$-adic Galois representations.

## 26.1 Inverse Galois Problem

A natural question, originally asked early in the 19th Century, is: Which finite groups can occur as the Galois group of a finite extension of the rationals $\mathbb{Q}$? Despite the simplicity of its statement, this question has not yet been answered completely. It is conjectured that the answer is 'any finite group', but this has neither been proved nor disproved. There are also generalisations; for example, let $K$ be any field and $G$ a given finite group. When is there a finite extension of $K$ with Galois group $G$? If so, we say that $G$ is *realisable over* $K$.

The topic has been widely studied, and numerous partial results are known, but they are rather disorganised. Malle and Matzat (1999) discusses them in depth. We summarise some representative ones, confining the discussion to the original case $K = \mathbb{Q}$.

### Cyclic and Abelian Groups

The cyclic group $\mathbb{Z}_n$ is realisable over $\mathbb{Q}$ for any $n$. The proof uses cyclotomic fields. Let $p$ be a prime such that $n$ divides $p - 1$; this exists by Dirichlet's Theorem on primes in an arithmetic progression, see for example Serre (1973). Let $\zeta$ be a primitive $p$th root of unity and consider $K = \mathbb{Q}(\zeta)$. The Galois group $G$ of $K/\mathbb{Q}$ is cyclic of order $p - 1$ by Theorem 21.8.

Since $n$ divides $p-1$, the group $G$ has a subgroup $H$ of order $(p-1)/n$. Let $L$ be its fixed field. By the Galois correspondence, the Galois group of $L/\mathbb{Q}$ is $G/H$, which is isomorphic to $\mathbb{Z}_n$. To find a polynomial with this Galois group requires a few more simple steps. The extension $L/\mathbb{Q}$ is simple by the Primitive Element Theorem 6.13, so $L = \mathbb{Q}(\alpha)$ for some $\alpha \in L$. Let $p$ be the minimal polynomial of $\alpha$ over $\mathbb{Q}$. Then the Galois group of $p$ is that of $L/\mathbb{Q}$, which is cyclic of order $n$.

DOI: 10.1201/9781003213949-26

**Example 26.1.** We find a polynomial over $\mathbb{Q}$ with Galois group $\mathbb{Z}_5$. Let $p = 11, n = 2$, and consider the primitive 11th root of unity $\zeta = \cos\frac{2\pi}{11} + i\sin\frac{2\pi}{11}$. The Galois group of $\mathbb{Q}(\zeta)/\mathbb{Q}$ is cyclic of order 10, generated by the map $\phi$ for which

$$\phi(\zeta^k) = \zeta^{-k}$$

Bearing in mind that $\zeta^{-1} = \zeta^{10}$, the fixed field $L$ of this group has basis over $\mathbb{Q}$ consisting of

$$\alpha = \zeta + \zeta^{10} \quad \beta = \zeta^2 + \zeta^9 \quad \gamma = \zeta^3 + \zeta^8 \quad \delta = \zeta^4 + \zeta^7 \quad \varepsilon = \zeta^5 + \zeta^6$$

These are the conjugates of $\zeta + \zeta^{10}$. By Lemma 5.5 the minimal polynomial $p$ of $\zeta + \zeta^{10}$ is irreducible over $\mathbb{Q}$. Its Galois group is that of $\mathbb{Q}(\zeta)/L$, which is $\mathbb{Z}_{10}/\mathbb{Z}_2 \cong \mathbb{Z}_5$.

We can compute $p$ using the identity $1 + \zeta + \cdots + \zeta^{10} = 0$. The calculation is quite lengthy. As a short cut, computer algebra shows that

$$p(t) = (t - \alpha)(t - \beta)(t - \gamma)(t - \delta)(t - \varepsilon) = t^5 + t^4 - 4t^3 - 3t^2 + 3t + 1$$

yielding an explicit polynomial with Galois group $\mathbb{Z}_5$ over $\mathbb{Q}$.

A simple generalisation of this method shows that any finite abelian group is realisable over $\mathbb{Q}$. See Exercise 26.1.

## Symmetric and Alternating Groups

We proved in Theorem 18.16 that the general $n$th degree polynomial 'over' any field $K$ (Definition 18.15) has Galois group $\mathbb{S}_n$. This in particular applies when $K = \mathbb{Q}$. However this polynomial is really one over $\mathbb{Q}(s_1, \ldots, s_n)$, where the $s_j$ are independent transcendental elements (the coefficients), so it does not answer the question for $\mathbb{Q}$. One approach is to find explicit polynomials over $\mathbb{Q}$ with Galois group $\mathbb{S}_n$. Another is to appeal to general theorems that demonstrate the existence of such polynomials without actually constructing them. It is not too hard to deal with the alternating group $\mathbb{A}_n$ by similar methods. These groups were first proved realisable over $\mathbb{Q}$ by Hilbert (1892). For full details see See Wikipedia (Inverse Galois problem).

## Soluble Groups

The next important step was taken by Scholz (1937) and Reichardt (1937), who proved that for any odd prime $p$, every finite $p$-group is realisable over $\mathbb{Q}$. Shafarevich (1954) extended this result to show that every finite soluble group is realisable over $\mathbb{Q}$. the proof had a gap which was corrected in 1989; see Neukirch, Schmidt and Wingberg (2000) Chapter IX. Shafarevich's proof does not construct an explicit polynomial with a given finite soluble group as Galois group.

### Simple Groups

At the opposite extreme to soluble groups are the (non-cyclic) finite simple groups. Thanks to the epic researches of finite group theorists in the second half of the 20th Century, the finite simple groups have been completely classified; for an accessible summary see Aschbacher (2004). They fall into a large number of families of groups of 'Lie type'—essentially, analogues over finite fields of the simple Lie groups, classified by Wilhelm Killing (with omissions) between 1888 and 1890, and simplified and completed by Élie Cartan in 1894—together with 26 curious exceptions known as *sporadic simple groups.*

All but one of the sporadic simple groups are known to be realisable over $\mathbb{Q}$, the sole exception being the Mathieu group $M_{23}$ of order $2^7.3^2.5.\Delta7.11.23 = 10,200,960$. whose status is unknown. This is one of the smaller sporadic simple groups; remarkably, Thompson (1984) proved some general theorems implying that (among others) the largest sporadic simple group is realisable over $\mathbb{Q}$. This group, known as the Fischer-Griess monster, has order

$$2^{46}.3^{20}.5{\cdot}7^6.11^2.13^3.17.19.23.29.31.41.47.59.71 \sim 8.10^{53}$$

The situation for the simple groups of Lie type is complicated. Many have been proved realisable over $\mathbb{Q}$, but many others have not. Zywina (2015) proves that the projective special linear groups $\mathbb{PSL}_2(\mathbb{F}_p)$ are realisable over $\mathbb{Q}$ for all primes $p \geq 5$.

The corresponding problem has been studied for many fields other than $\mathbb{Q}$. Further information can be found in Ranjbar and Ranjbar (2015) and Vila (1992).

## 26.2 Differential Galois Theory

We outline a version of Galois theory that applies to differential equations, developed initially by Vessiot (1892) and Picard (1896). Ritt (1932) developed the theory for differential equations in the context of complex functions, and Kolchin (1946) presented an algebraic approach formalising the ideas of Vessiot and Picard. For further details see Kaplansky (1957) and Magid (1994, 1999).

The basic idea is to axiomatise the operator $\mathrm{d}/\mathrm{d}x$ acting on functions of $x$:

**Definition 26.2.** A *differential field* is a field $K$ together with a *derivation.* This is a map $D : K \to K$ such that for all $k, l \in K$

$$\begin{aligned} D(k+l) &= D(k) + D(l) \\ D(kl) &= kD(l) + D(k)l \end{aligned} \tag{26.1}$$

The canonical example, and one of the most important for applications, arises when $K = \mathbb{C}(x)$, the field of rational complex functions, and $D$ is the usual derivative. The second equation in (26.1) is the *Leibniz formula* for products.

We can define *differential subfields* and *differential extensions* in the obvious way: the derivation is required to commute with the relevant inclusion maps. The *field of constants* of $K$ is the set of all $k \in K$ such that $D(k) = 0$; it is a subfield containing the prime subfield. A *differential automorphism* $\alpha$ of $K$ is an automorphism of $K$ such that the diagram

$$\begin{array}{ccc} K & \xrightarrow{\alpha} & K \\ \downarrow{\scriptstyle D} & & \downarrow{\scriptstyle D} \\ K & \xrightarrow{\alpha} & K \end{array} \tag{26.2}$$

commutes. That is, $\alpha(D(k)) = D(\alpha(k))$ for all $k \in K$.

**Definition 26.3.** If $M$ is a differential extension of $K$ then the *differential Galois group* of $M/K$ is the set of all differential automorphisms of $M$ that leave every element of $K$ fixed. It is obviously a group under composition, and unlike ordinary Galois theory, it is usually an infinite group.

So far all we have done is to tack the word 'differential' in front of everything. Even so, we can now set up the Galois correspondence. If $L$ is a differential field between $K$ and $M$ we define $L^*$ as the group of differential automorphisms of $M$ that fix every element of $L$. If $H$ is a subgroup of the differential Galois group $G = K^*$ of $M/K$ we define $H^\dagger$ to be the subfield of elements fixed by $H$, which is a differential subfield. The differential Galois group of $M/K$ is clearly a subgroup of the Galois group of $M/K$.

With these definitions the Galois correspondence does not quite work. We need one further ingredient:

**Definition 26.4.** A subgroup $H$ of $G$ is *closed* if $H^{\dagger *} = H$. A differential subfield $L$ with $K \subseteq L \subseteq M$ is *closed* if $L^{*\dagger} = L$.

It can then be proved that the maps $*$ and $\dagger$ define a bijection between closed subgroups and closed differential subfields. However, something is still lacking: we do not yet know what these closed subgroups and subfields look like.

Now we bring on the differential equations. For $y \in K$ we define

$$y' = D(y) \quad y'' = D^2(y) \quad \ldots \quad y^{(n)} = D^n(y)$$

Consider the 'homogeneous linear differential equation'

$$y^{(n)} + a_1 y^{(n-1)} + \cdots + a_{n-1}y' + a_n y = 0 \tag{26.3}$$

where $a_1, \ldots, a_n \in K$. Suppose that in some larger differential field, there are solutions $u_1, \ldots, u_{n+1}$ of (26.3). Then it can be proved that there exist constants $c_1, \ldots, c_{n+1}$ such that

$$c_1 u_1 + \cdots + c_{n+1} u_{n+1} = 0$$

where not all $c_j$ are zero. In other words, there are at most $n$ solutions of (26.3) that are linearly independent over constants.

**Definition 26.5.** A differential extension $M$ of $K$ is a *Picard-Vessiot extension* for equation (26.3) if:

(1) $M$ is the differential field generated by $K$ together with $u_1, \ldots, u_n$, where the $u_i$ satisfy (26.3) and are linearly independent over the field of constants.

(2) $M$ has the same field of constants as $K$.

In differential Galois theory, linear differential equations of the form (26.3) play the role of polynomials in standard Galois theory. The differential Galois group provides information about how the solutions $u_j$ can be expressed. So, just as we asked whether the quintic equation can be solved by radicals, and used properties of the Galois group $\mathbb{S}_5$ to prove that the answer is 'no' in general, we can use properties of the differential Galois group to decide whether (26.3) can be solved by specific methods. Two important methods are:

(1) *Adjoining an integral.* Adjoin $u$ such that $u' - a = 0$, for some $a \in K$.

(2) *Adjoining an exponential of an integral.* Adjoin $u$ such that $u' - au = 0$, for some $a \in K$.

To see where the terminology comes from, consider differential equations for real-valued functions $u$. Methods (1) and (2) are standard techniques in basic courses on differential equations.

Analogous to a radical extension in Galois theory, we have:

**Definition 26.6.** A differential field extension $M/K$ is a *Liouville extension* if there is a chain of differential fields

$$K = K_0 \subseteq K_1 \subseteq \cdots \subseteq K_n = M$$

such that each extension $K_{i+1}/K_i$ is either the adjunction of an integral or the adjunction of an exponential of an integral.

Elements of a Liouville extension are solutions of (26.3) that can be expressed using a series of integrals and exponentials of integrals. In other words, these are the solutions that can be expressed in an 'elementary' fashion; that is, in an explicit form using nothing worse than field operations, integrals, and exponentiation (in this sense).

Remarkably, it can then be proved that:

**Theorem 26.7.** *If $M$ is a Liouville extension of $K$, the differential Galois group of $M/K$ is soluble.*

For a concrete example we must introduce more structure on the differential Galois group. It can be considered as an *algebraic group* over the field of constants; that is, a group of matrices $(g_{ij})$ defined by a system of polynomial equations. For example, the equation $\det(g_{ij}) = 1$ defines the algebraic group of unimodular matrices. The matrices arise because differential field automorphisms induce linear transformations on the vector space of solutions of the differential equation (26.3), which is spanned by $u_1, \ldots, u_n$.

For a Picard-Vessiot extension it turns out that every intermediate field is closed, and the closed subgroups of the differential Galois group are its algebraic subgroups—those defined by systems of polynomial equations in the $g_{ij}$. Every algebraic group $\Gamma$ has a special normal subgroup $\Gamma^0$, the *connected component of the identity*. This is basically what it says on the tin: the connected component of the group that contains the identity element of the group, which in this context is the identity matrix. The phrase 'connected component' is topological, and in this instance it refers to a topology called the *Zariski topology* in which the closed sets are determined by zero-sets of polynomials. This topology is rather strange compared to the topologies usually encountered in analysis and metric spaces; for example, it generally lacks the Hausdorff property that any two distinct points can be surrounded by disjoint open sets.

We can now state a key theorem of differential Galois theory, closely analogous to the characterisation of polynomials that can be solved by radicals:

**Theorem 26.8.** *Suppose that $M/K$ is a Picard-Vessiot extension such that $K$ has characteristic zero and the field of contents $C$ is algebraically closed. Suppose that $M$ can be embedded in a differential field obtained from $K$ by a series of algebraic extensions, adjunctions of an integral, or adjunctions of the exponential of an integral. Then the connected component of the identity of the differential Galois group is soluble.*

*Conversely, if the component of the identity of the differential Galois group of $M/K$ is soluble, then $M$ can be obtained from $K$ by a finite (in particular, algebraic) extension followed by a Liouville extension.*

**Example 26.9.** Consider the differential equation

$$y'' - xy = 0 \tag{26.4}$$

over the field $\mathbb{C}(x)$ of rational complex functions. This is a homogeneous linear differential equation of order 2, and it can be proved that the differential Galois group is the group $\mathbb{SL}_2(\mathbb{C})$ of all unimodular $2 \times 2$ matrices over $\mathbb{C}$. That is, matrices

$$\begin{bmatrix} a & b \\ c & d \end{bmatrix} : \quad a, b, c, d \in \mathbb{C} \quad ad - bc = 1$$

The connected component of the identity in this group is not soluble (Exercises 26.2, 26.3), so (26.4) cannot be solved by starting with rational functions and performing algebraic operations, integrals, or exponentials of integrals. This means that (26.4) cannot be solved by any nice or even moderately nice formula involving the standard functions of analysis such as trigonometric functions, exponentials, or logarithms.

Nevertheless, equation (26.4), known as *Airy's equation*, has simple power-series solutions

$$1 + \frac{x^3}{2.3} + + \frac{x^6}{2.3.5.6} + \cdots$$
$$x + \frac{x^4}{3.4} + \frac{x^7}{3.4.6.7} + \cdots$$

Airy's equation is important in several areas of physics.

## 26.3 $p$-adic Numbers

Galois theory has important applications to algebraic number theory. This phrase can be read either as 'the algebraic theory of numbers', or as 'the theory of algebraic numbers', which are solutions of polynomial equations over $\mathbb{Q}$. In practice both interpretations apply, and boil down to much the same thing. At the current frontiers, the aim is to understand not just the Galois groups of specific finite extensions of $\mathbb{Q}$, as in this book, but all of them at once.

The main idea is to define $\overline{\mathbb{Q}}$ to be the algebraic closure of $\mathbb{Q}$, which contains as subfields all algebraic number fields, and study the Galois group $G_{\mathbb{Q}}$ of the extension $\overline{\mathbb{Q}}/\mathbb{Q}$, whose degree is infinite. The group $G_{\mathbb{Q}}$ is also infinite, but it has certain finiteness properties: technically is it a 'profinite' group, the projective limit of finite groups. The Galois correspondence remains valid, after a technical tweak: profinite groups have a natural topology, and we consider only subgroups of the Galois group that are closed in this topology. Not surprisingly, the subject has a strong topological flavour, so it often resembles complex or real analysis more than algebra. The entire area becomes highly technical, but it is also extremely powerful.

The central objects of study are *Galois representations* $\rho : G_{\mathbb{Q}} \to \mathbb{GL}_n(R)$ where $R$ is a topological ring. The main choices for $R$ are the complex numbers $\mathbb{C}$ and rings related to *p-adic numbers* for primes $p$. The $p$-adic numbers were introduced explicitly by Kurt Hensel in 1897, although some 'prehistory' can be detected in earlier number-theoretic research. Their construction resembles the way the real numbers $\mathbb{R}$ can be constructed from $\mathbb{Q}$ by considering convergent (more precisely, Cauchy) sequences of rationals; see for instance Stewart and Tall (2015a).

Abstractly, the construction of $\mathbb{R}$ from $\mathbb{Q}$ is achieved by forming the *completion* of $\mathbb{Q}$ with respect to the usual metric $d(q, r) = |q - r|$. The $p$-adic numbers arise in a similar manner, but with a very different metric. We define these numbers now.

Let $p$ be a prime. Any rational number $r$ can be written uniquely as

$$r = p^k a/b \tag{26.5}$$

where $a$ and $b$ are coprime and neither is divisible by $p$. For example if $p = 11$ and $r = 22/7$ then

$$r = 11.(2/7)$$

so $k = 1, a = 2, b = 7$. If instead $p = 7$ then $k = -1, a = 22, b = 1$.

The *p-adic absolute value* of $r$ is

$$|r|_p = p^{-k}$$

(when $r = 0$ we define $|0|_p = 0$). So when $r = 22/7$ we have $|r|_{11} = 11^{-1} = 1/11$. Similarly $|r|_7 = (7^{-1})^{-1} = 7$.

To interpret this definition it is useful to extend the usual notion of prime factorisation from $\mathbb{Z}$ to $\mathbb{Q}$. Every positive rational number $r$ can be written uniquely in the form

$$r = \prod_{p \text{ prime}} p^{m_p}$$

where the exponents $m_p \in \mathbb{Z}$; that is, they can be negative as well as positive. In this sense, the $p$-adic absolute value of $r$ is small when $k$ in (26.5) is large and positive; that is, $p$ occurs to a large power in the prime factorisation of $r$.

The rational numbers $\mathbb{Q}$ form a metric space with distance function

$$d(r, s) = |r - s|_p$$

In fact, this is an *ultrametric*, meaning that

$$|x + y|_p \leq \max(|x|_p, |y|_p) \tag{26.6}$$

which is a stronger condition than the usual triangle inequality $|x + y|_p \leq |x|_p + |y|_p$. The $p$-adic numbers $\mathbb{Q}_p$ are the *completion* of this metric space, which is constructed using equivalence classes of Cauchy sequences. For a more explicit construction, define a *p-adic series* to be a formal expression

$$a = \sum_{i=k}^{\infty} a_i p^i$$

where $k \in \mathbb{Z}$ and $p$ occurs to the power 0 in the prime factorisation of $a_i$. Two such series are *equivalent* if they have the same $k$, and if for every integer $n \geq k$ the difference

$$\sum_{i=k}^{n} (a_i - b_i) p^i$$

is a rational number $p^k\frac{a}{b}$ with $k > n$ where $a$ and $b$ are prime to $p$. The $p$-adic numbers are the equivalence classes for this equivalence relation, and the $p$-adic absolute value can be extended to these equivalence classes.

The $p$-adic numbers $\mathbb{Q}_p$ form a field with a metric, and it is complete in this metric. The importance of $p$-adic numbers in number theory goes back to their introduction by Kurt Hensel and is stated in the 'local-global principle' of Helmut Hasse: a polynomial equation in one or more variables with rational coefficients has a rational solution if and only if it has a solution in real numbers *and* in $p$-adic numbers for every prime $p$. Hasse proved that his principle holds for quadratic equations, but it is known to fail for polynomials in several indeterminates of higher degree. This idea can be seen as a major generalisation of a method we have used several times in earlier chapters: obtaining information about integer solutions of an equation by reducing it modulo $p$.

The study of the Galois group of $\overline{\mathbb{Q}}/\mathbb{Q}$ makes heavy use of representations over the $p$-adic numbers. The machinery relies strongly on $p$-adic cohomology, which is beyond the scope of this book. The results are central to modern algebraic number theory. For example, they are vital to Andrew Wiles's famous proof of Fermat's Last Theorem, which is sketched in Stewart and Tall (2015b) and other sources.

---

## EXERCISES

26.1 Arthur Cayley proved that every finite group $G$ is isomorphic to a subgroup of $\mathbb{S}_n$ where $n = |G|$. The proof is simple: for each $g \in G$ define $\pi_g : G \to G$ by $\pi_g h = gh$. Then $g \mapsto \pi_g$ is a monomorphism from $G$ to the group $\mathbb{S}_G$ of permutations of $G$.

Deduce that every finite group is isomorphic to the Galois group of some finite extension (not necessarily of $\mathbb{Q}$).

26.2* Let $\Gamma = \mathbb{SL}_2(\mathbb{C})$, the group of all $2 \times 2$ matrices

$$\begin{bmatrix} x & y \\ z & w \end{bmatrix} : \quad x, y, z, w \in \mathbb{C} \quad xy - zw = 1$$

over $\mathbb{C}$ of determinant 1. Show that $\Gamma^0 = \Gamma$. (Unpicking the meaning of the Zariski topology, you may assume that this is equivalent to proving that the defining polynomial equation $xw - yz - 1 = 0$ is irreducible in $\mathbb{C}[x, y, z, w]$.)

26.3* Prove that $\mathbb{SL}_2(\mathbb{C})$ is not soluble. (*Hint*: prove that the derived group $\Gamma' = \Gamma$.)

26.5 Let $r = 355/113$. Find the $p$-adic absolute value $|r|_p$ when:

(a) $p = 3$

(b) $p = 5$

(c) $p = 7$

(d) $p = 71$

(e) $p = 113$

26.5 Using a computer algebra package to calculate prime factors, if you wish, find the primes $p$ for which the $p$-adic absolute value $|r|_p$ is not equal to 1, when:

(a) $r = 12345$

(b) $r = 123456$

(c) $r = 1234567$

(d) $r = 12345678$

(e) $r = 123456789$

26.6 Prove that $|\cdot|_p$ is an ultrametric. That is, equation (26.6) is valid.

26.7 Look through the book to find examples where we use reduction modulo a prime to obtain information about polynomial equations over $\mathbb{Z}$ or $\mathbb{Q}$.

# References

## GALOIS THEORY

Artin, E. (1948) *Galois Theory*, Notre Dame University Press, Notre Dame.

Ayad, M. (2018) *Galois Theory and Applications: Solved Exercises and Problems*, World Scientific, Singapore.

Bastida, J.R. (1984) *Field Extensions and Galois Theory*, Addison-Wesley, Menlo Park.

Berndt, B.C., Spearman, B.K., and Williams, K.S. (2002) Commentary on a unpublished lecture by G.N. Watson on solving the quintic, *Mathematical Intelligencer* **24** number 4, 15–33.

Bewersdorrff, J. (2006) *Galois Theory for Beginners: A Historical Perspective*, American Mathematical Society, Providence.

Brown, K. (2010) Mathematics 6310: The Primitive Element Theorem, Lecture Notes, Cornell University. pi.math.cornell.edu/ kbrown/6310/primitive.pdf

Brzezinski, J. (2018) *Galois Theory Through Exercises*, Springer, Cham.

Cox, D.A. (2012) *Galois Theory*, 2nd ed., Wiley-Blackwell, Hoboken.

Edwards, H.M. (1984) *Galois Theory*, Springer, New York.

Fenrick, M.H. (1992) *Introduction to the Galois Correspondence*, Birkhäuser, Boston.

Garling, D.J.H. (1987) *A Course in Galois Theory*, Cambridge University Press, Cambridge.

Hadlock, C.R. (1978) *Field Theory and its Classical Problems*, Carus Mathematical Monographs **19**, Mathematical Assocation of America, Washington.

Hilbert, D. (1892) Über die Irreduzibilität ganzer rationaler Functionen mit ganzzähligen Koeffizienten, *Journal für die Reine und Angewandte Mathematik* **110** 104–129.

Howie, J.M. (2005) *Fields and Galois Theory*, Springer, Berlin.

Isaacs, M. (1985) Solution of polynomials by real radicals, *American Mathematical Monthly* **92** 571–575.

Jacobson, N. (1964) *Theory of Fields and Galois Theory*, Van Nostrand, Princeton.

Kaplansky, I. (1969) *Fields and Rings*, University of Chicago Press, Chicago.

King, R.B. (1996) *Beyond the Quartic Equation*, Birkhäuser, Boston.

Kuga, M. (2013) *Galois' Dream: Group Theory and Differential Equations*, Birkhäuser, Basel.

Lorenz, F. and Levy, S. (2005) *Algebra Volume 1: Fields and Galois Theory*, Springer, Berlin.

Malle, G. and Matzat, B.H. (1999) *Inverse Galois Theory*, Springer, Berlin.

Morandi, P. (1996) *Field and Galois Theory*, Graduate Texts in Mathematics **167**, Springer, Berlin.

Neukirch, J., Schmidt, A., and Wingberg, K. (2000) *Cohomology of Number Fields*, Grundlehren der Mathematischen Wissenschaften **23**, Springer, Berlin.

Newman, S.C. (2012) *A Classical Introduction to Galois Theory*, Wiley-Blackwell, Hoboken.

Postnikov, M.M. (2004) *Foundations of Galois Theory*, Dover, Mineola.

Ranjbar, F. and Ranjbar, S. (2015) Inverse Galois problem and significant methods, arXiv:1512.08708.

Reichardt, H. (1937) Konstruktion von Zahlkörpern mit gegebener Galoisgruppe von Primzahlpotenzordnung, *J. Reine Angew. Math.* **177** 1–5.

Rotman, J. (2013) *Galois Theory*, Springer, Berlin.

Scholz, A. (1937) Konstruktion algebraischer Zahlkörper mit beliebiger Gruppe von Primzahlpotenzordnung I, *Mathematsche Zeitschrift* **42** 161–188.

Shafarevich, I.R. (1954) Construction of fields of algebraic numbers with given solvable Galois group (in Russian), *Izvestiya Akademii Nauk SSSR*, Ser. Mat. **18** 525–578.

Tignol, J.-P. (2001) *Galois' Theory of Algebraic Equations*, World Scientific, Singapore.

Van der Waerden, B.L. (1953) *Modern Algebra* (2 vols), Ungar, New York.

Vila, N. (1992) On the inverse problem of Galois theory, *Publicacions Matemàtiques* **6** 1053–1073.

Weintraub, S.H. (2021) The theorem of the primitive element, *American Mathematical Monthly* **128** 753–754.

Wikpedia. Inverse Galois problem, en.wikipedia.org/wiki/Inverse_Galois_problem

---

# ADDITIONAL MATHEMATICAL MATERIAL

Adamchik, V.S. and Jeffrey, D.J. (2003) Polynomial transformations of Tschirnhaus, Bring and Jerrard, *ACM SIGSAM Bulletin* **37** 90–94.

Adams, J.F. (1969) *Lectures on Lie Groups*, University of Chicago Press, Chigago.

Anton, H. (1987) *Elementary Linear Algebra* (5th ed.), Wiley, New York.

Aschbacher, M. (2004) The status of the classification of finite simple groups, *Notices of the American Mathematical Society* **51** 736–740.

Chang, W.D. and Gordon, R.A. (2014) Trisecting angles in Pythagorean triangles, *American Mathematical Monthly* **121** 625–631.

de Bruijn, N. (1972) A solitaire game and its relation to a finite field, *Journal of Recreational Mathematics* **5** 133.

Dudley, U. (1987) *A Budget of Trisections*, Springer, New York.

Fieker, C. and Klüners, J. (2013) Computation of Galois groups of rational polynomials, arXiv:1211.3588.

Fraleigh, J.B. (1989) *A First Course in Abstract Algebra*, Addison-Wesley, Reading.

Geck, M. (2013) On the characterization of Galois extensions, *American Mathematical Monthly* **121** 637–639.

Gleason, A.M. (1988) Angle trisection, the heptagon, and the triskaidecagon, *American Mathematical Monthly* **95** 185–194.

Hardy, G.H. (1960) *A Course of Pure Mathematics*, Cambridge University Press, Cambridge.

Hardy, G.H. and Wright, E.M. (1962) *The Theory of Numbers*, Oxford University Press, Oxford.

Heath, T.L. (1956) *The Thirteen Books of Euclid's Elements* (3 vols) (2nd ed.), Dover, New York.

Herz-Fischler, R. (1998) *A Mathematical History of the Golden Number* (2nd ed.), Dover, Mineola.

Hulke, A. (1996) *Konstruktion transitiver Permutationsgruppen*, Dissertation, Rheinisch Westfälische Technische Hochschule, Aachen.

Humphreys, J.F. (1996) *A Course in Group Theory*, Oxford University Press, Oxford.

Kaplansky, I. (1957) *An Introduction to Differential Algebra*, Hermann, Paris.

Katz, B. (2013) Short proof of Abel's theorem that 5th degree polynomial equations cannot be solved, youtube.com/watch?v=zeRXVL6qPk4

Kolchin, E.R. (1946) The Picard–Vessiot theory of homogeneous linear ordinary differential equations, *Proceedings of the National Academy of Sciences of the USA* **32** 308–311.

Krantz, S.G. (2002) *Handbook of Logic and Proof Techniques for Computer Science*, Springer, New York.

Lang, S. (1983) *Fundamentals of Diophantine Geometry*, Springer, New York.

Lidl, R. and Niederreiter, H. (1986) *Introduction to Finite Fields and Their Applications*, Cambridge University Press, Cambridge.

Livio, M. (2002) *The Golden Ratio*, Broadway Books, New York.

Magid, A.R. (1994) *Lectures on Differential Galois Theory*, University Lecture Series **7** American Mathematical Society, Providence, R.I.

Magid, A.R. (1999) Differential Galois theory, *Notices of the American Mathematical Society* 1041–1049.

Neumann, P.M., Stoy, G.A., and Thompson, E.C. (1994) *Groups and Geometry*, Oxford University Press, Oxford.

Oldroyd, J.C. (1955) Approximate constructions for 7, 9, 11 and 13-sided polygons, *Eureka* **18**, 20.

Picard, É. (1896) *Traité d'Analyse*, Gauthier-Villars, Paris.

Ramanujan, S. (1962) *Collected Papers of Srinivasa Ramanujan*, Chelsea, New York.

Ritt, J.F. (1932) *Differential Equations from the Algebraic Standpoint*, American Mathematical Society, New York.

Salmon, G. (1885) *Lessons Introductory to the Modern Higher Algebra*, Hodges, Figgis, Dublin.

Serre, J-P. (1973) *A Course in Arithmetic*, Springer, New York.

Sharpe, D. (1987) *Rings and Factorization*, Cambridge University Press, Cambridge.

Soicher, L. and McKay, J. (1985) Computing Galois groups over the rationals, *Journal of Number Theory* **20** 273–281.

Stewart, I. (1977) Gauss, *Scientific American* **237** 122–131.

Stewart, I. and Tall, D. (2015a) *The Foundations of Mathematics* (2nd ed.), Oxford University Press, Oxford.

Stewart, I. and Tall, D. (2015b) *Algebraic Numbers and Fermat's Last Theorem* (4th ed.), CRC Press, Boca Raton FL.

Stewart, I. and Tall, D. (2018) *Complex Analysis* (2nd ed.), Cambridge University Press, Cambridge.

Thompson, J.G. (1984). Some finite groups which appear as gal$L/K$, where $K \subseteq \mathbb{Q}(\mu_n)$, *Journal of Algebra* **89** 437–499.

Thompson, T.T. (1983) *From Error-Correcting Codes Through Sphere-Packings to Simple Groups*, Carus Mathematical Monographs **21**, Mathematical Assocation of America, Washington DC.

Titchmarsh, E.C. (1960) *The Theory of Functions*, Oxford University Press, Oxford.

Vessiot, E. (1892) Sur l'intégration des équations différentielles linéaires, *Annales Scientifiques de l'École Normale Supérieure* **9** 197–280.

Zywina, D. (2015) The inverse Galois problem for $\mathbb{PSL}_2(\mathbb{F}_p)$, *Duke Mathematical Journal* **164** 2253–2292.

## HISTORICAL MATERIAL

Bell, E.T. (1965) *Men of Mathematics* (2 vols), Penguin, Harmondsworth, Middlesex.

Bertrand, J. (1899) La vie d'Évariste Galois, par P. Dupuy, *Bulletin des Sciences Mathématiques*, **23**, 198–212.

Bortolotti, E. (1925) L'algebra nella scuola matematica bolognese del secolo XVI, *Periodico di Matematica*, **5**(4), 147–84.

Bourbaki, N. (1969) *Éléments d'Histoire des Mathématiques*, Hermann, Paris.

Bourgne, R. and Azra, J.-P. (1962) *Écrits et Mémoires Mathématiques d'Évariste Galois*, Gauthier-Villars, Paris.

Cardano, G. (1931) *The Book of my Life*, Dent, London.

Clifford, W.K. (1968) *Mathematical Papers*, Chelsea, New York.

Coolidge, J.L. (1963) *The Mathematics of Great Amateurs*, Dover, New York.

Dalmas, A. (1956) *Évariste Galois, Révolutionnaire et Géomètre*, Fasquelle, Paris.

Dumas, A. (1967) *Mes Memoirs* (volume 4 chapter 204), Editions Gallimard, Paris.

Dupuy, P. (1896) La vie d'Évariste Galois, *Annales de l'École Normale*, **13**(3), 197–266.

Galois, E. (1897) *Oeuvres Mathématiques d'Évariste Galois*, Gauthier-Villars, Paris.

Gauss, C.F. (1966) *Disquisitiones Arithmeticae*, Yale University Press, New Haven.

Henry, C. (1879) Manuscrits de Sophie Germain, *Revue Philosophique* **631**.

Huntingdon, E.V. (1905) *Transactions of the American Mathematical Society* **6**, 181.

Infantozzi, C.A. (1968) Sur l'a mort d'Évariste Galois, *Revue d'Histoire des Sciences* **2**, 157.

Joseph, G.G. (2000) *The Crest of the Peacock*, Penguin, Harmondsworth.

Klein, F. (1913) *Lectures on the Icosahedron and the Solution of Equations of the Fifth Degree*, Kegan Paul, London.

Klein, F. (1962) *Famous Problems and other Monographs*, Chelsea, New York.

Kollros, L. (1949) *Évariste Galois*, Birkhäuser, Basel.

La Nave, F. and Mazur, B. (2002) Reading Bombelli, *Mathematical Intelligencer* **24** number 1, 12–21.

Midonick, H. (1965) *The Treasury of Mathematics* (2 vols), Penguin, Harmondsworth, Middlesex.

Neumann, P.M. (2011) *The Mathematical Writings of Évariste Galois*, European Mathematical Society, Zürich.

Richelot, F.J. (1832) De resolutione algebraica aequationis $x^{257} = 1$, sive de divisione circuli per bisectionam anguli septies repetitam in partes 257 inter se aequales commentatio coronata, *Journal für die Reine and Angewandte Mathematik* **9**, 1–26, 146–61, 209–30, 337–56.

Richmond, H.W. (1893) *Quarterly Journal of Mathematics* **26**, 206–7; and *Mathematische Annalen* **67** (1909), 459–61.

Rothman, A. (1982a) The short life of Évariste Galois, *Scientific American*, April, 112–20.

Rothman, A. (1982b) Genius and Biographers: The Fictionalization of Évariste Galois, *American Mathematical Monthly* **89** 84–106.

Tannery, J. (1908) (ed.) *Manuscrits d'Évariste Galois*, Gauthier-Villars, Paris.

Taton, R. (1947) Les relations d'Évariste Galois avec les mathématiciens de son temps. Cercle International de Synthèse, *Revue d'Histoire des Sciences* **1**, 114.

Taton, R. (1971) Sur les relations scientifiques d'Augustin Cauchy et d'Évariste Galois, *Revue d'Histoire des Sciences* **24**, 123.

---

## THE INTERNET

Websites come and go, and there is no guarantee that any of the following will still be in existence when you try to access them. Use a search engine, and look up Galois in Wikipedia.

Scans of the manuscripts:

www.bibliotheque-institutdefrance.fr/

The Évariste Galois archive.

galois-group.net/

Évariste Galois.

mathshistory.st-andrews.ac.uk/Biographies/Galois/

Évariste Galois postage stamp.
https://colnect.com/en/stamps/stamp/28999-Evariste_Galois_1811-1832-Famous_People_1984-France

Euclid the game: Interactive exploration of Euclidean geometry.
kasperpeulen.github.io/

Bright, C. Computing the Galois group of a polynomial.
cbright.myweb.cs.uwindsor.ca/reports/computing-galois-group.pdf

GAP data library containing all transitive subgroups of $\mathbb{S}_n$ for $n \leq 30$.
gap-system.org/Datalib/trans.html

Hulpke, A. Determining the Galois group of a rational polynomial.
math.colostate.edu/~hulpke/talks/galoistalk.pdf

Hulpke, A. Techniques for the computation of Galois groups.
math.colostate.edu/~hulpke/paper/gov.pdf

Fermat numbers.
fermatsearch.org

Mersenne primes.
isthe.com/chongo/tech/math/prime/mersenne.html

Pierpont primes.
en.wikipedia.org/wiki/Pierpont_prime

# Index

Printed in the United States
by Baker & Taylor Publisher Services